Springer Series in Statistics

Springer

New York
Berlin
Heidelberg
Barcelona
Budapest
Hong Kong
London
Milan
Paris
Santa Clara
Singapore
Tokyo

Springer Series in Statistics

Andersen/Borgan/Gill/Keiding: Statistical Models Based on Counting Processes.
Andrews/Herzberg: Data: A Collection of Problems from Many Fields for the Student and Research Worker.
Anscombe: Computing in Statistical Science through APL.
Berger: Statistical Decision Theory and Bayesian Analysis, 2nd edition.
Bolfarine/Zacks: Prediction Theory for Finite Populations.
Borg/Groenen: Modern Multidimensional Scaling: Theory and Applications
Brémaud: Point Processes and Queues: Martingale Dynamics.
Brockwell/Davis: Time Series: Theory and Methods, 2nd edition.
Daley/Vere-Jones: An Introduction to the Theory of Point Processes.
Dzhaparidze: Parameter Estimation and Hypothesis Testing in Spectral Analysis of Stationary Time Series.
Fahrmeir/Tutz: Multivariate Statistical Modelling Based on Generalized Linear Models.
Farrell: Multivariate Calculation.
Federer: Statistical Design and Analysis for Intercropping Experiments.
Fienberg/Hoaglin/Kruskal/Tanur (Eds.): A Statistical Model: Frederick Mosteller's Contributions to Statistics, Science and Public Policy.
Fisher/Sen: The Collected Works of Wassily Hoeffding.
Good: Permutation Tests: A Practical Guide to Resampling Methods for Testing Hypotheses.
Goodman/Kruskal: Measures of Association for Cross Classifications.
Gouriéroux: ARCH Models and Financial Applications.
Grandell: Aspects of Risk Theory.
Haberman: Advanced Statistics, Volume I: Description of Populations.
Hall: The Bootstrap and Edgeworth Expansion.
Härdle: Smoothing Techniques: With Implementation in S.
Hart: Nonparametric Smoothing and Lack-of-Fit Tests.
Hartigan: Bayes Theory.
Heyde: Quasi-Likelihood And Its Application: A General Approach to Optimal Parameter Estimation.
Heyer: Theory of Statistical Experiments.
Huet/Bouvier/Gruet/Jolivet: Statistical Tools for Nonlinear Regression: A Practical Guide with S-PLUS Examples.
Jolliffe: Principal Component Analysis.
Kolen/Brennan: Test Equating: Methods and Practices.
Kotz/Johnson (Eds.): Breakthroughs in Statistics Volume I.
Kotz/Johnson (Eds.): Breakthroughs in Statistics Volume II.
Kotz/Johnson (Eds.): Breakthroughs in Statistics Volume III.
Kres: Statistical Tables for Multivariate Analysis.
Küchler/Sørensen: Exponential Families of Stochastic Processes.
Le Cam: Asymptotic Methods in Statistical Decision Theory.

(continued after index)

Ludwig Fahrmeir   Gerhard Tutz

Multivariate Statistical Modelling Based on Generalized Linear Models

With Contributions by Wolfgang Hennevogl

With 45 Illustrations

 Springer

Ludwig Fahrmeir
Seminar für Statistik
Universität München
Ludwigstrasse 33
D-80539 München
Germany

Gerhard Tutz
Institut für Quantitative Methoden
Technische Universität Berlin
Franklinstr. 28/29
D-10587 Berlin
Germany

Mathematics Subject Classifications (1991): 62-02, 62-07, 62P10, 62P20

Library of Congress Cataloging-in-Publication Data
Fahrmeir, L.
    Multivariate statistical modelling based on generalized linear
    models   Ludwig Fahrmeir. Gerhard Tutz.
        p. cm. — (Springer series in statistics)
    Includes bibliographical references and index.
    ISBN 0-387-94233-5
    1. Multivariate analysis   2. Linear models (Statistics)
I. Tutz, Gerhard.   II. Title.   III. Series.
QA278.F34 1994
519.5′35—dc20
                                                    93-50900

Printed on acid-free paper.

Production coordinated by Henry Krell; manufacturing supervised by Vincent Scelta.
Photocomposed copy from the authors' TeX files.
Printed and bound by Edwards Brothers, Inc., Ann Arbor, MI.
Printed in the United States of America.

9 8 7 6 5 4 3 (Corrected third printing, 1997)

ISBN 0-387-94233-5 Springer-Verlag New York Berlin Heidelberg
ISBN 3-540-94233-5 Springer-Verlag Berlin Heidelberg New York   SPIN 10636358

Preface

Classical statistical models for regression, time series and longitudinal data provide well–established tools for approximately normally distributed variables. Enhanced by the availability of software packages these models dominated the field of applications for a long time. With the introduction of generalized linear models (GLM) a much more flexible instrument for statistical modelling has been created. The broad class of GLM's includes some of the classical linear models as special cases but is particularly suited for categorical discrete or nonnegative responses.

The last decade has seen various extensions of GLM's: multivariate and multicategorical models have been considered, longitudinal data analysis has been developed in this setting, random effects and nonparametric predictors have been included. These extended methods have grown around generalized linear models but often are no longer GLM's in the original sense. The aim of this book is to bring together and review a large part of these recent advances in statistical modelling. Although the continuous case is sketched sometimes, throughout the book the focus is on categorical data. The book deals with regression analysis in a wider sense including not only cross-sectional analysis but also time series and longitudinal data situations. We do not consider problems of symmetrical nature, like the investigation of the association structure in a given set of variables. For example, log-linear models for contingency tables, which can be treated as special cases of GLM's are totally omitted. The estimation approach that is primarily considered in this book is likelihood–based.

The book is aimed at applied statisticians, graduate students of statistics, and students and researchers with a strong interest in statistics and data analysis from areas like econometrics, biometrics and social sciences. It is written on an intermediate mathematical level with emphasis on basic ideas. Technical and mathematical details are often deferred to starred sections, and for rigorous proofs the reader is referred to the literature.

In preliminary versions of this book Wolfgang Hennevogl was the third author. A new job and its challenges reduced his involvement. Nevertheless he made valuable contributions, in particular to parts of Section 2.3, Section 4.2, Chapter 7, Section 8.3, Appendices A3, A4 and A5, and to many of the examples. In the final stage of the manuscript Thomas Kurtz made helpful contributions, by working out examples and Appendix B.

We are grateful to various colleagues and students in our courses for discussions and suggestions. Discussions with A. Agresti were helpful when the second author visited the University of Florida, Gainesville. We would like to thank Renate Meier-Reusch and Marietta Dostert for the skilful typing of the first version. Moreover, we thank Wolfgang Schneider, Clemens Biller, Martin Krauß, Thomas Scheuchenpflug and Michael Scholz for the preparation of later versions. Further we acknowledge the computational assistance of Christian Gieger, Arthur Klinger, Harald Nase and Stefan Wagenpfeil. We gratefully acknowledge support from Deutsche Forschungsgemeinschaft. For permission to use Tables 1.5, 3.12, 3.13 and 6.1 we are grateful to the Royal Statistical Society and Biometrika Trust.

We hope you will enjoy the book.

München and Berlin, 26.02.1994

Ludwig Fahrmeir
Gerhard Tutz
Wolfgang Hennevogl

Since the first printing of this book several very readable books and other important work have been published. We have included some remarks at the end of each chapter.

München and Berlin, 28.05.1997

Ludwig Fahrmeir
Gerhard Tutz

Contents

Preface   **v**

List of Examples   **xv**

List of Figures   **xix**

List of Tables   **xxiii**

1 Introduction   **1**
   1.1 Outline and examples   2
   1.2 Remarks on notation   13
   1.3 Further reading   13

2 Modelling and analysis of cross–sectional data:
a review of univariate generalized linear models   **15**
   2.1 Univariate generalized linear models   16
     2.1.1 Data   16
       Coding of covariates   16
       Grouped and ungrouped data   17
     2.1.2 Definition of univariate generalized linear models . .   18
     2.1.3 Models for continuous responses   22
       Normal distribution   22
       Gamma distribution   23
       Inverse Gaussian distribution   23

	2.1.4	Models for binary and binomial responses	24
		Linear probability model	25
		Probit model	25
		Logit model	26
		Complementary log–log model	26
		Binary models as threshold models of latent linear models	27
		Parameter interpretation	29
		Overdispersion	34
	2.1.5	Models for counted data	35
		Log–linear Poisson model	35
		Linear Poisson model	35
2.2	Likelihood inference		37
	2.2.1	Maximum likelihood estimation	37
		Log–likelihood, score function and information matrix	38
		Computation of the MLE by iterative methods	40
		Uniqueness and existence of MLE's*	41
		Asymptotic properties	42
		Discussion of regularity assumptions*	43
		Additional scale or overdispersion parameter	44
	2.2.2	Hypothesis testing and goodness–of–fit statistics	45
		Goodness–of–fit statistics	48
2.3	Some extensions		52
	2.3.1	Quasi–likelihood models	52
		Basic models	52
		Variance functions with unknown parameters	55
		Nonconstant dispersion parameter	55
	2.3.2	Bayes models	57
	2.3.3	Nonlinear and nonexponential family regression models*	60
2.4	Further developments		62

3 Models for multicategorical responses: multivariate extensions of generalized linear models   63

3.1	Multicategorical response models		64
	3.1.1	Multinomial distribution	64
	3.1.2	Data	65
	3.1.3	The multivariate model	66
	3.1.4	Multivariate generalized linear models	68
3.2	Models for nominal responses		70
	3.2.1	The principle of maximum random utility	70
	3.2.2	Modelling of explanatory variables: choice of design matrix	71
3.3	Models for ordinal responses		73
	3.3.1	Cumulative models: the threshold approach	75

Cumulative logistic model or proportional odds model   76
Grouped Cox model or proportional hazards model .   78
Extreme–maximal–value distribution model   79
3.3.2   Extended versions of cumulative models   79
3.3.3   Link functions and design matrices for cumulative
models   80
3.3.4   Sequential models   84
Generalized sequential models   87
Link functions of sequential models   90
3.3.5   Strict stochastic ordering*   90
3.3.6   Two–step models   91
Link function and design matrix for two–step models   94
3.3.7   Alternative approaches*   95
3.4   Statistical inference   96
3.4.1   Maximum likelihood estimation   97
Numerical computation   98
3.4.2   Testing and goodness–of–fit   99
Testing of linear hypotheses   99
Goodness–of–fit statistics   99
3.4.3   Power–divergence family* 101
Asymptotic properties under classical "fixed cells"
assumptions 103
Sparseness and "increasing–cells" asymptotics 103
3.5   Multivariate models for correlated responses 104
3.5.1   Conditional models 105
Asymetric models 105
Symmetric models 108
3.5.2   Marginal models 110
Statistical inference 113

4   Selecting and checking models   119
4.1   Variable selection 119
4.1.1   Selection criteria 120
4.1.2   Selection procedures 122
All–subsets selection 122
Stepwise backward and forward selection 122
4.2   Diagnostics 124
4.2.1   Diagnostic tools for the classical linear model 125
4.2.2   Generalized hat matrix 126
4.2.3   Residuals and goodness–of–fit statistics 130
4.2.4   Case deletion 138
4.3   General tests for misspecification* 140
4.3.1   Estimation under model misspecification 142
4.3.2   Hausman–type tests 144
Hausman tests 144

Information matrix test 145
4.3.3   Tests for non–nested hypotheses 146
Tests based on artificial nesting 146
Generalized Wald and score tests 147

**5 Semi– and nonparametric approaches to regression
analysis   151**
5.1 Smoothing techniques for continuous responses 152
5.1.1   Simple neighbourhood smoothers 152
5.1.2   Spline smoothing 153
Cubic smoothing splines 153
Regression splines 155
5.1.3   Kernel smoothing 156
Relation to other smoothers 158
Bias–variance trade–off 158
5.1.4   Selection of smoothing parameters* 160
5.2 Kernel smoothing with multicategorical response 162
5.2.1   Kernel methods for the estimation of discrete
distributions 162
5.2.2   Smoothed categorical regression 167
5.2.3   Choice of smoothing parameters* 172
5.3 Spline smoothing in generalized linear models 175
5.3.1   Cubic spline smoothing with a single covariate . . . 175
Fisher scoring for generalized spline smoothing* . . 176
Choice of smoothing parameter 177
5.3.2   Generalized additive models 180
Fisher scoring with backfitting* 181
5.4 Further developments 185

**6 Fixed parameter models for time series and
longitudinal data   187**
6.1 Time series 188
6.1.1   Conditional models 188
Generalized autoregressive models 188
Quasi–likelihood models and extensions 191
6.1.2   Statistical inference for conditional models 194
6.1.3   Marginal models 200
Estimation of marginal models 202
6.2 Longitudinal data 204
6.2.1   Conditional models 205
Generalized autoregressive models, quasi–likelihood
models 205
Statistical inference 206
Transition models 207
Subject–specific approaches and conditional likelihood 208

| | 6.2.2 | Marginal models | 211 |
| | | Statistical inference | 213 |

7 Random effects models   **219**

7.1	Linear random effects models for normal data	221
	7.1.1	Two–stage random effects models	221
		Random intercepts	222
		Random slopes	223
		Multi–level models	224
	7.1.2	Statistical inference	224
		Known variance–covariance components	225
		Unknown variance–covariance components	225
		Derivation of the EM–algorithm*	227
7.2	Random effects in generalized linear models	228
7.3	Estimation based on posterior modes	233
	7.3.1	Known variance–covariance components	234
	7.3.2	Unknown variance–covariance components	235
	7.3.3	Algorithmic details*	235
		Fisher scoring for given variance–covariance components	235
		EM–type algorithm	237
7.4	Estimation by integration techniques	238
	7.4.1	Maximum likelihood estimation of fixed parameters	238
	7.4.2	Posterior mean estimation of random effects	240
	7.4.3	Algorithmic details*	241
		Direct maximization	241
		Indirect maximization	243
		Posterior mean estimation	247
7.5	Examples	249
7.6	Marginal estimation approach to random effects models	252
7.7	Further approaches	254

8 State space models   **257**

8.1	Linear state space models and the Kalman filter	258
	8.1.1	Linear state space models	258
	8.1.2	Statistical inference	263
		Linear Kalman filtering and smoothing	263
		Kalman filtering and smoothing as posterior mode estimation*	265
		Unknown hyperparameters	267
		EM–algorithm for estimating hyperparameters*	268
8.2	Non–normal and nonlinear state space models	269
	8.2.1	Dynamic generalized linear models	270
		Categorical time series	271
	8.2.2	Nonlinear and nonexponential family models*	274

8.3   Non–normal filtering and smoothing   275
    8.3.1   Posterior mode estimation   276
        Generalized extended Kalman filter and smoother* .   277
        Gauss–Newton and Fisher–scoring filtering and
            smoothing*   279
        Estimation of hyperparameters*   281
        Some applications   281
    8.3.2   Posterior mean estimation   286
        A Gibbs sampling approach*   287
        Integration–based approaches*   290
8.4   Longitudinal data   293
    8.4.1   State space modelling of longitudinal data   293
    8.4.2   Filtering and smoothing   295
        Generalized Kalman filter and smoother for
            longitudinal data*   296
8.5   Further developments   303

9   Survival models   **305**
9.1   Models for continuous time   305
    9.1.1   Basic models   305
        Exponential distribution   306
        Weibull distribution   307
    9.1.2   Parametric regression models   307
        Location–scale models for log T   308
        Proportional hazards models   308
        Linear transformation models and binary regression
            models   309
    9.1.3   Censoring   310
        Random censoring   310
        Type I censoring   311
    9.1.4   Estimation   312
        Exponential model   312
        Weibull model   313
9.2   Models for discrete time   314
    9.2.1   Life table estimates   315
    9.2.2   Parametric regression models   318
        The grouped proportional hazards model   318
        A generalized version: the model of Aranda–Ordaz .   320
        The logistic model   321
        Sequential model and parameterization of the
            baseline hazard   321
    9.2.3   Maximum likelihood estimation   322
    9.2.4   Time–varying covariates   325
        Internal covariates*   328
        Maximum likelihood estimation*   329

9.3   Discrete models for multiple modes of failure 331
      9.3.1 Basic models 331
      9.3.2 Maximum likelihood estimation 333
9.4   Smoothing in discrete survival analysis 337
      9.4.1 Dynamic discrete time survival models 337
             Posterior mode smoothing 338
      9.4.2 Kernel smoothing 340
      9.4.3 Further developments 344

Appendix A   **345**
A.1   Exponential families and generalized linear models 345
A.2   Basic ideas for asymptotics 350
A.3   EM–algorithm 355
A.4   Numerical integration 357
A.5   Monte Carlo methods 363

Appendix B Software for fitting generalized linear models   **367**

References   **379**

Author Index   **417**

Subject Index   **423**

3.3	Job expectation	74
3.4	Breathing test results	82
3.5	Job expectation	83
3.6	Tonsil size	85
3.7	Tonsil size	88
3.8	Breathing test results	89
3.9	Rheumatoid arthritis	91
3.10	Rheumatoid arthritis	93
3.11	Caesarean birth study	100
3.12	Reported happiness	107
3.13	Visual impairment	116
4.1	Credit–scoring	124
4.2	Vaso–constriction	127
4.3	Job expectation	130
4.4	Vaso–constriction	133
4.5	Job expectation	134
4.6	Vaso–constriction	140
4.7	Job expectation	140
4.8	Credit–scoring	148
5.1	Motorcycle data	159
5.2	Memory	165
5.3	Vaso–constriction data	168
5.4	Unemployment data	170
5.5	Rainfall data	177
5.6	Vaso–constriction data	183
6.1	Polio incidence in USA	197
6.2	Polio incidence in USA	203
6.3	IFO business test	209
6.4	Ohio children	215
7.1	Ohio children data	228
7.2	Bitterness of white wines	228
7.3	Ohio children data	249
7.4	Bitterness of white wines	250
8.1	Rainfall data	281
8.2	Advertising data	283
8.3	Phone calls	284

8.4   Rainfall data 292

8.5   Business test 298

9.1   Duration of unemployment 317

9.2   Duration of unemployment 325

9.3   Duration of unemployment 335

9.4   Head and neck cancer 341

List of Figures

1.1   Number of occurrences of rainfall in the Tokyo area for each
      calendar day during 1983–1984   9
1.2   Monthly number of polio cases in USA from 1970 to 1983 .   10

2.1   The gamma distribution: $G(\mu = 1, \nu)$ 24
2.2   Response functions for binary responses   26
2.3   Response functions for binary responses adjusted to the lo-
      gistic function (that means linear transformation yielding
      mean zero and variance $\pi^2/3$).   28
2.4   Log–likelihood (—) and quadratic approximation (- - -) for
      Wald test and slope for score test   47

3.1   Densities of the latent response for two subpopulations with
      different values of x (logistic, extreme–minimal–value, ex-
      treme–maximal–value distributions)   77

4.1   Index plot of h_{ii} for vaso–constriction data   129
4.2   Index plot of $\text{tr}(H_{ii})$ and $\det(H_{ii})$ for grouped job expecta-
      tion data   131
4.3   Index plot of $r_i^P, r_{i,s}^P$ and r_i^D for vaso–constriction data . . .   134
4.4 $N(0, 1)$–probability plot of $r_{i,s}^P$ for vaso–constriction data . .   135

4.5   Relative frequencies and response curve of the fitted cumu-
      lative logistic model (responses are "don't expect adequate
      employment," "not sure," "expect employment immediately
      after getting the degree") 136
4.6   Index plot of $(r_{i,s}^P)'(r_{i,s}^P)$ for grouped job expectation data . 137
4.7 $\chi^2(2)$–probability plot of $(r_{i,s}^P)'(r_{i,s}^P)$ for grouped job expec-
      tation data 137
4.8   Index plot of $c_{i,1}$ and c_i for vaso–constriction data 139
4.9   Index plot of $c_{i,1}$ and c_i for grouped job expectation data . 141

5.1   Smoothed estimates for motorcycle data showing time $(x -
      axis)$ and head acceleration $(y - axis)$ after a simulated im-
      pact. Smoothers are running lines (top left), running medi-
      ans (top right), Epanechnikov kernel (bottom left) and cubic
      splines (bottom right) 159
5.2   Memory of stressful event for nominal and ordinal kernel
      (relative frequencies given as points) 166
5.3   Kernel–based estimate of the nonoccurrence of vaso–
      constriction in the skin 169
5.4   Response surface for the nonoccurrence of vaso–constriction
      based on a fitted logit model 169
5.5   Categorical kernel regression (—) and sequential logit model
      (- - -) for unemployment data 171
5.6   Smoothed probability of rainfall $\lambda = 4064$ 178
5.7   Smoothed probability of rainfall $\lambda = 32$ 179
5.8   Generalized cross–validation criterion, with logarithmic scale
      for λ 179
5.9   Nonoccurrence of vaso–constriction of the skin smoothed by
      an additive model with $\lambda_1 = \lambda_2 = 0.001$ 184
5.10 Nonoccurrence of vaso–constriction of the skin smoothed by
      an additive model with $\lambda_1 = \lambda_2 = 0.003$ 184

6.1   Monthly number of polio cases in USA from 1970 to 1983 . 198
6.2   Predicted polio incidence $\hat{\mu}_t$ based on a log–linear AR(1=5)–
      model fit 199

8.1   Number of occurrences of rainfall in the Tokyo area for each
      calendar day during 1983–1984 281
8.2   Smoothed probabilites $\hat{\pi}_t$ of daily rainfall, obtained by gen-
      eralized Kalman (- - - -) and Gauss–Newton smoothing (—) 282
8.3   Smoothed trend (lower line) and advertising effect 283
8.4   Advertising data and fitted values (—) 284
8.5   Number of phone calls for half–hour intervals 285
8.6   Observed and fitted (———) values 286

8.7 Smoothed probabilities $\hat{\pi}_t$ of daily rainfall, obtained by posterior mean (———) and posterior mode (- - - -) smoothing   292
8.8 Covariate effects for model 1.   299
8.9 Time–varying effect of increasing demand for models 1 (———) and 2 (- - - -)   300
8.10 Trend parameters for models 1 (———) and 2 (- - - -) . . .   301
8.11 Seasonal effect of threshold 1, model 2   302
8.12 Seasonal effect of threshold 2, model 2   302

9.1 Life table estimate for unemployment data   318
9.2 Estimated survivor function for unemployment data   319
9.3 Life table estimate for duration of unemployment with causes, full–time job or part–time job   335
9.4 Cubic–linear spline fit for head and neck cancer data (- - -) and posterior mode smoother (—) with ± standard deviation confidence bands   343
9.5 Smoothed kernel estimate for head and neck cancer data . .   343

List of Tables

1.1 Data on infection from 251 births by Caesarean section . .   3
1.2 Cellular differentiation data   4
1.3 Grouped data for job expectation of students of psychology
     in Regensburg   5
1.4 Breathing results of Houston industrial workers   6
1.5 Visual impairment data   7
1.6 Presence and absence of respiratory infection   11

2.1 Simple exponential families with dispersion parameter . . .   21
2.2 Grouped data on infection   30
2.3 Cellular differentiation data   36
2.4 Logit model fit to Caesarean birth study data   48
2.5 Logit model fit to credit–scoring data   50
2.6 Log–linear model fits to cellular differentiation data based
     on Poisson–likelihoods   51
2.7 Log–linear model fits to cellular differentiation data based
     on quasi–likelihoods   56

3.1 Breathing results of Houston industrial workers   74
3.2 Grouped data for job expectation of students of psychology
     in Regensburg   75
3.3 Estimates of cumulative models for breathing test data
     (p-values in each second column)   82
3.4 Cumulative model for job expectation data   84

3.5   Tonsil size and Streptococuss pyogenes   85
3.6   Fits for tonsil size data   88
3.7   Sequential logit models for the breathing test data (p–values
       in brackets)   89
3.8   Clinical trial of a new agent and an active control   92
3.9   Analysis of clinical trial data on rheumatoid arthritis   93
3.10   Cross classification of sex, reported happiness and years of
       schooling   107
3.11   Estimates for the cross classification of sex, reported happi-
       ness and years of schooling   108
3.12   Visual impairment data (from Liang et al., 1992)   116
3.13   Estimation results for visual impairment data   117

4.1   Logit model fit to credit–scoring data   124
4.2   Vaso–constriction data   128
4.3   Logit model fit to vaso–constriction data   129

5.1   Discrete ordinal kernel functions   164
5.2   Unemployment in categories $y = 1$ (1–5 months), $y = 2$
       (7–12 months), $y = 3$ (> 12 months)   172
5.3   Loss functions and empirical loss for observation (y, x) . . .   174

6.1   Monthly number of poliomyelitis cases in USA for 1970 to
       1983   198
6.2   Log–linear AR(5)–model fit to polio data   199
6.3   Log–linear AR(1)–model fits to polio data   200
6.4   Marginal model fit for polio data   204
6.5   Variables and questions of the IFO business test   209
6.6   Estimates of main effects   210
6.7   Presence and absence of respiratory infection   216
6.8   Marginal logit model fits for Ohio children data   217
6.9   Main effects model fits for Ohio children data   217

7.1   Bitterness of wine data (Randall, 1989)   228
7.2   Random intercept logit model for Ohio children data (effect
       coding of smoking and age, standard deviations in parentheses)250
7.3   Estimation results of bitterness of wine data   251

8.1   Contingency table for 55 firms of branch "Steine und Erden"   298

9.1   Duration of unemployment   326
9.2   Estimates of cause–specific logistic model for duration of un-
       employment data   336
9.3   Head and neck cancer (Efron, 1988)   342

1
Introduction

Classical statistical models for regression, time series and longitudinal data analysis are generally useful in situations where data are approximately Gaussian and can be explained by some linear structure. These models are easy to interpret and the methods are theoretically well understood and investigated. However, the underlying assumptions may be too stringent and applications of the methods may be misleading in situations where data are clearly non–normal, such as categorical or counted data. Statistical modelling aims at providing more flexible model–based tools for data analysis.

Generalized linear models have been introduced by Nelder and Wedderburn (1972) as a unifying family of models for nonstandard cross–sectional regression analysis with non–normal responses. Their further development had a major influence on statistical modelling in a wider sense, and they have been extended in various ways and for more general situations. This book brings together and reviews a large part of recent advances in statistical modelling that are based on or related to generalized linear models. Throughout the text the focus is on discrete data. The topics include models for ordered multicategorical responses, multivariate correlated responses in cross–sectional or repeated measurements situations, nonparametric approaches, random effects, autoregressive–type and dynamic or state space extensions for non–normal time series and longitudinal data, and discrete time survival data.

To make the book accessible for a broader readership it is written at an intermediate mathematical level. The emphasis is on basic ideas and on explanation of methods from the viewpoint of an applied statistician.

It is helpful to be familiar with linear regression models. Although some concepts like that of design matrices and coding of qualitative variables are introduced, they are not described at length. Of course, knowledge of matrix calculus and basic theory of estimation and testing are necessary prerequisites. Extension of linear regression for contionuous responses to nonparametric approaches, random coefficient models and state spacing models are reviewed in compact form at the beginning of corresponding chapters of this book. For a more detailed and deeper study of methods for continuous data, the reader should consult the literature cited there. Mathematical and technical details of the theory are kept at a comparably simple level. For detailed and rigorous proofs of some results, for example asymptotics, the reader is referred to the literature. Moreover, some extensions and a number of mathematical and algorithmic details are deferred to starred sections. Knowledge of these sections is not neccessary to understand the main body of this text. Real data examples and applications from various fields such as economics, social science and medicine illustrate the text. They should encourage statisticians, students and researchers working in these fields to use the methods for their own problems, and they should stimulate mathematically interested readers for a deeper study of theoretical foundations and properties.

1.1   Outline and examples

The following survey of contents is illustrated by some of the examples that will be used in later chapters.

Modelling and analysis of univariate cross–sectional data (Chapter 2)

Chapter 2 gives a review of univariate generalized linear models and some basic extensions such as quasi–likelihood models. These models are useful for cross–sectional parametric regression analysis with non–normal response variables as in Examples 1.1, 1.2 and 1.3 where responses are binary or counts.

Example 1.1: Caesarean birth study

Table 1.1, kindly provided by Prof. R. E. Weissenbacher, Munich, Klinikum Großhadern, shows data on infection from births by Caesarean section. The response variable of interest is the occurrence or nonoccurrence of infection, with two types (I,II) of infection. Three dichotomous covariates, which may influence the risk of infection, were considered: Was the Caesarean section planned or not? Have risk-factors, such as diabetes, overweight and others, been present? Were antibiotics given as a prophylaxis? The aim is to analyze effects of the covariates on the risk of infection, especially whether antibiotics can decrease the risk of infection. If one ignores the two types

TABLE 1.1. Data on infection from 251 births by Caesarean section

	Caesarean planned			Not planned		
	Infection			Infection		
	I	II	non	I	II	non
Antibiotics						
Risk-factors	0	1	17	4	7	87
No risk-factors	0	0	2	0	0	0
No antibiotics						
Risk-factors	11	17	30	10	13	3
No risk-factors	4	4	32	0	0	9

of infection the response variable is binary (infection yes/no); otherwise it is three-categorial.   □

Example 1.2: Credit–Scoring

In credit business, banks are interested in information whether prospective consumers will pay back their credit as agreed upon or not. The aim of credit-scoring is to model or predict the probability that a consumer with certain covariates ("risk factors") is to be considered as a potential risk. The data set, which will be analyzed later, consists of 1000 consumer's credits from a German bank. For each consumer the binary variable y "creditability" ($y = 0$ for credit–worthy, $y = 1$ for not credit–worthy) is available. In addition, 20 covariates that are assumed to influence creditability were collected, for example,

- running account, with categories no, medium and good,

- duration of credit in months,

- amount of credit in "Deutsche Mark,"

- payment of previous credits, with categories good and bad,

- intended use, with categories private and professional.

The original data set is reproduced in Fahrmeir and Hamerle (1984, Appendix C) and is available on electronic file from the authors. (Some information on the distribution of covariates is also given in Example 2.2).

We will analyse the effect of covariates on the variable "creditability" by a binary regression model. Other tools used for credit-scoring are discriminance analysis, classification trees and neural networks.   □

TABLE 1.2. Cellular differentiation data

Number y of cells differentiating	Dose of TNF(U/ml)	Dose of IFN(U/ml)
11	0	0
18	0	4
20	0	20
39	0	100
22	1	0
38	1	4
52	1	20
69	1	100
31	10	0
68	10	4
69	10	20
128	10	100
102	100	0
171	100	4
180	100	20
193	100	100

Example 1.3: Cellular differentiation

The effect of two agents of immuno–activating ability that may induce cell differentiation was investigated by Piegorsch, Weinberg and Margolin (1988). As response variable the number of cells that exhibited markers after exposure was recorded. It is of interest if the agents TNF (tumor necrosis factor) and IFN (interferon) stimulate cell differentiation independently or if there is a synergetic effect. The count data are given in Table 1.2.   □

Generalized linear models extend classical linear regression for approximately normal responses to regression for non–normal responses, including binary responses or counts as in the preceding examples. In their original version, generalized linear models assume that the mean $E(y|x)$ of the response given the covariates is related to a linear predictor $\eta = \beta_0 + \beta_1 x_1 + \ldots + \beta_p x_p = \beta_0 + x'\beta$ by a response or link function h in the form

$$E(y|x) = h(\beta_0 + x'\beta). \qquad (1.1.1)$$

Due to the distributional assumptions the variance function $\text{var}(y|x)$ is then determined by choice of the specific distribution of y given x. Quasi–likelihood models and nonlinear or nonexponential family models allow to weaken these assumptions within a parametric likelihood–based framework. Most of the material of this chapter is fundamental for subsequent chapters.

TABLE 1.3. Grouped data for job expectation of students of psychology in Regensburg

Observation number	Age in years	Response categories 1	2	3	n_i
1	19	1	2	0	3
2	20	5	18	2	25
3	21	6	19	2	27
4	22	1	6	3	10
5	23	2	7	3	12
6	24	1	7	5	13
7	25	0	0	3	3
8	26	0	1	0	1
9	27	0	2	1	3
10	29	1	0	0	1
11	30	0	0	2	2
12	31	0	1	0	1
13	34	0	1	0	1

Models for multicategorical data (Chapter 3)

Model (1.1.1) is appropriate for unidimensional responses, like counts or binary 0-1-variables. However, in the Caesarean section study (Example 1.1) one may distinguish between different kinds of infection, getting a *multicategorical* response with categories "no infection," "infection type I" and "infection type II." These response categories are not strictly ordered and therefore may be treated as unordered. In many cases the response of interest is given in ordered categories as in the following examples.

Example 1.4: Job expectation

In a study on the perspectives of students, psychology students at the University of Regensburg have been asked if they expect to find an adequate employment after getting their degree. The response categories were ordered with respect to their expectation. The responses were "don't expect adequate employment" (category 1), "not sure" (category 2) and "immediately after the degree" (category 3). The data given in Table 1.3 show the response and the covariate age.   □

Example 1.5: Breathing test results

Forthofer and Lehnen (1981) considered the effect of age and smoking upon breathing test results for workers in industrial plants in Texas. The test results are given on an ordered scale with categories "normal," "borderline" and "abnormal." It is of interest how age and smoking status are connected to breathing test results. Table 1.4 contains the data.   □

TABLE 1.4. Breathing results of Houston industrial workers

		Breathing test results		
Age	Smoking status	Normal	Borderline	Abnormal
< 40	Never smoked	577	27	7
	Former smoker	192	20	3
	Current smoker	682	46	11
40–59	Never smoked	164	4	0
	Former smoker	145	15	7
	Current smoker	245	47	27

Multicategorial responses cannot be considered as unidimensional. If the categories are labeled with numbers like $1, \ldots, k$ the response looks like a univariate variable. But one has to keep in mind that the numbers are mere labels, especially if the categories are not ordered. Therefore one has to consider a separate variable for each category in a multivariate modelling framework. Chapter 3 deals with multivariate approaches where the response variable is a vector $y = (y_1, \ldots, y_q)$. Instead of one predictor $\beta_0 + x'\beta$ one has q predictors $\beta_{01} + x'\beta_1, \ldots, \beta_{0q} + x'\beta_q$, that determine the response y_j in the form

$$E(y_j|x) = h_j(\beta_{01} + x'\beta_1, \ldots, \beta_{0q} + x'\beta_q)$$

where h_j is a link function for the jth component of y. In analogy to the unidimensional model (1.1.1) one has

$$E(y|x) = h(\beta_{01} + x'\beta_1, \ldots, \beta_{0q} + x'\beta_q)$$

with $h = (h_1, \ldots, h_q) : \mathbb{R}^q \to \mathbb{R}^q$ denoting the multidimensional link function.

This extension is particularly helpful in modelling multicategorical responses based on the multinomial distribution. Then y_j represents observations in category j. For unordered response categories the q predictors $\beta_{01} + x'\beta_1, \ldots, \beta_{0q} + x'\beta_q$ usually are necessary. However, when the response is in ordered categories the use of the order information yields simpler models where the number of parameters may be reduced by assuming $\beta_1 = \ldots = \beta_q$. A large part of this chapter considers ordinal response variables like the test results in breathing tests (Example 1.5) or expectations of students (Example 1.4).

In the preceding examples, we still considered the situation of *one* response variable, though multicategorial. The truly multivariate case arises if a vector of correlated responses is observed for each unit as in the following example.

TABLE 1.5. Visual impairment data
(reproduced from Liang, Zeger and Qaqish, 1992)

Visual impairment	White 40–50	White 51–60	White 61–70	White 70+	Black 40–50	Black 51–60	Black 61–70	Black 70+	Total
Left eye									
Yes	15	24	42	139	29	38	50	85	422
No	617	557	789	673	750	574	473	344	4777
Right eye									
Yes	19	25	48	146	31	37	49	93	448
No	613	556	783	666	748	575	474	226	4751

Example 1.6: Visual impairment
In a visual impairment study (Liang, Zeger and Qaqish, 1992) binary responses y_1 and y_2 for both eyes of each individual of a sample population were recorded, indicating whether or not an eye was visually impaired. Covariates include age in years, race (white or black) and education in years. The main objective is to analyze the influence of race and age on visual impairment, controlling for education. The response variable "visual impairment" is bivariate, with two correlated binary components y_1, y_2, and an appropriate analysis has to take care of this. □

Similar situations occur, e.g., in twin and family studies or in dentistry. In other applications, the response vector consists of different variables, e.g., different questions in an interview. Section 3.5 surveys recent developments in regression analysis for correlated mainly binary responses.

Selecting and checking models (Chapter 4)

This chapter discusses some topics concerning the specification of models. The first topic is the selection of variables, which is important when one has 20 variables like in the credit–scoring example. Procedures and selection criteria that help to reduce the number of explanatory variables are considered. In Section 4.2 diagnostic tools are given within the framework of multivariate generalized linear models. These tools help to identify observations that are influential or determine the goodness–of–fit of the model in an extreme way. Residual analysis may show if the lack of fit is due to some specific observations (outliers) or if the model is inappropriate for most of the observations. In Section 4.3 an approach is outlined that does not investigate single observations but is based on tests that should reflect the appropriateness of a model. We do not consider tests for specific deviations, e.g., for the correct link function, but general specification tests that may be used in principle in any maximum likelihood type regression problem.

Semi– and nonparametric approaches (Chapter 5)

Chapter 5 gives a brief introduction to semi- and nonparametric approaches. Nonparametric smoothing techniques are considered when one may keep faith with the data by low smoothing or one may produce a very smooth estimate showing low fluctuations. The principle is not to use a parametric model but to let the data decide the functional form with a minimum of restrictions. Smoothing techniques are very flexible and allow one to look at data with varying degrees of smoothing.

Example 1.7: Rainfall data

Figure 1.1 displays the number of occurrences of rainfall in the Tokyo area for each calendar day during the years 1983–1984 (Kitagawa, 1987). This is an example of a discrete time series. To compare it to similar data of other areas, or of other years, and to see some seasonal yearly pattern, it will be useful to estimate a smooth curve, representing the probability of rainfall on each calendar day.   □

Simple regression smoothers like kernel smoothing are considered in Section 5.1 and extended to the case of categorial responses in Section 5.2. Alternative extensions of generalized linear models leading to additive models are based on the following approach: Instead of specifying the linear predictor $x'\beta$, covariates may enter the model by a smooth function $f(x)$ yielding

$$E(y|x) = h\big(f(x)\big)$$

with known response function. By splitting up the influence of covariates one gets generalized additive models that have the form

$$E(y|x) = h\big(f_1(x_1) + \ldots + f_p(x_p)\big),$$

where each component enters the model by a smooth function. Techniques of this type will give a picture for rainfall data that shows the pattern more clearly and is much more comfortable to look at.

Fixed parameter models for time series and longitudinal data (Chapter 6)

Chapter 6 extends the modelling approach of Chapter 2 and Chapter 3, which are mainly appropriate for (conditionally) independent observations $(y_i, x_i), i = 1, \ldots, n$ in a cross section, to time series data $(y_t, x_t), t = 1, \ldots, T$, as in Example 1.8, and to longitudinal data, where many time series are observed simultaneously as in Examples 1.9 and 1.10.

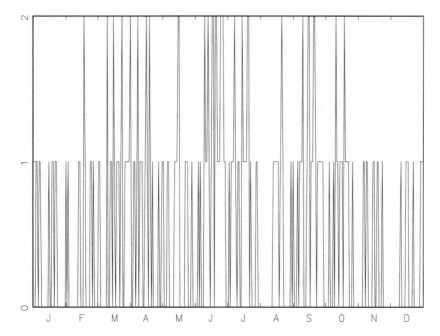

FIGURE 1.1. Number of occurrences of rainfall in the Tokyo area for each calendar day during 1983–1984

Example 1.8: Polio incidence
Figure 1.2 displays the monthly number of Polio cases in USA from 1970 to 1983 (Zeger, 1988a). This is a time series of count data, and analyzing it by traditional methods of time series analysis may give false conclusions. Looking at the plot in Figure 1.2, there seems to be some seasonality and, perhaps, a decrease in the rate of polio infection. Therefore one might try to model the data with a log–linear Poisson model with trend and seasonal terms as covariates. However, in the time series situation one has to take into account dependence among observations. This can be accomplished by introducing past observations as additional covariates, as in common autoregressive models. Another possibility is to use a marginal modelling approach. □

Example 1.9: IFO business test
The IFO institute for economic research in Munich collects categorical monthly data of firms in various industrial branches. The monthly questionnaire contains questions on the tendency of realizations and expectations of variables like production, orders in hand and demand. Answers are categorical, most of them trichotomous with categories like increase, decrease, or no change. Thus for each firm the data form a categorical time series. Considering all firms within a certain branch we have categorical panel or

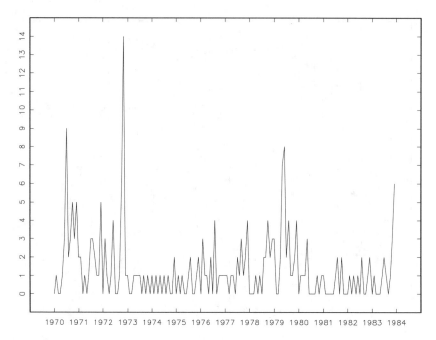

FIGURE 1.2. Monthly number of polio cases in USA from 1970 to 1983

longitudinal data. Based on such data, one may, for example, analyze the
dependency of current production plans on demand and orders at hand. □

Example 1.10: Ohio children
Within the Harvard Study of Air Pollution and Health 537 children (Laird,
Beck and Ware, 1984) were examined annually from age 7 to age 10 on the
presence or absence of respiratory infection. Thus we have four repeated
measurements on one child or, in contrast to Example 1.9, many "short"
binary time series. The only available covariate is mother's smoking status
(regular smoker, nonregular smoker) at the beginning of the study. One
of the primary goals is to analyze the influence of mother's smoking on
children's respiratory disease. Responses of one child, however, are likely to
be correlated. This dependence should be taken into account in an adequate
way. The data are given in Table 1.6 where "1" stand for infection and "0"
for no infection.   □

We distinguish between conditional or autoregressive–type models and
marginal models. In conditional models, the independence assumption is
dropped by including past observations $y_{t-1}, y_{t-2}, \ldots$ as additional covariates in generalized linear or quasi–likelihood models. This kind of modelling is quite analogous to autoregressive models for Gaussian time series,
that means conditional densities $f(y_t|y_{t-1}, y_{t-2}, \ldots, x_t)$ are parametrically

TABLE 1.6. Presence and absence of respiratory infection

Mother did not smoke					Mother smoked				
Age of child				Frequency	Age of child				Frequency
7	8	9	10		7	8	9	10	
0	0	0	0	237	0	0	0	0	118
0	0	0	1	10	0	0	0	1	6
0	0	1	0	15	0	0	1	0	8
0	0	1	1	4	0	0	1	1	2
0	1	0	0	16	0	1	0	0	11
0	1	0	1	2	0	1	0	1	1
0	1	1	0	7	0	1	1	0	6
0	1	1	1	3	0	1	1	1	4
1	0	0	0	24	1	0	0	0	7
1	0	0	1	3	1	0	0	1	3
1	0	1	0	3	1	0	1	0	3
1	0	1	1	2	1	0	1	1	1
1	1	0	0	6	1	1	0	0	4
1	1	0	1	2	1	1	0	1	2
1	1	1	0	5	1	1	1	0	4
1	1	1	1	11	1	1	1	1	7

modelled. In certain applications, e.g., as in Example 1.10, the marginal effect of covariates x_t on y_t is of primary interest, whereas correlation of observations is regarded as a nuisance or only of secondary importance. Then it is more reasonable to base inference on marginal models for $f(y_t|x_t)$, but taking into account correlation as in the work of Liang and Zeger (1986 and later).

Random effects models (Chapter 7)

This chapter deals with random effects models for non–normal data. Such data appear in cross–sectional and longitudinal studies, when one has repeated measurements $y_{it}, t = 1, \ldots T_i$, from individual i. As an example consider the Ohio children data (Example 1.10). The presence or absence of respiratory infection is stated at four ages. Since children may respond differently to the smoking behaviour of their mother one may have so–called unobserved heterogeneity. Random effects models take into account unobserved heterogeneity which arises from covariate effects varying from one individual to another or is due to omitted covariates. In analogy to linear random effects models for normal data the models are defined as two–stage models or, equivalently, mixed models. In the first stage it is assumed that each of the repeated measurements $y_{it}, t = 1, ..., T_i$, on individual i follows a (possibly autoregressive) generalized linear model with individual–specific unknown effects b_i. In the second stage between–individual variation is introduced by assuming individual effects to be i.i.d. among individuals. Estimation of parameters and random effects, however, is more complicated than in linear random effects models. Several approaches based on posterior modes and means are discussed and applied.

State space models (Chapter 8)

Chapter 8 surveys state space or dynamic approaches for analyzing non–normal time series or longitudinal data. State space models relate time series observations y_t to unobserved "states," like trend and seasonal components or time–varying effects of covariates. Estimation of "states" (filtering and smoothing) is a primary goal of inference. For approximately normal data, linear state space models and the famous Kalman filter provide a unifying framework to analyze structural time series models (see, e.g., Harvey, 1989). Work on a dynamic generalized model began much more recently. While formulation of non–Gaussian models is in analogy to the linear Gaussian case, the filtering and smoothing problem becomes harder. Posterior mode and mean approaches are described in this chapter and the methods are applied to smoothing the rainfall data of Example 1.7 as well as to the IFO business test data (Example 1.9).

Survival models (Chapter 9)

The objective here is to identify the factors that determine survival or transition. The case of continuous time is very briefly considered; only models that may be estimated in a way similar to generalized linear models are sketched. The focus in this chapter is on models for discrete time, which is a case often found in practice when only weeks or months of survival are recorded (compare the following example). Some of the parametric approaches are strongly related to the models considered in ordinal modelling (Chapter 3). In addition to the parametric models nonparametric smoothing techniques are briefly introduced.

Example 1.11: Duration of unemployment

The data set comprises 1669 unemployed persons who are observed over several years in the socio–economic panel in Germany (Hanefeld, 1987). Time is measured in months. The focus of interest is how explanatory variables like sex, education level or nationality influence the transition from unemployment to employment. Since the follow–up of persons over time is incomplete one has to deal with the problem of censored observations. □

1.2   Remarks on notation

When writing vectors within the text most often the transposed sign $"'"$ will be omitted. A combination of vectors $x_1, \ldots x_n$ reads simply $x = (x_1, \ldots, x_n)$ instead of $x' = (x_1', \ldots, x_n')$. When considering square roots of matrices we use the T as transposed sign. So $A^{1/2}(A^{T/2})$ denotes the left (the corresponding right) square root of matrix A such that $A^{1/2}A^{T/2} = A$ holds. The inverse matrices are denoted by $A^{-1/2} = (A^{1/2})^{-1}$ and $A^{-T/2} = (A^{T/2})^{-1}$.

1.3   Further reading

Several topics that are strongly related to generalized linear models but are treated extensively in books are only sketched here or simply omitted. McCullagh and Nelder (1989) is a standard source of information about generalized linear models. In particular the univariate case is considered very extensively.

Since the focus here is on regression modelling contingency table analysis is totally omitted. Agresti (1984, 1990) considers very thoroughly among other topics models for contingency tables with unordered or ordered categories. Sources for log–linear models are also Christensen (1990) and Bishop, Fienberg and Holland (1975). Whittaker (1990) is a book to consult when one is interested in graphical models. Santner and Duffy (1989) consider cross–classified data and univariate discrete data. A survey of exact inferences for discrete data is given by Agresti (1992).

Non– and semiparametric smoothing approaches are treated extensively in Hastie and Tibshirani (1990). Härdle (1990 a,b) demonstrates the developments in applied nonparametric regression with continuous response. Interesting monographs that have been published recently or are about to be published deal extensively with longitudinal data and repeated measurements (Jones, 1993; Lindsey, 1993; Diggle, Liang and Zeger, 1994).

Further sources of information in the area of generalized linear models are the Proceedings of GLIM and the International Workshop on Statistical Modelling (Gilchrist, Francis and Whittaker, 1985; Decarli, Francis, Gilchrist and Seeber, 1989; Van der Heijden, Jansen, Francis and Seeber, 1992; Fahrmeir, Francis, Gilchrist and Tutz, 1992).

2
Modelling and analysis of cross–sectional data: a review of univariate generalized linear models

The material in this chapter provides an introduction to univariate generalized linear models and serves as a basis for the following chapters, which contain extensions, e.g., to multivariate, nonparametric, random effects or dynamic models. It is not intended to replace a deeper study of detailed expositions like in McCullagh and Nelder (1989) or, with focus on the GLIM package, in Aitkin, Anderson, Francis and Hinde (1989). Shorter introductions are given, e.g., by Dobson (1989), Firth (1991) and Fahrmeir and Kredler (1984). Collett (1991) focuses on modelling of binary data and discusses practical aspects in more detail.

After an introductory section on the type of data that will be considered, Section 2.1 gives the general definition of generalized linear models and describes some important members. Binary and Poisson regression models are of particular relevance for the applications in this chapter as well as later. Binary regression models, such as the logit model, are appropriate for analyzing the relationship of a binary response with explanatory variables. In Example 1.1, the response is "infection" or "no infection"; in Example 1.2 the response is "creditworthy" or "not creditworthy." Poisson regression models are useful if the response is a count like the number of cells differentiating as in Example 1.3.

The common tools for estimation, testing and simple goodness–of–fit criteria are contained in Section 2.2. Section 2.3 presents extensions to quasi–likelihood models for univariate responses and shortly considers Bayes approaches and nonlinear and nonexponential family models.

2.1 Univariate generalized linear models

2.1.1 Data

Consider the common situation of cross–sectional regression analysis, where a univariate variable of primary interest, the response or dependent variable y, has to be explained by a vector $x = (x_1, \ldots, x_m)$ of covariates or explanatory variables. The data

$$(y_i, x_i), \qquad i = 1, \ldots, n \qquad (2.1.1)$$

consist in observations on (y, x) for each unit or individual i of a cross section of size n. Responses can be continuous real variables as common in classical linear models, nonnegative, e.g., duration or income, counts as in Example 1.3, or binary as in Examples 1.1 and 1.2. Covariates may be quantitative (metrical), such as duration of credit in months, amount of credit in Example 1.2, dose of TNF or IFN in Example 1.3, or qualitative (ordered or unordered categorical), such as the dichotomous covariates "Caesarian birth" (planned or not), "presence of risk factors" (yes/no), "antibiotics" (given or not) in Example 1.1 or "running account" (categories no, medium, good) in Example 1.2. Covariates may be deterministic, i.e., known values or experimental conditions, or they may be stochastic , i.e., observations of a random vector x.

Coding of covariates

As in common linear regression qualitative covariates have to be coded appropriately. A categorical covariate (or factor) x with k possible categories $1, \ldots, k$ will generally be coded by a "dummy vector" with $q = k - 1$ components $x^{(1)}, \ldots, x^{(q)}$. If $0 - 1$ dummies are used, which is shortly referred to as "dummy coding", then $x^{(j)}$ is defined by

$$x^{(j)} = \begin{cases} 1, & \text{if category } j \text{ is observed} \\ 0, & \text{else} \end{cases} \qquad j = 1, \ldots, q.$$

If the kth category, the "reference category," is observed, then x is the zero vector.

An alternative coding scheme, which is referred to as "effect coding", is defined by

$$x^{(j)} = \begin{cases} 1, & \text{if category } j \text{ is observed,} \\ -1, & \text{if category } k \text{ is observed,} \\ 0, & \text{else} \end{cases} \qquad j = 1, \ldots, q.$$

In the case of effect coding, the reference category k is given by the vector $(-1, \ldots, -1)$ instead of the zero vector. Other types of coding may sometimes be more useful for the purpose of interpretation, but will not be considered in this text.

Grouped and ungrouped data

In (2.1.1), it was implicitly assumed that the data are *ungrouped*, i.e., (y_i, x_i) is the original observation on unit i of the cross section. In this case, each covariate vector $x_i = (x_{i1}, \ldots, x_{im})$ corresponds exactly to one unit i and to the ith row of the data matrix X:

$$
\text{Unit } 1 \\
\vdots \\
\text{Unit } i, \qquad y = \begin{bmatrix} y_1 \\ \vdots \\ y_i \\ \vdots \\ y_n \end{bmatrix}, \qquad X = \begin{bmatrix} x_{11} & \cdots & x_{1m} \\ \vdots & & \vdots \\ x_{i1} & \cdots & x_{im} \\ \vdots & & \vdots \\ x_{n1} & \cdots & x_{nm} \end{bmatrix}. \\
\vdots \\
\text{Unit } n
$$

The cellular differentiation data (Table 1.2) are an example of ungrouped data. If some of the covariate vectors or rows of X have identical covariate values $(x_{i1}, \ldots, x_{im})$, the data may be *grouped*: after relabeling the index, only rows x_i with different combinations of covariate values appear in the data matrix X, together with the number n_i of repetitions and the arithmetic mean $\bar{y}_i$ of the individual responses on the same vector x_i of covariates. Thus grouped data are of the form

$$
\text{Group } 1 \\
\vdots \\
\text{Group } i, \quad \begin{bmatrix} n_1 \\ \vdots \\ n_i \\ \vdots \\ n_g \end{bmatrix}, \quad \bar{y} = \begin{bmatrix} \bar{y}_1 \\ \vdots \\ \bar{y}_i \\ \vdots \\ \bar{y}_g \end{bmatrix}, \quad X = \begin{bmatrix} x_{11} & \cdots & x_{1m} \\ \vdots & & \vdots \\ x_{i1} & \cdots & x_{im} \\ \vdots & & \vdots \\ x_{g1} & \cdots & x_{gm} \end{bmatrix} \\
\vdots \\
\text{Group } g
$$

where g is the number of different covariates x_i in the data set, n_i denotes the number of units with equal covariate vector x_i, and $\bar{y}_i$ denotes the arithmetic mean. Equivalently, instead of the arithmetic mean, the total sum of responses could be used. In particular if individual responses are binary, coded by $0 - 1$ dummies, it is common to use absolute instead of relative frequencies and to display them in the form of contingency tables. It is seen that ungrouped data are a special case of grouped data for $n_1 = \ldots = n_g = 1$. Grouping of ungrouped raw data is generally possible and can lead to a considerable amount of data reduction if *all covariates are categorical*, as, e.g., in Example 1.1. In this example, individual responses in the raw data set were binary (infection yes or no), whereas Table 1.1 shows the grouped data. With *metrical covariates* it can often be impossible to group raw data, as, e.g., in Example 1.3, or there are only very few individuals with identical covariate values that can be grouped, as in Example 1.2.

For the following main reasons it is important to distinguish between grouped and ungrouped data:

(i) Some statistical procedures are meaningful only for grouped data, in particular in connection with goodness–of–fit tests, residuals and influence analysis.

(ii) Asymptotic results for grouped data can be obtained under $n_i \to \infty$ for all $i = 1, \ldots, g$, or under $n \to \infty$, without necessarily requiring $n_i \to \infty$. The latter type of asymptotics, which is more difficult to deal with, is appropriate for ungrouped data or grouped data with small n_i.

(iii) From the computational point of view, considerable savings in computing times and memory requirements can be achieved by grouping the data as far as possible.

2.1.2 Definition of univariate generalized linear models

The *classical* linear model for ungrouped normal responses and deterministic covariates is defined by the relation

$$y_i = z_i'\beta + \epsilon_i, \qquad i = 1, \ldots, n,$$

where z_i, the design vector, is an appropriate function of the covariate vector x_i and β is a vector of unknown parameters. For a vector of metric variables the simplest form of z_i is $z_i' = (1, x_i')$, for a vector of qualitative variables or a mixture of metric and qualitative variables dummy variables have to be included. The errors ϵ_i are assumed to be normally distributed and independent,

$$\epsilon_i \sim N(0, \sigma^2).$$

We rewrite the model in a form that leads to generalized linear models in a natural way: The observations y_i are independent and normally distributed,

$$y_i \sim N(\mu_i, \sigma^2), \qquad i = 1, \ldots, n, \qquad (2.1.2)$$

with $\mu_i = E(y_i)$. The mean μ_i is given by the linear combination $z_i'\beta$,

$$\mu_i = z_i'\beta, \qquad i = 1, \ldots, n. \qquad (2.1.3)$$

If covariates are stochastic, we assume the pairs (y_i, x_i) to be independent and identically distributed. Then the model is to be understood conditionally, i.e., (2.1.2) is the conditional density of y_i given x_i, and the y_i are conditionally independent.

This remark applies also to the following definition of generalized linear models, where the preceding assumptions are relaxed in the following way:

1. *Distributional assumption:*
   Given x_i, the y_i are (conditionally) independent, and the (conditional) distribution of y_i belongs to a simple exponential family with

(conditional) expectation $E(y_i \mid x_i) = \mu_i$ and, possibly, a common scale parameter ϕ, not depending on i.

2. *Structural assumption:*
   The expectation μ_i is related to the linear predictor $\eta_i = z_i'\beta$ by

   $$\mu_i = h(\eta_i) = h(z_i'\beta) \qquad \text{resp.} \quad \eta_i = g(\mu_i)$$

   where

   h is a known one–to–one, sufficiently smooth response function;

   g is the link function, i.e., the inverse of h;

   β is a vector of unknown parameters of dimension p; and

   z_i is a design vector of dimension p, which is determined as an appropriate function $z_i = z(x_i)$ of the covariates.

Thus, a specific generalized linear model is fully characterized by three components:

– *the type of the exponential family,*

– *the response or link function, and*

– *the design vector.*

Some important models, in particular for binary responses, are considered in more detail in the following subsections. Therefore, only some general remarks on the three components are given here.

(i) Exponential families and some of its properties are described in more detail in Appendix A1. For univariate generalized linear models as considered in this chapter, the densities of responses y_i can always be written as

$$f(y_i|\theta_i, \phi, \omega_i) = \exp\left\{ \frac{y_i\theta_i - b(\theta_i)}{\phi}\omega_i + c(y_i, \phi, \omega_i) \right\}, \qquad (2.1.4)$$

where

θ_i is the so–called natural parameter,

ϕ is an additional scale or dispersion parameter ,

$b(\cdot)$, and $c(\cdot)$ are specific functions corresponding to the type of exponential family, and

ω_i is a weight with $\omega_i = 1$ for ungrouped data ($i = 1, \ldots, n$) and $\omega_i = n_i$ for grouped data ($i = 1, \ldots, g$) if the *average* is considered as response (or $\omega_i = 1/n_i$ if the *sum* of individual responses is considered).

Important members are the normal, the binomial, the Poisson, the gamma and the inverse Gaussian distributions. Their characteristics are expressed in exponential family terms in Table 2.1. The natural parameter θ is a function of the mean μ, i.e., $\theta_i = \theta(\mu_i)$, which is uniquely determined by the specific exponential family through the relation $\mu = b'(\theta) = \partial b(\theta)/\partial \theta$. Moreover, the variance of y is of the form

$$\text{var}(y_i|x_i) \;=\; \sigma^2(\mu_i) \;=\; \phi v(\mu_i)/\omega_i \qquad (2.1.5)$$

where the variance function $v(\mu)$ is uniquely determined by the specific exponential family through the relation $v(\mu) = b''(\theta) = \partial^2 b(\theta)/\partial \theta^2$. Thus, specification of the mean structure by $\mu = h(z'\beta)$ implies a certain variance structure. However, many results and procedures remain valid under appropriate modifications if the mean and the variance function are specified separately, thereby dropping the exponential family assumption and considering quasi–likelihood models (see Section 2.3.1).

(ii) Choice of appropriate response or link functions depends on the specific exponential family, i.e., on the type of response, and on the particular application. For each exponential family there exists a so–called natural or canonical link function. Natural link functions relate the natural parameter directly to the linear predictor:

$$\theta = \theta(\mu) = \eta = z'\beta, \qquad (2.1.6)$$

i.e., $g(\mu) \equiv \theta(\mu)$. The natural link functions can thus be determined from Table 2.1, e.g.,

$$\begin{aligned}
\eta &= \mu & \text{for the normal,} \\
\eta &= \log \mu & \text{for the Poisson,} \\
\eta &= \log\left[\tfrac{\mu}{1-\mu}\right] & \text{for the Bernoulli}
\end{aligned}$$

distribution. Natural link functions lead to models with convenient mathematical and statistical properties. However, this should not be the main reason for choosing them, and non–natural link functions may be more appropriate in a particular application.

(iii) Concerning the design vector, nothing new has to be said compared to linear models: In most cases a constant, corresponding to the "grand mean," is added so that z is of the form $z = (1, w)$. Metrical covariates can be incorporated directly or after appropriate transformations like $\log(x), x^2, \ldots$ etc. Categorical covariates, ordered or unordered, have to be coded by a dummy vector as described in Section 2.1.1: For the unidimensional covariate $x \in \{1, \ldots, k\}$ a meaningful linear predictor is

$$\eta = \beta_0 + x^{(1)}\beta_1 + \ldots + x^{(q)}\beta_q$$

where $x^{(1)}, \ldots, x^{(q)}$ are dummy variables as given in Section 2.1.1.

TABLE 2.1. Simple exponential families with dispersion parameter

$$f(y|\theta,\phi,\omega) \;=\; \exp\left\{ \frac{y\theta - b(\theta)}{\phi}\omega + c(y,\phi,\omega) \right\}$$

(a) Components of the exponential family

Distribution		$\theta(\mu)$	$b(\theta)$	ϕ
Normal	$N(\mu,\sigma^2)$	μ	$\theta^2/2$	σ^2
Bernoulli	$B(1,\pi)$	$\log(\pi/(1-\pi))$	$\log(1+\exp(\theta))$	1
Poisson	$P(\lambda)$	$\log\lambda$	$\exp(\theta)$	1
Gamma	$G(\mu,\nu)$	$-1/\mu$	$-\log(-\theta)$	ν^{-1}
Inverse Gaussian	$IG(\mu,\sigma^2)$	$1/\mu^2$	$-(-2\theta)^{1/2}$	σ^2

(b) Expectation and variance

Distribution	$E(y)=b'(\theta)$	var. fct. $b''(\theta)$	$\mathrm{var}(y)=b''(\theta)\phi/\omega$
Normal	$\mu=\theta$	1	σ^2/ω
Bernoulli	$\pi=\frac{\exp(\theta)}{1+\exp(\theta)}$	$\pi(1-\pi)$	$\pi(1-\pi)/\omega$
Poisson	$\lambda=\exp(\theta)$	λ	λ/ω
Gamma	$\mu=-1/\theta$	μ^2	$\mu^2\nu^{-1}/\omega$
Inverse Gaussian	$\mu=(-2\theta)^{-1/2}$	μ^3	$\mu^3\sigma^2/\omega$

Derivations are denoted by $b'(\theta) = \partial b(\theta)/\partial\theta, b''(\theta) = \partial^2 b(\theta)/\partial\theta^2$. The weight ω is equal to 1 for individual ungrouped observations. For grouped data, y denotes the average of individual responses, the densities are scaled and the weight ω equals the group size (i.e., the number of repeated observations in a group).

If the original vector x comprises more than one qualitative component the linear predictor will contain dummy variables for all the components and possibly products of dummy variables (interactions). In addition to "main effects," the "interaction effects" that are of interest are obtained by adding the product of two or more covariates to the design vector. Metric covariates are multiplied directly, for categorical covariates corresponding dummy vectors have to be multiplied. As in linear models, one has to be aware of the problem of multicollinearity or aliasing, to avoid problems of parameter identifiability.

(iv) A last remark concerns the relationship between models for grouped and ungrouped data. Suppose that ungrouped data (y_i, x_i), $i = 1, \ldots, n$, are modelled by a generalized linear model with response and variance function

$$E\left(y_i | x_i\right) = \mu_i = h(z_i' \beta), \qquad \text{var}\left(y_i | x_i\right) = \phi v(\mu_i).$$

If data are grouped as in Section 2.1.1, with $\bar{y}_i$ denoting the arithmetic mean of n_i observations on the same covariate resp. design value $z_i = z(x_i)$, then the distribution of $\bar{y}_i$ is within the exponential family again, with the same mean structure

$$E\left(\bar{y}_i | x_i\right) = h(z_i' \beta)$$

as for ungrouped data, but with variance

$$\text{var}\left(\bar{y}_i | x_i\right) = \frac{\phi v(\mu_i)}{n_i};$$

see Appendix A1. So the definition of generalized linear models applies to grouped data as well, if the variance function is properly adjusted by $1/n_i$, or, equivalently, by defining the weights ω_i in (2.1.4) as $\omega_i = n_i$. Note that in Table 2.1 the response y refers to the average over n_i observations if observations are grouped. Then ω is the number of grouped observations instead of $\omega = 1$ as for the ungrouped case.

2.1.3 Models for continuous responses

Normal distribution

Assuming a normal distribution choice of the natural link function leads to the classical linear normal model

$$\mu = \eta = z' \beta.$$

Sometimes a nonlinear relationship

$$\mu = h(\eta)$$

will be more appropriate, e.g.,

$$h(\eta) = \eta^2, \qquad h(\eta) = \log \eta \qquad \text{or} \qquad h(\eta) = \exp \eta.$$

This type of nonlinear normal regression can easily be handled within the framework of generalized linear models.

Gamma distribution

The gamma distribution is useful for regression analysis of (nearly) continuous non–negative variables, such as life spans, insurance claims, amount of rainfall, etc. In terms of its mean $\mu > 0$ and the shape parameter $\nu > 0$, the density is given by

$$f(y|\mu,\nu) \;=\; \frac{1}{\Gamma(\nu)} \left(\frac{\nu}{\mu}\right)^{\nu} y^{\nu-1} \exp\left(-\frac{\nu}{\mu} y\right), \qquad y \geq 0.$$

From Table 2.1 we have

$$\mathrm{var}(y) = \sigma^2(\mu) = \phi\mu^2,$$

with $\phi = 1/\nu$. The shape parameter determines the form of the density. For $0 < \nu < 1$, $f(y)$ decreases monotonically, with $f(y) \to \infty$ for $y \to 0$ and $f(y) \to 0$ for $y \to \infty$. For $\nu = 1$, the exponential distribution is obtained as a special case. For $\nu > 1$, the density is zero at $y = 0$, has a mode at $y = \mu - \mu/\nu$ and is positively skewed. Figure 2.1 displays the density for $\nu = 0.5$, 1.0, 2.0 and 5.0 ($\mu = 1$). It can be seen that the densities are all positively skewed, but a normal limit is attained as $\nu \to \infty$.

The natural response function is the reciprocal

$$\mu = \eta^{-1}.$$

Since $\mu > 0$, the linear predictor is restricted to $\eta = z'\beta > 0$, implying restrictions on β. Two other important response functions are the identity

$$h(\eta) = \eta = \mu$$

and the exponential response function

$$h(\eta) = \exp(\eta) = \mu,$$

or, equivalently, the log–link

$$g(\mu) = \log(\mu) = \eta.$$

Inverse Gaussian distribution

This distribution can be applied for nonsymmetric regression analysis and for lifetimes. Detailed expositions are given in Folks and Chhikara (1978) and Jorgensen (1982).

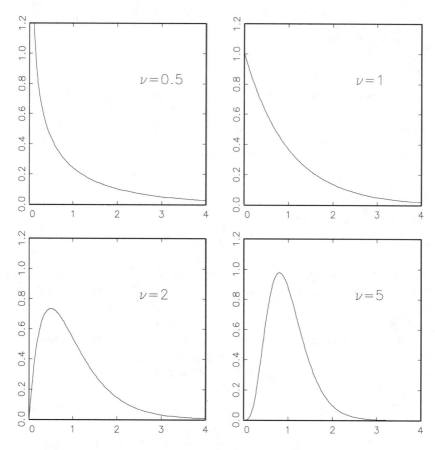

FIGURE 2.1. The gamma distribution: $G(\mu = 1, \nu)$

2.1.4   Models for binary and binomial responses

Consider first the case of ungrouped binary responses, coded by 0 and 1. Given the covariate x, a binary variable y is completely determined by its response probability

$$E(y|x) \;=\; P(y = 1|x) \;=\; \pi,$$

implying

$$\mathrm{var}(y|x) \;=\; \pi(1 - \pi).$$

If binary data are grouped as in Section 2.1.1, we let $\bar{y}$ denote the relative frequency of observed 1s for the, say m, independent binary observations with the same covariate vector x. The absolute frequencies $m\bar{y}$ are binomially distributed with

$$E(m\bar{y}|x) = m\pi, \qquad \mathrm{var}(m\bar{y}|x) = m\pi(1 - \pi).$$

The relative frequencies $\bar{y}$ are *scaled binomial*, i.e., they take the corresponding values $0, 1/m, 2/m, \cdots, 1$ with the same binomial probabilities as $m\bar{y}$, and

$$E(\bar{y}|x) = \pi, \qquad \text{var}(\bar{y}|x) = \frac{\pi(1-\pi)}{m}.$$

Thus, for grouped binary data, response functions are the same as for individual binary responses, while the variance function has to be divided by m, i.e., $\omega = m$.

If the individual (ungrouped) response (given x) is binomially distributed with $y \sim B(m, \pi)$ we will usually consider the scaled binomial or relative frequency y/m as response. Since y/m may be understood as an average of individual independent binary observations, this case may be treated as occurring from grouped observations. Then the variance

$$\text{var}(y/m) = \pi(1-\pi)/m$$

is the same as for grouped binary observations when grouping is done over m observations. Models for binary and binomial responses are determined by relating the response probability π to the linear predictor $\eta = z'\beta$ via some response function $\pi = h(\eta)$ resp. link function $g(\pi) = \eta$. The following models are most common.

Linear probability model

In analogy to linear normal models, π is related directly to η by the identity link

$$\pi = \eta = z'\beta.$$

Though easy to interpret, this model has a serious drawback: Since π is a probability, $z'\beta$ has to be in $[0,1]$ for all possible values of z. This implies severe restrictions on β.

This disadvantage is avoided by the following models. They relate π to the linear predictor η by

$$\pi = F(\eta) \tag{2.1.7}$$

where F is a strictly monotonous distribution function on the whole real axis, so that no restrictions on η and on β have to be imposed.

Probit model

It is defined by

$$\pi = \Phi(\eta) = \Phi(z'\beta),$$

where Φ is the standard normal distribution function. It imposes no restrictions on η, however it is computationally more demanding when computing likelihoods.

Logit model

It corresponds to the natural link function

$$g(\pi) = \log \left\{ \frac{\pi}{1-\pi} \right\} = \eta,$$

with the logistic distribution function

$$\pi = h(\eta) = \frac{\exp(\eta)}{1 + \exp(\eta)}$$

as the resulting response function. The logistic distribution function also has support on the entire real axis and is symmetric, but it has somewhat heavier tails than the standard normal. Apart from π values near 0 or 1, which correspond to the tails, fits by probit or logit models are generally quite similar (see Figure 2.2). Since the logistic function is easier to compute than the standard normal, logit models are often preferred in practice.

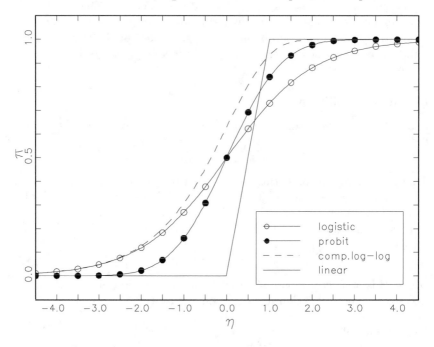

FIGURE 2.2. Response functions for binary responses

Complementary log–log model

It has the link function

$$g(\pi) = \log\left(-\log(1-\pi)\right),$$

and the extreme–minimal–value distribution function

$$h(\eta) = 1 - \exp\big(-\exp(\eta)\big)$$

as the corresponding response function. It is nonsymmetric, close to the logistic function for small π, but with considerably less heavy right tail.

The four response functions corresponding to the linear probability, probit, logit and complementary log–log models are displayed in Figure 2.2, where the values of η are plotted against π.

Figure 2.2 suggests that the response functions are quite different. However, one should keep in mind that the predictor η is linear in the form $\eta = z'\beta$, for simplicity $\eta = \beta_0 + x\beta_1$. So if instead of the distribution function F the transformed distribution function $\tilde{F}(u) = F(\frac{u-\mu}{\sigma})$ is used the models

$$\pi = F(\beta_0 + x\beta_1) \quad \text{and} \quad \pi = \tilde{F}(\tilde{\beta}_0 + x\tilde{\beta}_1)$$

are equivalent by setting $\tilde{\beta}_0 = \sigma\beta_0 + \mu$ and $\tilde{\beta}_1 = \sigma\beta_1$. Therefore models should be compared for the appropriate scaling of the predictor η. This may be done by transforming F so that the mean and variance according to different distribution functions are the same. For the functions used earlier mean and variance are given by

Response function F	Mean	Variance
linear	0.5	1/12
probit	0	1
logistic	0	$\pi^2/3$
compl. log–log	-0.5772	$\pi^2/6$

where $\pi = 3.14159$. Figure 2.3 displays the four response functions with η adjusted to have the logistic mean and variance for all four response functions.

In contrast to Figure 2.2 the logistic and probit function, which now have variance $\pi^2/3$, are nearly identical. Therefore fits of probit and logit models are generally quite similar after the adjustment of η, which is most often implicitly done in estimation. The complementary log–log function, however, is steeper than the logistic and probit function even after the adjustment. Thus, for small values η it approaches 0 more slowly, and as η approaches infinity it approaches 1 faster than the logistic and adjusted probit functions.

Binary models as threshold models of latent linear models

As presented so far, binary response models seem to be ad hoc specifications having some useful properties. However, all these models can be derived as

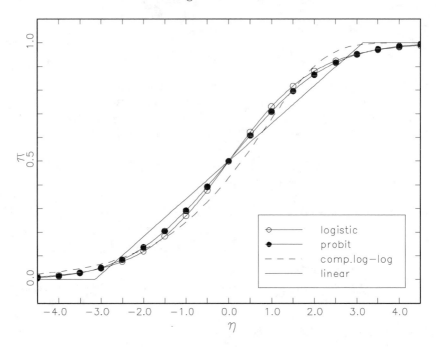

FIGURE 2.3. Response functions for binary responses adjusted to the logistic function (that means linear transformation yielding mean zero and variance $\pi^2/3$).

threshold models, where binary responses y are based on a latent continuous variable $\tilde{y}$ that obeys a linear model

$$\tilde{y} = \alpha_0 + w'\alpha + \sigma\epsilon,$$

where ϵ is distributed according to $F(\cdot)$, e.g. a logistic or standard normal distribution function, and σ is a scale parameter. The relation between y and $\tilde{y}$ is given by

$$y = \left\{ \begin{array}{ll} 1, & \tilde{y} \leq \tau \\ 0, & \tilde{y} > \tau, \end{array} \right.$$

with a threshold value τ. From this assumption one gets

$$P(y = 1) = P(\alpha_0 + w'\alpha + \sigma\epsilon \leq \tau) = F\left(\frac{\tau - \alpha_0 - w'\alpha}{\sigma}\right).$$

Defining

$$\beta = \left(\frac{\tau - \alpha_0}{\sigma}, \frac{\alpha'}{\sigma}\right)', \qquad z' = (1, -w')$$

the general model (2.1.7) is obtained. It should be noted that covariate effects α of the underlying linear model can be identified only up to the common but generally unknown factor $1/\sigma$, and that the grand mean parameter α_0 cannot be identified at all if τ is unknown.

Parameter interpretation

If we base the binary model on a latent linear model as above, covariate effects $\beta = (\beta_1, \beta_2, \ldots)$ may be interpreted with respect to this latent model. The preceding discussion shows that only relative values, e.g., β_1/β_2, but not absolute values β_1, β_2 are meaningful.

Parameter interpretation becomes more difficult for direct interpretation of covariate effects on the binary response y. For the logistic model, we have a linear model for the "log odds," i.e., the logarithm of the odds $\pi/(1-\pi)$ of a response $y = 1$. So interpretation of covariate effects on the log odds is the same as on the expectation $\mu = E(y)$ in the linear model. In particular in a medical context, the odds is often called "relative risk." For a linear predictor of the form $\eta = \beta_0 + \beta_1 x_1 + \beta_2 x_2$ say, we have

$$\frac{\pi}{1-\pi} = \exp(\beta_0)\exp(\beta_1 x_1)\exp(\beta_2 x_2)$$

so that (exponentials of) covariate effects have a multiplicative effect on the relative risk. However, this kind of interpretation is not possible for other models. Generally it seems best to break up interpretation in two stages:

1. Interpret covariate effects on the linear predictor $\eta = z'\beta$ in the same way as in linear models.

2. Transform this linear effect on η into a nonlinear effect on π with the aid of a graph of the response function

$$\pi = h(\eta)$$

as given in Figures 2.2 and 2.3.

Example 2.1: Caesarean birth study
Recall Example 1.1 where the response variable of interest is the occurrence or nonoccurrence of infection following birth by Caesarean section. Ignoring the two different types of infection, we form a binary response variable y with $y = 0$ if there is no infection and $y = 1$ if there is infection of type I or II. The risk of infection will be modelled with respect to the following three covariates, which are all dichotomous:

NOPLAN : whether the Caesarean was not planned or planned,
FACTOR : presence of one or more risk factors, such as diabetes, overweight, early labour pain and others,
ANTIB   : whether antibiotics were given as prophylaxis.

TABLE 2.2. Grouped data on infection

| | Caesarean planned | | Not planned | |
| | Infection | | Infection | |
	yes	no	yes	no
Antibiotics				
Risk factors	1	17	11	87
No risk factors	0	2	0	0
No antibiotics				
Risk factors	28	30	23	3
No risk factors	8	32	0	9

Table 2.2 summarizes the data of 251 births and is obtained from Table 1.1 by combining the two types I and II of infection.

We model the probability of infection by a logit model. All covariates are 0–1–coded (NOPLAN=1 for "the Caesarean was not planned," ANTIB=1 for "there were antibiotics given," FACTOR=1 for "there were risk factors present"). Including only three covariates (and no interactions) in the explanatory linear term, the model can be explicitly written in the form

$$\log \frac{P(\text{infection})}{P(\text{no infection})} = \beta_0 + \beta_1 \text{NOPLAN} + \beta_2 \text{FACTOR} + \beta_3 \text{ANTIB}.$$

The estimates of the covariate effects are given by

covariate	1	NOPLAN	FACTOR	ANTIB
effect	-1.89	1.07	2.03	-3.25

The estimates can be interpreted using the model formula from above. Antibiotics lower the log odds of infection. Since the logarithm is strictly monotone, we can directly conclude that antibiotics lower the relative risk of infection itself. Similary, risk–factors and the fact that a Caesarean was not planned do increase the relative risk of infection.

For a better interpretation of the parameter values we can rewrite the model in the form of relative risks

$$\frac{P(\text{infection})}{P(\text{no infection})} = \exp(\beta_0) \cdot \exp(\beta_1 \text{NOPLAN}) \cdot \exp(\beta_2 \text{FACTOR})$$
$$\cdot \exp(\beta_3 \text{ANTIB}).$$

It can now be easily seen that when the Caesarean was not planned the relative risk of infection increases by a factor of $\exp(\hat{\beta}_1)=\exp(1.07)=2.92$.

Additionally, with risk factors present the relative risk of infection is $\exp(\hat{\beta}_2) = 7.6$ times the relative risk without risk factors present.

Note, however, that all these values are only valid if the model fitted is accurate. For comparison we also fit a probit model with the same covariates. Parameter estimates are now -1.09, 0.61, 1.20 and -1.90. Although absolute values are quite different, relative values of effects are nearly the same as for the logit model.   □

Example 2.2: Credit–scoring

If banks give a credit to a client, they are interested in estimating the risk that the client will not pay back the credit as agreed upon by contract. The aim of credit–scoring systems is to model or predict the probability that a client with certain covariates ("risk factors") is to be considered as a potential risk. We will analyze the effect of covariates on the binary response "creditability" by a logit model. Other tools currently used in credit scoring are (linear) discriminance analysis, classification and regression trees, and neural networks.

The data set consists of 1000 consumer's credits from a south German bank. The response variable of interest is "creditability," which is given in dichotomous form ($y = 0$ for creditworthy, $y = 1$ for not creditworthy). In addition, 20 covariates that are assumed to influence creditability were collected. The raw data are recorded in Fahrmeir and Hamerle (1984, see p. 334 ff. and p. 751 ff.) and are available on electronic file. Fahrmeir and Hamerle (1984, p. 285–86) used a logit model to analyze a subset of these data containing only the following 7 covariates, which are partly metrical and partly categorical:

X1 running account, trichotomous with categories "no running account" (=1), "good running account"(=2), "medium running account" ("less than 200 DM"= 3 = reference category)

X3 duration of credit in months, metrical

X4 amount of credit in DM, metrical

X5 payment of previous credits, dichotomous with categories "good," "bad" (= reference category)

X6 intended use, dichotomous with categories "private" or "professional" (= reference category)

X7 and X8 are dummies for sex and marital status with reference category "male" resp. "living alone."

Since there are only three clients with the same covariate values, it is not possible to present the data in grouped form. Individual data of clients look like this:

Client	Y	X1	X3	X4	X5	X6	X7	X8
No.1	1	1	18	1049	0	1	0	0
No.23	0	3	36	2348	1	1	0	0

It is important to note that the data come from a *stratified sample:* 300 clients were drawn from the stratum defined by $y = 1$ (not creditworthy), while 700 clients are from the stratum $y = 0$. Within these strata, the empirical distributions of the covariates $X1$, $X3$, $X4$, $X5$ are represented below (relative frequencies in percents):

X1: account	1	y	0
no account	45.0		19.9
good	15.3		49.7
medium	39.7		30.2

X3: duration in months	1	y	0
≤ 6	3.00		10.43
$6 < \ldots \leq 12$	22.33		30.00
$12 < \ldots \leq 18$	18.67		18.71
$18 < \ldots \leq 24$	22.00		22.57
$24 < \ldots \leq 30$	6.33		5.43
$30 < \ldots \leq 36$	12.67		6.84
$36 < \ldots \leq 42$	1.67		1.17
$42 < \ldots \leq 48$	10.67		3.14
$48 < \ldots \leq 54$	.33		.14
$54 < \ldots \leq 60$	2.33		1.00

X4: amount of credit in DM	1	y	0
≤ 500	1.00		2.14
$500 < \ldots \leq 1000$	11.33		9.14
$1000 < \ldots \leq 1500$	17.00		19.86
$1500 < \ldots \leq 2500$	19.67		24.57
$2500 < \ldots \leq 5000$	25.00		28.57
$5000 < \ldots \leq 7500$	11.33		9.71
$7500 < \ldots \leq 10000$	6.67		3.71
$10000 < \ldots \leq 15000$	7.00		2.00
$15000 < \ldots \leq 20000$	1.00		.29

X5: previous credit	1	y	0
good	82.33		94.85
bad	17.66		5.15

X6: intended use	1	y	0
private	57.53		69.29
professional	42.47		30.71

Looking at these empirical distributions, we see that variable $X1$ is distinctly different for $y = 1$ and 0, while variable $X4$ does not differ very much for the two strata. Indeed it was found out in a variable selection procedure (Example 4.1) that variable $X4$ has no significant influence on Y; the same is true for $X7$. Both variables therefore have been neglected in the following analysis.

The probability of being "not creditworthy" is assumed to follow the logit model

$$\pi = P(y = 1|x) = \frac{\exp(z'\beta)}{1 + \exp(z'\beta)},$$

with design vector $z' = (1, X1[1], X1[2], X3, X5, X6, X8)$. The dummy $X1[1]$ stands for "no running account," $X1[2]$ for "good running account." All qualitative covariates are $(0-1)$–coded with reference categories as described above. The maximum likelihood estimates of the covariate effects β are given by:

Covariate	1	X1[1]	X1[2]	X3	X5	X6	X8
Effect	0.026	0.617	-1.32	0.039	-0.988	-0.47	-0.533

Before interpreting the results, we remark the following: Although the data come from a sample stratified with respect to the binary response, covariate effects are estimated correctly, see Anderson (1972).

However, instead of the intercept term β_0 one estimates $\beta_0 + \log p(1)/N(1) - \log p(0)/N(0)$, where $N(1) = 300$, $N(2) = 700$ are the (fixed) sample sizes in the stratified sample, while $p(1)$, $p(0)$ are the corresponding prior probabilities in the population. If the latter are known or can be estimated, the intercept term can be adjusted to obtain a consistent estimate of β_0.

First let us consider the effect of the metrical covariate "duration of credit" (X3). Due to the positive effect of X3 an increasing credit duration increases the probability of being "not creditworthy." The effect of the (0–1)–coded qualitative covariates on the creditability may be interpreted with respect to the chosen reference category. As an example consider the effect of "running account" (X1). For persons with "no running account" (X1[1]) the probability of being "not creditworthy" is higher than for those with "medium running account" (effect is 0 by definition) and for those with "good running account" (X1[2]). Vice versa, for persons with "good running account" the probability of being "creditworthy" is higher than for those with "medium running account" and "no running account." Furthermore, persons "living alone" (X8, effect is 0 by definition) are less "creditworthy" in probability than others, and persons who intend the "private use" (X6) of the credit are also more "creditworthy" in probability than those who intend to use the credit in a professional way (effect is 0 by definition). □

Overdispersion

A phenomen often observed in applications is that the actual variance of binary data is larger than explained by the nominal variance of the (scaled) binomial model. This phenomen is called "overdispersion" or "extra bino-mial variation." There are a number of causes why overdispersion may be observed. Two main reasons are: *unobserved heterogeneity* not taken into account by covariates in the linear predictor, and *positive correlation be-tween individual binary responses*, e.g., when individual units belong to a cluster, as, e.g., members of a household. In the latter situation, the simple formula for the sum of independent binary variables leading to the bino-mial variance is no longer valid, since positive correlations also contribute to the variance of the sum or the average. This type of overdispersion can only be caused if the local sample size $n_i > 1$. Unobserved heterogeneity, not modelled by the linear predictor, also leads to positive correlation (see Chapter 7), so that the same effect is observed. The simplest way to account for overdispersion is to introduce a multiplicative *overdispersion* parameter $\phi > 1$ and to assume that

$$\text{var}(y_i|x_i) = \phi \, \frac{\pi_i(1 - \pi_i)}{n_i} \, .$$

Since only $\pi_i = E(y_i|x_i)$ and $\text{var}(y_i|x_i)$ are actually needed in the estima-tion of effects β and of ϕ, statistical inference may be formally carried out as if ϕ were the nuisance parameter occurring in an exponential family dis-tribution, e.g., as in the gamma distribution; see the next section. However, the introduction of a multiplicative overdispersion parameter leads in fact already to the simplest form of a quasi–likelihood model (Section 2.3.1): Although there may exist distributions with variance $\phi\pi_i(1 - \pi_i)/n_i$, such as the beta–binomial model (see e.g. Collet, 1991, Chapter 6), a genuine likelihood is not actually needed for the purpose of inference. For reasons of robustness it may be preferable to specify only the mean and the vari-ance function. Note hereby that simply allowing $\phi \neq 1$ in the exponential family form of the binomial distribution in Table 2.1 will not give a normed density function.

More complex but explicit approaches to deal with overdispersion are contained in Section 3.5, where correlated binary responses are consid-ered, and in Chapter 7, where unobserved heterogeneity is modelled by introducing random effects in the linear predictor. A nice discussion of overdispersion is given in Collett (1991, Chapter 6). More recently, tests for overdispersion have been proposed by Dean (1992).

2.1.5   Models for counted data

Counted data appear in many applications, e.g., as the number of certain events within a fixed period of time (insurance claims, accidents, deaths, births, ...) or as the frequencies in each cell of a contingency table. Under certain circumstances, such data may be approximately modelled by models for normal data, or, if only some small values 0, 1, ..., q are observed, by models for multicategorical data (Chapter 3). Generally the Poisson distribution or some modification should be the first choice.

Log–linear Poisson model

It relates μ and η by the natural link

$$\log(\mu) = \eta = z'\beta, \qquad \mu = \exp(\eta).$$

The dependence of μ on the design vector is multiplicative, which is a sensible assumption in many applications. If all covariates are categorical and appropriate interaction effects are included in z, this leads to log–linear modelling of frequencies in higher–dimensional contingency tables. We do not discuss these models here, but refer the reader to Bishop et al. (1975), McCullagh and Nelder (1989), Agresti (1990), and Chapter 10 in Fahrmeir and Hamerle (1984).

Linear Poisson model

The direct relation

$$\mu = z'\beta$$

may be useful if covariates are assumed to be additive. Since $z'\beta$ has to be non–negative for all z, restrictions on β are implied.

If y is exactly Poisson–distributed, its variance equals its expectation:

$$\mathrm{var}(y|x) \;=\; \mu.$$

For similar reasons as for binomial data, overdispersion can be present in count data. For count data, as a rule, a nuisance parameter should be included so that

$$\mathrm{var}(y|x) \;=\; \sigma(\mu) \;=\; \phi\mu.$$

Again, as for binary data, more complex models that account for extra–variation in the data are available, see, e.g., Breslow (1984), Cameron and Trivedi (1986), and Chapter 7.

Example 2.3: Cellular differentiation

In a biomedical study of the immuno–activating ability of two agents, TNF (tumor necrosis factor) and IFN (interferon), to induce cell differentiation, the number of cells that exhibited markers of differentiation after exposure

to TNF and/or IFN was recorded. At each of the 16 dose combinations of
TNF/INF 200 cells were examined. The number y of cells differentiating
in one trial and the corresponding dose levels of the two factors are given
in Table 2.3, which is reproduced from Piegorsch, Weinberg and Margolin
(1988).

TABLE 2.3. Cellular differentiation data

Number y of cells differentiating	Dose of TNF(U/ml)	Dose of IFN(U/ml)
11	0	0
18	0	4
20	0	20
39	0	100
22	1	0
38	1	4
52	1	20
69	1	100
31	10	0
68	10	4
69	10	20
128	10	100
102	100	0
171	100	4
180	100	20
193	100	100

An important scientific question is whether the two agents stimulate cell differentiation synergistically or independently. Example 2.6 treats this problem within a log–linear Poisson model of the form

$$\mu = E(y|\text{TNF}, \text{IFN}) = \exp(\beta_0 + \beta_1 \text{TNF} + \beta_2 \text{IFN} + \beta_3 \text{TNF} * \text{IFN})$$

where $E(y|\text{TNF}, \text{IFN})$ denotes the expected number of cells differentiating
after exposure to TNF and IFN. The synergistic effect between TNF and
IFN is represented by the influence of the two–factor interaction TNF $*$
IFN. The following estimates were obtained:

$$\hat{\beta}_0 = 3.436, \quad \hat{\beta}_1 = 0.016, \quad \hat{\beta}_2 = 0.009, \quad \hat{\beta}_3 = -0.001.$$

Therefore it seems doubtful whether there is synergistic effect. A more
refined analysis that relaxes the assumption of Poisson distributed counts
y is given in Example 2.7.   □

2.2   Likelihood inference

Regression analysis with generalized linear models is based on likelihoods. This section contains the basic inferential tools for parameter estimation, hypothesis testing and goodness–of–fit tests, whereas more detailed material on model choice and model checking is deferred to Chapter 4. The methods rely on the genuine method of maximum likelihood, i.e., it is assumed that the model is completely and correctly specified in the sense of the definition in Section 2.1.2. In many applications, this assumption may be too idealistic. Quasi–likelihood models, where only the mean and the variance function are to be specified, are considered in Section 2.3.1.

2.2.1   Maximum likelihood estimation

Given the sample $y_1, \ldots, y_i, \ldots$, together with the covariates $x_1, \ldots, x_i, \ldots$, or design vectors $z_1, \ldots, z_i, \ldots$, a maximum likelihood estimator (MLE) of the unknown parameter vector β in the model $\mathrm{E}(y_i|x_i) = \mu_i = h(z_i'\beta)$ is obtained by maximizing the likelihood. To treat the cases of individual data $(i = 1, \ldots, n)$ and of grouped data $(i = 1, \ldots, g)$ simultaneously, we omit n or g as the upper limit in summation signs. Thus, sums may run over i from 1 to n or from 1 to g, and weights ω_i have to be set equal to 1 for individual data and equal to n_i for grouped data.

We first assume that the *scale* parameter ϕ is known. Since ϕ appears as a factor in the likelihood, we may set $\phi = 1$ in this case without loss of generality if we are only interested in a point estimate of β. Note, however, that ϕ (or a consistent estimate) is needed for computing variances of the MLE. Consistent estimation of an unknown ϕ by a method of moments, which is carried out in a subsequent step, is described at the end of this subsection. The parameter ϕ may also be considered as an *overdispersion* parameter, and it may be treated *formally* in the same way as a scale parameter. (Note, however, that only the mean $\mu_i = h(z_i'\beta)$ and the variance function are then properly defined so that one has to start with the expression for the score function $s(\beta)$ instead of with the log–likelihood $l(\beta)$ in (2.2.1)).

To avoid additional complexities concerning parameter identifiability, it is assumed from now on that the "grand" design matrix

$$Z = (z_1, \ldots, z_i, \ldots)' \qquad \text{has (full) rank} \quad p,$$

or, equivalently,

$$\sum_i z_i z_i' = Z'Z \qquad \text{has rank} \quad p.$$

Log–likelihood, score function and information matrix

According to (2.1.4), the log–likelihood contribution of observation y_i is, up to an additive constant,

$$l_i(\theta_i) = \log f(y_i|\theta_i, \phi, \omega_i) = \frac{y_i\theta_i - b(\theta_i)}{\phi}\omega_i.$$

The function $c(y_i, \phi, \omega_i)$, which does not contain θ_i, has been omitted. Inserting the relation $\theta_i = \theta(\mu_i)$ between the natural parameter and the mean, as given in Table 2.1 and Appendix A1, this contribution becomes a function of μ_i:

$$l_i(\mu_i) = \frac{y_i\theta(\mu_i) - b(\theta(\mu_i))}{\phi}\omega_i,$$

using l as a generic symbol for log–likelihoods. For example, in the case of binary responses ($\mu_i = \pi_i$) one obtains the well–known form

$$l_i(\pi_i) = y_i \log \pi_i + (1 - y_i) \log(1 - \pi_i).$$

For (scaled) binomial responses, i.e., relative frequencies $\bar{y}_i$ and repetition number n_i we have

$$l_i(\pi_i) = n_i(\bar{y}_i \log \pi_i + (1 - \bar{y}_i) \log(1 - \pi_i)).$$

In the case of Poisson responses ($\mu_i = \lambda_i$) one has

$$l_i(\lambda_i) = y_i \log \lambda_i - \lambda_i.$$

Inserting the mean structure $\mu_i = h(z_i'\beta)$ finally provides

$$l_i(\beta) = l_i(h(z_i'\beta))$$

as a function of β. Since $y_1, \ldots, y_i, \ldots$ are observations of independent random variables, the log–likelihood of the sample is the sum of the individual contributions:

$$l(\beta) = \sum_i l_i(\beta). \tag{2.2.1}$$

Its first derivative is the p–dimensional score function

$$s(\beta) = \frac{\partial l}{\partial \beta} = \sum_i s_i(\beta).$$

The individual score function contributions are

$$s_i(\beta) = z_i D_i(\beta) \sigma_i^{-2}(\beta)[y_i - \mu_i(\beta)], \tag{2.2.2}$$

where

$$\mu_i(\beta) = h(z_i'\beta), \qquad \sigma_i^2(\beta) = v(h(z_i'\beta))\phi/\omega_i$$

and

$$D_i(\beta) = \frac{\partial h(z_i'\beta)}{\partial \eta} \qquad (2.2.3)$$

is the first derivative of the response function $h(\eta)$ evaluated at $\eta_i = z_i'\beta$. The parameter ϕ may be interpreted here as a scale parameter in the likelihood or as an overdispersion factor for the variance function.

The expected Fisher information matrix is

$$F(\beta) = \text{cov } s(\beta) = \sum_i F_i(\beta), \qquad (2.2.4)$$

with

$$F_i(\beta) = z_i z_i' w_i(\beta)$$

and the weight functions

$$w_i(\beta) = D_i^2(\beta)\,\sigma_i^{-2}(\beta). \qquad (2.2.5)$$

The observed information matrix is the matrix

$$F_{obs}(\beta) = -\frac{\partial^2 l(\beta)}{\partial\beta\,\partial\beta'}$$

of negative second derivatives. Its explicit form will not be needed in the sequel, but is given in Appendix A1. Observed and expected information matrices are related by

$$F(\beta) = E\big(F_{obs}(\beta)\big).$$

For natural link functions, score functions and Fisher information matrices simplify to

$$s(\beta) = \frac{1}{\phi}\sum_i w_i z_i\big[y_i - \mu_i(\beta)\big], \qquad F(\beta) = \frac{1}{\phi}\sum_i w_i v\big(\mu_i(\beta)\big) z_i z_i'.$$

Moreover, expected and observed Fisher information are identical,

$$F(\beta) = F_{obs}(\beta).$$

For some purposes, matrix notation is convenient. For the more general grouped data case one has

$$y = \begin{bmatrix} y_1 \\ \vdots \\ y_g \end{bmatrix}, \quad \mu(\beta) = \begin{bmatrix} \mu_1(\beta) \\ \vdots \\ \mu_g(\beta) \end{bmatrix}, \quad \Sigma(\beta) = \begin{bmatrix} \sigma_1^2(\beta) & & 0 \\ & \ddots & \\ 0 & & \sigma_g^2(\beta) \end{bmatrix},$$

$$D(\beta) = \begin{bmatrix} D_1(\beta) & & 0 \\ & \ddots & \\ 0 & & D_g(\beta) \end{bmatrix},$$

$$W(\beta) \;=\; \begin{bmatrix} w_1(\beta) & & 0 \\ & \ddots & \\ 0 & & w_g(\beta) \end{bmatrix}$$

and one obtains

$$s(\beta) = Z'\,D(\beta)\,\Sigma^{-1}(\beta)\,\big[y - \mu(\beta)\big], \qquad F(\beta) = Z'\,W(\beta)\,Z.$$

For canonical link functions one obtains

$$s(\beta) = \frac{1}{\phi}Z'\Omega\big[y - \mu(\beta)\big]; \qquad F(\beta) = \frac{1}{\phi}Z'\Omega V(\beta)Z$$

with $\Omega = \mathrm{diag}(\omega_i)$, $V(\beta) = \mathrm{diag}(v(\mu_i))$.

Numerical computation of the MLE by iterative methods

Generally, MLE's $\hat\beta$ are not computed as global maximizers of $l(\beta)$, but as solutions of the likelihood equations

$$s(\hat\beta) = 0, \tag{2.2.6}$$

which correspond to local maxima, i.e., with $F_{obs}(\hat\beta)$ positive definite. For many important models, however, the log–likelihood $l(\beta)$ is concave so that local and global maxima coincide. For strictly concave log–likelihoods, the MLE is even unique whenever it exists. Existence means that there is at least one $\hat\beta$ within the admissible parameter set B such that $l(\hat\beta)$ is a global or local maximum. Some additional information on the important questions of existence and uniqueness is given later.

The likelihood equations are in general nonlinear and have to be solved iteratively. The most widely used iteration scheme is Fisher scoring or iteratively weighted least–squares. Starting with an initial estimate $\hat\beta^{(0)}$, Fisher scoring iterations are defined by

$$\hat\beta^{(k+1)} \;=\; \hat\beta^{(k)} + F^{-1}\big(\hat\beta^{(k)}\big)\,s\big(\hat\beta^{(k)}\big), \qquad k = 0, 1, 2, \ldots.$$

Note that the dispersion parameter ϕ cancels out in the term $F^{-1}\big(\hat\beta^{(k)}\big)\cdot s\big(\hat\beta^{(k)}\big)$. As a simple initial estimate $\hat\beta^{(0)}$, one may compute the unweighted least squares estimate for the data set $(g(y_i), z_i)$, $i = 1, \ldots, n$, thereby slightly modifying observations y_i for which the link function is undefined (e.g., $g(y_i) = \log(y_i)$ for $y_i = 0$). Iterations are stopped if some termination criterion is reached, e.g., if

$$\frac{\big\|\hat\beta^{(k+1)} - \hat\beta^{(k)}\big\|}{\big\|\hat\beta^{(k)}\big\|} \;\le\; \varepsilon$$

for some prechosen small number $\varepsilon > 0$.

If one defines the "working observation vector"

$$\tilde{y}(\beta) = \big(\tilde{y}_1(\beta), \ldots, \tilde{y}_n(\beta)\big)',$$
$$\tilde{y}_i(\beta) = z_i'\beta + D_i^{-1}(\beta)\left[y_i - \mu_i(\beta)\right],$$

then the Fisher scoring iterations may be expressed by

$$\hat{\beta}^{(k+1)} = \big(Z'\,W^{(k)}\,Z\big)^{-1}\,Z'\,W^{(k)}\,\tilde{y}^{(k)},$$

where $W^{(k)}$, $\tilde{y}^{(k)}$ means evaluation of W and $\tilde{y}$ at $\beta = \hat{\beta}^{(k)}$. This form can be interpreted as iteratively weighted least squares, and it has the advantage that a number of results in linear and nonlinear least squares estimation can be used after appropriate modifications.

Of course other iterative schemes may be applied to solve the likelihood equations. The Newton–Raphson scheme is obtained from Fisher scoring if expected information $F(\beta)$ is replaced by observed information $F_{obs}(\beta)$. However, $F(\beta)$ is easier to evaluate and always positive semidefinite. Quasi–Newton methods are often better alternatives than the simple Newton–Raphson scheme.

In defining the scoring iterations, we have tacitly assumed that $F(\beta)$ is nonsingular, i.e., positive definite, for the sequence of iterates $\hat{\beta}^{(k)}$. Since full rank of $Z'Z$ is assumed, this is the case if (most of) the weights $w_i(\beta)$ are positive. In this case, iterations will usually stop after a few iterations near a maximum. If they diverge, i.e., if successive differences $\|\hat{\beta}^{(k+1)} - \hat{\beta}^{(k)}\|$ increase, this can indicate a bad initial estimate or, more often, nonexistence of an MLE within the admissible parameter set B. If $B = \mathrm{I\!R}^p$ this means that at least one component of $\hat{\beta}^{(k)}$ tends to infinity. In the following, some results for discrete response models are discussed more formally.

Uniqueness and existence of MLE's*

The questions whether MLE's exist, whether they lie in the interior of the parameter space and whether they are unique have been treated by various authors. Results of Haberman (1974, log–linear and binomial models), Wedderburn (1976, normal, Poisson, gamma and binomial models), Silvapulle (1981) and Kaufmann (1988) (binomial and multicategorical models) are based on concavity of the log–likelihood. We restrict discussion to binomial and Poisson models; for other models we refer to Wedderburn (1976) and see also Fahrmeir and Kredler (1984).

Consider the general binary model (2.1.7), with distribution function F as a response function. If F is a continuous distribution function on the real line without constancy intervals, then, without further restrictions, the admissible parameter set is $B = \mathrm{I\!R}^p$. In this case existence means that there is a finite $\hat{\beta}$ where $l(\beta)$ attains its maximum. Furthermore let F be such that $\log F$ and $\log(1 - F)$ are strictly concave. This is fulfilled, e.g., for the logit, probit and double exponential models. Then, for full rank of Z, existence and uniqueness are equivalent to the condition that the equality/inequality system

$$y_i z_i' \beta \geq 0, \qquad (1 - y_i) z_i' \beta \leq 0, \qquad \text{for all } i$$

has only the trivial solution $\beta = 0$. Though easy to formulate, this condition can be difficult to check in practice. (Note that according to our convention, y_i is the relative frequency of ones observed for the design value z_i.) A sufficient but rather strong condition is that the matrix

$$\sum_i y_i(1 - y_i) z_i z_i' = Z' \operatorname{diag}\big(y_i(1 - y_i)\big)$$

has full rank. This condition cannot be fulfilled for purely binary responses.

If F is a continuous distribution with $0 < F(x) < 1$ only for a subinterval (a, b) of the real line, then the admissible set

$$B = \{\beta : a \leq z_i' \beta \leq b \text{ for all } i\}$$

is restricted, but still convex. Existence now means that there is a $\hat{\beta} \in B$ maximizing $l(\beta)$. If $\log F$ resp. $\log(1 - F)$ is strictly concave on $(a, b]$ resp. $[a, b)$, then a unique MLE always exists, if Z has full rank. The most prominent model of this type is the linear probability one.

For the log–linear Poisson model, the admissible parameter space is $B = \mathbb{R}^p$. A finite and unique MLE exists if the matrix

$$\sum_i y_i z_i z_i' \qquad \text{has full rank p.} \tag{2.2.7}$$

For the linear Poisson model the admissible parameter space is given by

$$B = \{\beta : z_i' \beta \geq 0 \text{ for all } i\},$$

and a non–negative MLE exists if Z has full rank. It is unique if and only if (2.2.7) holds. Let us conclude with some general remarks: If Z has full rank, $F(\beta) = F_{obs}(\beta)$ is always positive definite for models with canonical link or response function. Conditions for existence are often difficult to check. In practice, it may be easier to start ML iterations to see whether divergence or convergence occurs. Nonexistence may be overcome by larger sample sizes, since the asymptotic theory guarantees asymptotic existence, or by Bayes estimation with a strictly concave informative prior for β, see Section 2.3.2.

Asymptotic properties

Inferential methods for GLM's rely on asymptotic properties of ML estimators. Under "regularity assumptions," discussed informally later, the following properties hold.

Asymptotic existence and uniqueness:
The probability that $\hat\beta$ exists and is (locally) unique tends to 1 for $n \to \infty$.
Consistency:
If β denotes the "true" value, then for $n \to \infty$ we have $\hat\beta \to \beta$ in probability (weak consistency), or with probability 1 (strong consistency).
Asymptotic normality:
The distribution of the (normed) MLE becomes normal for $n \to \infty$, or, somewhat more informally, for large n

$$\hat\beta \overset{a}{\sim} N\big(\beta, F^{-1}(\hat\beta)\big),$$

i.e., $\hat\beta$ is approximately normal with approximate (or "asymptotic") covariance matrix

$$\mathrm{cov}(\hat\beta) \overset{a}{=} A(\hat\beta) = F^{-1}(\hat\beta),$$

where $A(\beta) := F^{-1}(\beta)$ is the inverse of the Fisher matrix. In case of an unknown scale parameter ϕ, all results remain valid if it is replaced by a consistent estimate $\hat\phi$.

Furthermore the MLE is asymptotically efficient compared to a wide class of other estimators.

Discussion of regularity assumptions*

With respect to the underlying regularity assumptions and the complexity of proofs, one can distinguish three types of asymptotics:

(i) Asymptotic theory for grouped data assumes a fixed number g of groups and $n_i \to \infty$, $i = 1, \ldots, g$ such that $n_i/n \to \lambda_i$ for fixed "proportions" $\lambda_i > 0$, $i = 1, \ldots, n$. For applications with finite sample size this means that there is a sufficient number of repeated observations for each design vector z_i. This assumption will normally be violated in the presence of continuous covariates.

(ii) Standard asymptotic theory for ungrouped data requires only that the total sample size n tends to infinity, but as a typical regularity assumption it is required that

$$\frac{F(\beta)}{n} \quad \text{has a positive definite limit,} \quad (2.2.8)$$

together with additional moment conditions. This type of asymptotic analysis is standard in the sense that it is rather near to the case of i.i.d. observations. Therefore results of this type are often stated without proof under "mild regularity conditions." However, the convergence condition (2.2.8) induces convergence conditions on the covariates itself. It is commonly fulfilled for stochastic regressors in the population case, where (y_i, x_i) are i.i.d. drawings from the joint distribution of (y, x). In planned experiments (2.2.8) can be fulfilled if regressors are similarly scattered as stochastic regressors. However, (2.2.8) is typically violated for trending or growing regressors.

(iii) General asymptotic theory requires only divergence of $F(\beta)$, i.e.,

$$A(\beta) = F^{-1}(\beta) \to 0 \qquad (2.2.9)$$

together with additional continuity properties of $F(\beta)$, or conditions on the moments of responses, on the sequence of covariates, etc. Condition (2.2.9) seems to be an indispensable requirement. It guarantees that information in the data increases with the sample size. Such general results are given in Haberman (1977, for natural link functions) and in Fahrmeir and Kaufmann (1985). Proofs require additional effort, in particular reducing general conditions on $F(\beta)$ to conditions on the sequence of covariates in particular models. As a "sample" of more specific conditions, some results of Fahrmeir and Kaufmann (1986) for the logit and probit models, as well as the linear and log–linear Poisson model are reviewed.

For bounded regressors, i.e., $\|z_n\| < c$ for all n, divergence of $Z'Z$, or, equivalently

$$(Z'Z)^{-1} \to 0 \qquad \text{for} \quad n \to \infty \qquad (2.2.10)$$

implies all the asymptotic properties. In the classical linear model, condition (2.2.10) alone is necessary and sufficient for (weak and strong) consistency and for asymptotic normality. Although bounded regressors cover a large number of situations, there are applications where growing regressors will be of interest, e.g., to model certain trends in a longitudinal analysis or in dose response experiments. Under a slight sharpening of (2.2.9) it can be shown that

$$\|z_n\| \;=\; 0(\log n) \qquad \text{for the logit model and the log–linear Poisson model,}$$

$$\|z_n\| \;=\; 0(\log n)^{\frac{1}{2}} \qquad \text{for the probit model}$$

$$\|z_n\| \;=\; 0(n^\alpha) \qquad \text{for the linear Poisson model, with some } \alpha > 0,$$

are sharp upper bounds for the admissible growth of regressors, assuring asymptotic properties. Compared to linear models, where, e.g., exponential growth is admissible, growth of regressors is much more restricted, e.g., to sublogarithmic growth in the logit model.

Additional scale or overdispersion parameter

If the scale or an overdispersion parameter ϕ is unknown, it can be consistently estimated by

$$\hat{\phi} = \frac{1}{g - p} \sum_{i=1}^{g} \frac{(y_i - \hat{\mu}_i)^2}{v(\hat{\mu}_i)/n_i}, \qquad (2.2.11)$$

with $\hat{\mu}_i = h(z_i'\hat{\beta})$ and $v(\hat{\mu}_i)/n_i$ as the estimated expectation and variance function of y_i, after grouping the data as far as possible. In all expressions where ϕ occurs, e.g., in $F(\hat{\beta})$, it is replaced by its estimate $\hat{\phi}$ to obtain correct standard errors.

Note that for linear Gaussian regression the moment estimate (2.2.11) reduces to the well–known estimate for σ^2 from the sum of squared residuals.

2.2.2   Hypothesis testing and goodness–of–fit statistics

Most of the testing problems for β are linear hypotheses of the form

$$H_0 : C\beta = \xi \qquad \text{against} \quad H_1 : C\beta \neq \xi, \tag{2.2.12}$$

where the matrix C has full row rank $s \leq p$. An important special case is

$$H_0 : \beta_r = 0 \qquad \text{against} \quad H_1 : \beta_r \neq 0, \tag{2.2.13}$$

where β_r is a subvector of β. This corresponds to testing the submodel defined by $\beta_r = 0$ against the full model. In the following it is assumed that unknown scale or overdispersion parameters ϕ are replaced by consistent estimates.

The *likelihood ratio statistic*

$$\lambda = -2\{l(\tilde{\beta}) - l(\hat{\beta})\}$$

compares the unrestricted maximum $l(\hat{\beta})$ of the (log–)likelihood with the maximum $l(\tilde{\beta})$ obtained for the restricted MLE $\tilde{\beta}$, computed under the restriction $C\beta = \xi$ of H_0. If the unrestricted maximum $l(\hat{\beta})$ is significantly larger than $l(\tilde{\beta})$, implying that λ is large, H_0 will be rejected in favour of H_1. A likelihood ratio test of (2.2.13), i.e. testing of a submodel defined by $\beta_r = 0$ requires new scoring iterations for the submodel, whereas considerable more effort is required to estimate $\tilde{\beta}$ under the general H_0 of (2.2.12). In case of an unknown scale parameter ϕ, all results remain valid if it is replaced by a consistent estimate $\hat{\phi}$.

Note that the likelihood ratio statistic is not properly defined for overdispersion models where the distributional assumptions are not fully given. As an approximation, however, it is common to work with the usual log-likelihoods for binomial or Poisson models, additionally divided by the estimate $\hat{\phi}$ in (2.2.11), which one obtains from the larger model or, if several nested models are compared, from some maximal model containing both models under concern.

The Wald test and the score test are computationally attractive quadratic approximations of the likelihood ratio statistic. The *Wald statistic*

$$w = \left(C\hat{\beta} - \xi\right)' \left[C\, F^{-1}(\hat{\beta})\, C'\right]^{-1} \left(C\hat{\beta} - \xi\right)$$

uses the weighted distance between the unrestricted estimate $C\hat{\beta}$ of $C\beta$ and its hypothetical value ξ under H_0. The weight is determined by the inverse of the asymptotic covariance matrix $CF^{-1}(\hat{\beta})C'$ of $C\hat{\beta}$. The Wald statistic is useful if the unrestricted MLE has already been computed, as e.g., in subset selection procedures. The *score statistic*

$$u = s'\left(\tilde{\beta}\right) F^{-1}\left(\tilde{\beta}\right) s\left(\tilde{\beta}\right)$$

is based on the following idea: The score function $s(\beta)$ for the unrestricted model is the zero vector if it is evaluated at the unrestricted MLE $\hat{\beta}$. If $\hat{\beta}$ is replaced by the MLE $\tilde{\beta}$ under H_0, $s(\tilde{\beta})$ will be significantly different from zero if H_0 is not true. The distance between $s(\tilde{\beta})$ and zero is measured by the score statistic u, with the inverse information matrix $F^{-1}(\tilde{\beta})$ acting as a weight. The score test is of advantage if a restricted model has already been fitted and is to be tested against a more complex model as, e.g., in a forward selection procedure. No new scoring iterations for computing the MLE $\hat{\beta}$ of the larger model are required.

An advantage of the Wald and score statistics is also that they are properly defined for models with overdispersion, since only first and second moments are involved.

Let $A(\beta) = F^{-1}(\beta)$ denote the inverse information matrix. For the special hypothesis (2.2.13), the statistics reduce to

$$w = \hat{\beta}'_r \, \hat{A}_r^{-1} \, \hat{\beta}_r, \tag{2.2.14}$$

and

$$u = \tilde{s}'_r \, \tilde{A}_r \, \tilde{s}_r$$

where A_r is the submatrix of $A = F^{-1}$ corresponding to the elements of β_r, s_r is the corresponding subvector of s, and " $\hat{}$ " or " $\tilde{}$ " means evaluation at $\beta = \hat{\beta}$ or $\beta = \tilde{\beta}$.

For the special case where β_r consists of only a scalar component of β, the Wald statistic is the square of the "t–value"

$$t_r = \frac{\hat{\beta}_r}{\sqrt{\hat{a}_{rr}}},$$

the standardized estimate of β_r, with $\hat{a}_{rr} = \text{var}(\hat{\beta}_r)$ the rth diagonal element of the (estimated) asymptotic covariance matrix $A(\hat{\beta}) = F^{-1}(\hat{\beta})$ of $\hat{\beta}$.

Under H_0 the three test statistics are asymptotically equivalent and have the same limiting χ^2–distribution with s degrees of freedom,

$$\lambda, \, w, u \overset{a}{\sim} \chi^2(s), \tag{2.2.15}$$

under similar general conditions, which ensure asymptotic properties of the MLE (Fahrmeir, 1987a). Critical values or p–values for the testing procedure are determined according to this limiting χ^2–distribution. In particular p–values corresponding to the squared t–values t_r^2 of effects β_r, are computed from the $\chi^2(1)$ distribution.

For finite sample size n, the quality of approximation of the distribution of λ, ω, u to the limiting $\chi^2(s)$–distribution depends on n and on the form of the log–likelihood function. As an example consider the linear Poisson model $\mu = \beta$ for a single observation y that takes the value 3. The quality of approximation can be seen from Figure 2.4 where the special case of testing $H_0 : \beta = \beta_0$ against $H_1 : \beta \neq \beta_0$ is treated. In this case $\tilde{\beta} = \beta_0$. Furthermore, $\hat{\beta} = 3$ denotes the MLE. The likelihood ratio statistic takes the vertical log–likelihood distance $l(\hat{\beta}) - l(\beta_0)$ as a measure of evidence for or against H_1. The Wald statistic is based on the quadratic log–likelihood approximation $\tilde{l}$, which is obtained by a second–order Taylor series expansion around $\hat{\beta}$. It takes into account the distance $\tilde{l}(\hat{\beta}) - \tilde{l}(\beta_0)$. The larger the horizontal distance $(\hat{\beta} - \beta_0)$ the larger is the discrepancy between the likelihood ratio and Wald statistic as long as the log–likelihood is non-quadratic. The score statistic looks at the (squared) slope of $l(\beta)$ at β_0, weighted by the inverse of the curvature, which should be near to zero if β_0 is near to $\hat{\beta}$. If the log–likelihood is a quadratic function of β, then all three test statistics coincide. For large sample size n, asymptotic theory shows that $l(\beta)$ becomes approximately quadratic, so that λ, w and u will tend to be close to each other. For medium or small n, however, differences can become more serious. A more detailed discussion is presented in Buse (1982).

The hypotheses considered so far form linear subspaces. In some applications, other hypotheses may be of interest, e.g., inequality restrictions. Already in the classical linear normal model the null distribution of the LR

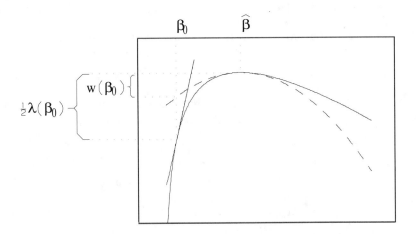

FIGURE 2.4. Log–likelihood (—) and quadratic approximation (- - -) for Wald test and slope for score test.

statistic is then a mixture of χ^2–distributions (e.g., Gourieroux, Holly and Montfort, 1982; Wolak, 1987; Kaufmann, 1989). Following Wolak (1989), such results should carry over to generalized linear models under appropriate regularity assumptions.

Goodness–of–fit statistics

Two summary statistics often used to assess the adequacy of a model are the *Pearson statistic*

$$\chi^2 = \sum_{i=1}^{g} \frac{(y_i - \hat{\mu}_i)^2}{v(\hat{\mu}_i)}$$

and the *deviance*

$$D = -2\phi \sum_{i=1}^{g} \left\{ l_i(\hat{\mu}_i) - l_i(y_i) \right\},$$

where $\hat{\mu}_i$, $v(\hat{\mu}_i)$ are the estimated mean and variance function, and $l_i(y_i)$ is the individual log–likelihood where μ_i is replaced by y_i (the maximum likelihood achievable). For both cases data should be grouped as far as possible. If grouped data asymptotics apply, i.e., if the number of observations is sufficiently large in all groups, both statistics are approximately $\phi\chi^2(g - p)$–distributed where p is the number of estimated coefficients. Then the statistics may be used for formally testing the goodness–of–fit of a model. However, if n is large and n_i remains small, in particular $n_i = 1$, formal use of these test statistics becomes dangerous; see McCullagh and Nelder (1989), pp.118–119. Large values of χ^2 or D cannot necessarily provide evidence for lack of fit.

Model checking has to be supported by additional formal and informal diagnostic tools, see Chapter 4.

Example 2.4: Caesarian birth study (Example 2.1 continued)
In Example 2.1, a logit model was fitted for the risk of infection.

TABLE 2.4. Logit model fit to Caesarian birth study data

	$\hat{\beta}$	$\sqrt{var\hat{\beta}}$	t
1	-1.89	0.41	-4.61
NOPLAN	1.07	0.43	2.49
FACTOR	2.03	0.46	4.41
ANTIB	-3.25	0.48	-6.77

In addition to the point estimates $\hat{\beta}$ of effects, already given and interpreted there, estimated standard deviations and t–values are given in Table 2.4. For the deviance, the value 10.997 was obtained, with 3 degrees of freedom. The model is rejected for $\alpha = 5\%$, indicating a bad fit. Let us check the model by comparing observed relative frequencies to fitted probabilities (in parenthesis).

	Caesarian planned		Not planned	
Antibiotics				
Risk factors	0.06	(0.04)	0.11	(0.11)
No risk factors	0.00	(0.01)		
No antibiotics				
Risk factors	0.48	(0.53)	0.88	(0.77)
No risk factors	0.20	(0.13)	0.00	(0.30)

This shows that the fit is extremly bad in the group defined by Caesarian not planned/no antibiotics/no risk. Let us try to improve the fit by including interaction terms. The following table shows, in the last column, the differences of deviances to the main effect model if the interaction terms NOPLAN*ANTIB and FACTOR*ANTIB are included.

	Deviance	df	Difference to main effect model
main effect model	10.997	3	.
Main effects + NOPLAN*ANTIB	10.918	2	0.07839
Main effects + FACTOR*ANTIB	10.974	2	0.02272

This indicates that these interactions may be omitted. A problem occurs for the inclusion of the interaction NOPLAN*FACTOR. Some program packages yield estimates for this model, with very large values for the NO-PLAN and NOPLAN*FACTOR effects, but without any warning. Actually in the first print of this book such misleading estimates have been given (thanks to J.-L. Foulley for his hints). In fact, no finite ML estimate exists, i.e., parameter estimates for NOPLAN and NOPLAN*FACTOR are going to infinity, since data are too sparse with responses for the cell NO-PLAN=1, FACTOR=0 and ANTIB=0 all in the "no" category. Instead of using robustified estimates, a common and simple method to deal with such a situation is to add a fictious datum to this cell, with 0.5 as response category and 0.5 in the other one. Doing this, estimated effects and standard deviations for this model are

Covariate	1	NOPLAN	FACTOR	ANTIB	NOPLAN*FACTOR
$\hat{\beta}$	-1.39	-1.56	1.36	-3.83	3.41
$\sqrt{var(\hat{\beta})}$	0.40	1.50	0.47	0.60	1.61

The deviance for this model is 0.955, with 2 degrees of freedom, providing strong evidence for including the interaction NOPLAN*FACTOR. Although some parameter estimates change considerably, one can easily see that there is only little change in the linear predictor for most of the covariate combinations. For example, the contribution of NOPLAN and FACTOR to the linear predictor in the main effect model was 3.10 for NOPLAN=1 and FACTOR=1, whereas it is 3.21 in the model including the interaction NOPLAN*FACTOR. However, there is a distinct change for NOPLAN=1 and FACTOR=0, leading to improved fitted probabilities. Note that omission of NOPLAN suggested by the standard deviation, would impair fit and deviance.   □

	Caesarian planned		Not planned	
Antibiotics				
Risk factors	0.06	(0.02)	0.11	(0.12)
No risk factors	0.00	(0.01)		
No antibiotics				
Risk factors	0.48	(0.49)	0.88	(0.86)
No risk factors	0.20	(0.20)	0.05	(0.05)

Example 2.5: Credit–scoring (Example 2.2 continued)
Consider the logit model of Example 2.2 where the response variable was the consumers' creditability. Table 2.5 gives the MLE's $\hat{\beta}_r$, the standard errors $\sqrt{var(\hat{\beta}_r)}$, the standardized MLE's $t_r = \hat{\beta}_r/\sqrt{var(\hat{\beta}_r)}$ and their p–values for each component β_r of the parameter vector β. Using a 0.05

TABLE 2.5. Logit model fit to credit–scoring data

	$\hat{\beta}$	$\sqrt{var(\hat{\beta})}$	t	p–value
1	0.026	0.316	0.082	0.933
X1[1]	0.617	0.176	3.513	0.0
X1[2]	-1.320	0.202	-6.527	0.0
X3	0.039	0.006	6.174	0.0
X5	-0.988	0.253	-3.910	0.0
X6	-0.470	0.160	-2.940	0.003
X7	-0.533	0.160	-3.347	0.001

significance level the hypothesis $H_0 : \beta_r = 0$ is rejected for all parameters with the exception of the grand mean "1."

In addition, the logit model yields deviance 1017.35 and $\chi^2 = 1006.53$ with 991 degrees of freedom. Assuming an approximate $\chi^2(991)$–distribution the model is not rejected, as can be seen from the p–values 0.387 (deviance) and 0.277 (χ^2). However, the assumption of a $\chi^2(991)$–distribution may be dangerous, even approximately, since the fit is based on binary observations that cannot be grouped because of the metrical covariate X3. □

Example 2.6: Cellular differentiation (Example 2.3 continued)
Recall the log–linear model of Example 2.3, where the logarithm of the expected number y of cells differentiating depends linearly on the main effects TNF, IFN and on the interaction between these two factors. The first column of Table 2.6 contains the MLE's $\hat{\beta}$, which are based on a Poisson likelihood with nuisance–parameter ϕ set equal to one. Concerning the effect of the interaction between TNF and IFN the p–value (in brackets) would suggest a high significance. However, deviance 142.4 and $\chi^2 = 140.8$ at 12 degrees of freedom indicates a high level of extravariation or overdispersion that is not explained by the fitted Poisson model. Since the counts y are rather large the asymptotic $\chi^2(12)$–distribution of deviance and χ^2 seem to be justified, so that the fitted Poisson model has to be rejected.

TABLE 2.6. Log–linear model fits to cellular differentiation data based on Poisson–likelihoods

	Poisson, $\phi = 1$		Poisson, $\hat{\phi} = 11.734$	
1	3.436	(.0)	3.436	(.0)
TNF	.016	(.0)	.016	(.0)
IFN	.009	(.0)	.009	(.0)
TNF*IFN	-.001	(.0)	-.001	(.22)

To take into account overdispersion the nuisance parameter ϕ has to be estimated by (2.2.11). Since the likelihood equations (2.2.6) do not depend on ϕ, the estimated nuisance–parameter $\hat{\phi} = 11.734$ does not affect the MLE's so that $\hat{\beta}$ is the same for ϕ set to one and $\hat{\phi} = 11.734$. The asymptotic variance–covariance matrix $\text{cov}(\hat{\beta}) = F^{-1}(\hat{\beta})$, however, depends on ϕ and has to be corrected with $\hat{\phi} = 11.734$. Due to this correction the Wald statistics $\hat{\beta}_r/\sqrt{\text{var}(\hat{\beta}_r)}, r = 0, \ldots, 3$ change, as well as the p–values (given in brackets). In contrast to the fit where ϕ was set to one (column 1), the interaction effect is no longer significant. This result is also obtained by using a quasi–likelihood approach (see Example 2.7). □

2.3   Some extensions

The original class of generalized linear models and related techniques for statistical inference have been modified and extended in several ways, further enhancing its flexibility and potential in applications. In this section we first describe two approaches for univariate cross–sectional models that also play an important role in later chapters where responses are correlated (Section 3.5, Chapters 6, 7 and 8): quasi–likelihood and Bayes models. As a further generalization nonlinear or nonexponential family models are shortly addressed in Subsection 2.3.3.

2.3.1   Quasi–likelihood models

One of the basic assumptions in the definition of generalized linear models is that the true density of the responses belongs to a specific exponential family, e.g., normal, binomial, Poisson, gamma, etc. Apart from the normal family, choice of the mean structure $\mu = h(z'\beta)$ implies a certain variance structure $v(\mu) = v(h(z'\beta))$. For example, in a linear Poisson model $\mu = z'\beta$ implies $v(\mu) = \mu = z'\beta$. If this is not consistent with the variation of the data, it was proposed to introduce an additional overdispersion parameter to account for extravariation, so that $var(y) = \phi\mu = \phi z'\beta$. The resulting score function, with typical contribution $z(\phi z'\beta)^{-1}(y - z'\beta)$, is no longer the first derivative form from a Poisson likelihood but from some "quasi–likelihood." Quasi–likelihood models allow dropping the exponential family assumption and separating the mean and variance structure. No full distributional assumptions are necessary; only first and second moments have to be specified. Under appropriate conditions, parameters can be estimated consistently, and asymptotic inference is still possible under appropriate modifications.

Basic models

Wedderburn (1974), McCullagh (1983) and McCullagh and Nelder (1983, 1989) assume that the mean and variance structure are correctly specified by

$$E(y|x) \; = \; \mu \; = \; h(z'\beta), \qquad var(y|x) \; = \; \sigma^2(\mu) \; = \; \phi v(\mu), \qquad (2.3.1)$$

where $v(\mu)$ is a variance function, generally defined separately and without reference to some exponential family, and ϕ is the dispersion parameter. Then a "quasi–likelihood" $Q(\beta, \phi)$ or an extended version (Nelder and Pregibon, 1987; McCullagh and Nelder, 1989) can be constructed such that $\partial Q/\partial\beta$ has the form (2.2.2) of a score function, with $\sigma^2(\mu) = \phi v(\mu)$ given by (2.3.1). For the original definition of quasi–likelihood, an equivalent genuine likelihood exists if there is a simple exponential family with the same variance function (Morris, 1982). In a closely related approach

Gourieroux, Montfort and Trognon (1984) assume only a correctly specified mean structure. Estimation is based on a "pseudo"–exponential family
model that need not contain the true distribution. Consequently the true
variance function will be different from the variance function appearing in
the score function corresponding to the quasi–model.

In the following, both approaches are considered in a unifying way. It is
supposed that the *mean is correctly specified* by $\mu = h(z'\beta)$ as in (2.3.1),
but *the true variance*

$$\text{var}(y|x) = \sigma_0^2(x)$$

may be different from $\sigma^2(\mu) = \phi v(\mu)$ in (2.3.1), which is used as a *"working" variance function*. The basic assumption of independent responses is
maintained in this section. Estimation is based on the *quasi–score function*
or *generalized estimating function* (GEE)

$$s(\beta) = \sum_i z_i \, D_i(\beta) \, \sigma_i^{-2}(\beta) \left[y_i - \mu_i(\beta)\right], \tag{2.3.2}$$

where $\mu_i(\beta) = h(z_i'\beta)$ is the correctly specified mean, and $D_i(\beta)$ is the
first derivative of h evaluated at $\eta_i = z_i'\beta$ as in Section 2.2.1. However,
$\sigma_i^2(\beta) = \phi v(\mu_i(\beta))$ is now a "working" variance: The variance function $v(\mu)$
may in principle be specified freely as in the approach of Wedderburn (1976)
and others, or it is implied by a "working" pseudo–likelihood model $l(\beta)$ as
in Gourieroux et al. (1984). For reasons of efficiency, the "working" variance
should be close to the true variance, which means in accordance with the
variability of the data. However, consistent estimation of β is possible with
any choice. Therefore, specification of $v(\mu)$ will be a compromise between
simplicity and loss of efficiency.

A global Q(uasi)MLE would be a global maximizer of an associated
quasi–likelihood $l(\beta)$. As in the case of correctly specified generalized linear models, we consider only local QMLE's, i.e., roots of the quasi–score
function where the matrix $\partial s(\beta)/\partial \beta'$ is negative definite. Global and local
QMLE's may be different, but for many models of interest they coincide.

Negative first derivatives $-\partial s(\beta)/\partial \beta'$ have the same form as
$-\partial^2 l(\beta)/\partial \beta \, \partial \beta'$ in Section 2.2.1 resp. Appendix A1. Moreover,

$$F(\beta) = E\left(-\frac{\partial s(\beta)}{\partial \beta'}\right) = \sum_i z_i z_i' w_i(\beta), \tag{2.3.3}$$

where the weights are given as in (2.2.5), with $\sigma_i^2(\beta)$ as the "working"
variances. In (2.3.3) and in the following, E, cov, var, etc., are to be understood with respect to the true but incompletely or incorrectly specified data–generating probability mechanism P. Equation (2.3.3) follows
from Appendix A1, since $E(y_i) - \mu_i(\beta) = 0$. However, the expected quasi–
information $F(\beta)$ is generally different from cov $s(\beta)$ so that (2.2.4) will

not hold. In fact

$$V(\beta) = \text{cov } s(\beta) = \sum_i z_i \, z_i' \, D_i^2(\beta) \, \frac{\sigma_{oi}^2}{\sigma_i^4(\beta)}, \qquad (2.3.4)$$

where $\sigma_{0i}^2 = \sigma_0^2(x_i)$ is the true (conditional) variance $\text{var}(y_i|x_i)$. Equation (2.3.4) follows from (2.3.2) using $Es(\beta) = 0$, $E((y_i - \mu_i(\beta))^2|x) = \text{var}(y_i|x_i)$, and the assumption of independent observations. Comparing (2.3.3) and (2.3.4), it is seen that generally $F(\beta) \neq V(\beta)$. However, if the variance structure is correctly specified, i.e., if

$$\sigma_0^2(x) = \phi v(\mu), \qquad \mu = h(z'\beta)$$

as assumed by Wedderburn and others, then $F(\beta) = \text{cov}(s(\beta)) = V(\beta)$ holds again.

Asymptotic properties of QMLE's can be obtained under appropriate regularity conditions, similar to those in Section 2.2.1. For standard asymptotic theory we refer to some of the work cited above; nonstandard results can be obtained along the lines of Fahrmeir (1990). Under appropriate assumptions, a (local) QMLE $\hat{\beta}$ as a root of the estimating equation (2.3.2) exists asymptotically, and it is consistent and asymptotically normal,

$$\hat{\beta} \overset{a}{\sim} N\big(\beta, \hat{F}^{-1}\hat{V}\hat{F}^{-1}\big),$$

with estimates $\hat{F} = F(\hat{\beta})$ and

$$\hat{V} = \sum_i z_i z_i' D_i^2(\hat{\beta}) \frac{\big[y_i - h(z_i'\hat{\beta})\big]^2}{\sigma_i^4(\hat{\beta})}$$

for $V(\beta)$ and $F(\beta)$.

Compared with the corresponding result for completely specified models, it is seen that essentially only the asymptotic covariance matrix $\hat{\text{cov}}(\hat{\beta})$ has to be corrected to the "sandwich"–matrix $\hat{A} = \hat{F}^{-1}\hat{V}\hat{F}^{-1}$. Thus, the quasi–likelihood approach allows consistent and asymptotically normal estimation of β under quite weak assumptions, however, with some loss of efficiency due to the corrected asymptotic covariance matrix. To keep this loss small, the "working" variance structure should be as close as possible to the true variance structure.

Tests of linear hypotheses of the form

$$H_0 : C\beta = \xi \qquad \text{against} \quad H_1 : C\beta \neq \xi$$

are still possible by appropriately modified Wald and score statistics. For example, the *modified Wald statistic* is

$$w_m = (C\hat{\beta} - \xi)' \, [C\hat{A}C']^{-1} \, (C\hat{\beta} - \xi)$$

with the corrected covariance matrix $\hat{A} = \hat{F}^{-1}\hat{V}\hat{F}^{-1}$ instead of $\hat{F}$ as in the common Wald statistic w. It has a limiting χ^2–distribution with $r = \text{rank}(C)$ degrees of freedom. Results of this kind are contained in White (1982) for the i.i.d. setting and they carry over without major difficulties to the present framework. In contrast to testing in completely specified models, "quasi–likelihood ratio test statistics" are in general not asymptotically equivalent to modified Wald or score statistics and may not have a limiting χ^2–distribution (see, e.g., Foutz and Srivastava, 1977, in the i.i.d. setting).

Variance functions with unknown parameters

Up to now the variance function was assumed to be a known function $v(\mu)$ of the mean. Several authors, e.g., Nelder and Pregibon (1987), relaxed this requirement by allowing unknown parameters θ in the variance function, so that

$$\text{var}(y|x) = \phi v(\mu; \theta).$$

For example, a useful parameterized variance function is obtained by considering powers of μ:

$$v(\mu; \theta) = \mu^\theta.$$

The values $\theta = 0, 1, 2, 3$ correspond to the variance functions of the normal, Poisson, gamma, and inverse Gaussian distributions.

For fixed θ, a QMLE $\hat{\beta}$ (and an estimate $\hat{\phi}$) can be obtained as before. Estimation of θ, given β and ϕ, can, e.g., be carried out by some method of moments. Cycling between the two steps until convergence gives a joint estimation procedure. Asymptotic results for $\hat{\beta}$ remain valid if θ is replaced by a consistent estimate $\hat{\theta}$.

Nonconstant dispersion parameter

A further extension is quasi–likelihood models where both the mean μ and the dispersion parameter are modelled as functions of separate linear predictors (Pregibon, 1984, Nelder and Pregibon, 1987, Efron, 1986, McCullagh and Nelder, 1989, ch.10, Nelder, 1992):

$$\mu = h(z'\beta), \qquad \phi = \phi(w'\theta), \qquad \text{var}(y) = \phi v(\mu).$$

Here w is a vector of covariates affecting the dispersion parameter. A two–step estimating procedure is proposed by cycling between the generalized estimating equation for β, holding the current iterate $\hat{\phi} = \phi(w'\hat{\theta})$ fixed, and a second generalized estimating equation for θ, holding $\hat{\beta}$ and $\hat{\mu} = h(z'\hat{\beta})$ fixed. This second GEE is obtained by differentiating Nelder and Pregibon's extended quasi–likelihood. Alternatively, joint estimation of β and θ is in principle possible by the general techniques for fitting likelihood and quasi–likelihood models described by Gay and Welsch (1988); see also Section 2.3.3. For consistent estimation of β and θ it is required that not only the mean but also the dispersion parameter be correctly specified.

Example 2.7: Cellular differentiation (Examples 2.3, 2.6 continued)
In Example 2.3 a log–linear model was proposed to analyze the synergistic effect of TNF and IFN on μ, the expected number y of cells differentiating after exposure to TNF and/or IFN. Example 2.6 gives the estimation results, which are based on Poisson–likelihood fits involving the variance structure $\sigma^2(\mu) = \text{var}(y|TNF, IFN) = \phi\mu$. However, a comparison of sample mean and sample variance for each group of counts having the same dose level of TNF reveals that a variance structure proportional to μ is less adequate:

		Dose of TNF		
	0	1	10	100
$\bar{x}$	22	45.25	74	161.5
s^2	107.5	300.7	1206.5	1241.25

The same holds for IFN. The sample moments lead one to suppose that the variance structure has the form $\sigma^2(\mu) = \text{var}(y|TNF, IFN) = \phi\mu^2$ or $\mu + \theta\mu^2$, where the last one corresponds to the variance of the negative binomial distribution. QMLE's $\hat{\beta}$, which are based on the log–linear mean structure proposed in Example 2.3 and on three alternative variance structures $\sigma^2(\mu)$, are given together with their p–values (in brackets) and moment estimators for the dispersion parameter ϕ or θ in Table 2.7. Parameter estimates and p–values for the variance assumptions being quadratic in the mean μ are nearly identical and do not differ very much from the results, which are based on a variance depending linearly on μ. However, in contrast to the fit, which is based on a Poisson model with unknown nuisance–parameter (see the second column of Table 2.6), the p–value of

TABLE 2.7. Log–linear model fits to cellular differentiation data based on quasi–likelihoods

	$\sigma^2(\mu) = \phi\mu$		$\sigma^2(\mu) = \phi\mu^2$		$\sigma^2(\mu) = \mu + \theta\mu^2$	
1	3.436	(0.0)	3.394	(0.0)	3.395	(0.0)
TNF	0.016	(0.0)	0.016	(0.0)	0.016	(0.0)
IFN	0.009	(0.0)	0.009	(0.003)	0.009	(0.003)
TFN*IFN	-0.001	(0.22)	-0.001	(0.099)	-0.001	(0.099)
$\hat{\phi}$	11.734		0.243		—	
$\hat{\theta}$	—		—		0.215	

the interaction TNF*IFN only moderately supports the presence of a synergistic effect between TNF and IFN. Moreover, the Poisson assumption seems to be less appropriate due to the discrepancy between estimated means and variances.   □

2.3.2   Bayes models

Bayesian variants of GLM's have been used by a number of authors to incorporate additional information about the parameter vector β in the original GLM. External information about the unknown parameters is modelled by a parametric prior density. Thus the parameters of the data density (2.1.4) are not considered unknown constants but random variables.

Bayes models on β assume that β is a random vector with prior density $p(\beta)$, and Bayes estimators are based on the posterior density $p(\beta|Y = (y_1, \ldots, y_n))$ of β given the data. Bayes' theorem relates prior and posterior densities by

$$p(\beta|Y) = \frac{L(\beta|Y)\, p(\beta)}{\int L(\beta|Y)\, p(\beta)\, d\beta}, \qquad (2.3.5)$$

where $L(\beta|Y)$ is the likelihood of the data. Marginal posterior densities for components of β, posterior means and covariance matrices, etc., can be obtained from the posterior by integration (resp. summation in the discrete case). For example

$$E(\beta|Y) = \int \beta\, p(\beta|Y)\, d\beta \qquad (2.3.6)$$

is the *posterior mean*, which is an optimal estimator of β for quadratic loss functions, and

$$\text{cov}(\beta|Y) = \int \big(\beta - E(\beta|Y)\big)\big(\beta - E(\beta|Y)\big)' p(\beta|Y)\, d\beta \qquad (2.3.7)$$

is the associated *posterior covariance matrix*, which is a measure for the precision of the posterior mean estimate. So, at a first glance, the Bayesian paradigm seems to be easily implemented. However, exact analytic solutions of the preceding integrations are available only for some special models, e.g., for the normal linear model. For most of the important models, e.g., binomial logit models with at least one covariate, no conjugate priors that would allow convenient analytic treatment exist. Therefore, implementation of the Bayesian estimation approach via (2.3.6), (2.3.7) requires numerical or Monte Carlo integration. Since the integrals have the dimension of β, which may be high–dimensional, this is not a trivial task. A number of techniques have been proposed and discussed, and research in this area is still in progress. We refer the reader, e.g., to Naylor and Smith (1982), Zellner and Rossi (1984), Smith et al. (1985), West and Harrison (1989, ch.13) for methods such as Gauss–Hermite integration or Monte

Carlo integration, and, e.g., to Gelfand and Smith (1990), for the application of "Gibbs sampling." Appendix A.5 gives a short presentation of this techniques, which will also be used in Chapters 7 and 8.

Posterior mode estimation is an alternative to full posterior analysis or posterior mean estimation, which avoids numerical integrations. It has been proposed by a number of authors, e.g., by Leonard (1972), Laird (1978), Stiratelli, Laird and Ware (1984), Zellner and Rossi (1984), Duffy and Santner (1989) and Santner and Duffy (1989). The posterior mode estimator $\hat{\beta}_p$ maximizes the posterior density p or equivalently the log posterior likelihood

$$l_p(\beta|Y) = l(\beta) + \log p(\beta), \tag{2.3.8}$$

where $l(\beta)$ is the log–likelihood of the generalized linear model under consideration. If a normal prior

$$\beta \sim N(\alpha, Q), \qquad Q > 0$$

is chosen, (2.3.8) specializes to

$$l_p(\beta|Y) = l(\beta) - \frac{1}{2}(\beta - \alpha)'Q^{-1}(\beta - \alpha), \tag{2.3.9}$$

dropping terms that are constant with respect to β. The criterion (2.3.9) is a penalized likelihood, with the penalty $(\beta - \alpha)'Q^{-1}(\beta - \alpha)$ for deviations from the prior parameter α. The addition of such a concave log prior to $l(\beta)$ also helps to avoid problems of nonexistence and nonuniqueness for ML estimators. Computation can be carried out iteratively, e.g., by Fisher scoring. For normal priors, $s(\beta)$ is modified to

$$s_p(\beta) = \frac{\partial l_p(\beta|Y)}{\partial \beta} = s(\beta) - Q^{-1}(\beta - \alpha),$$

and $F(\beta) = -E(\partial^2 l(\beta|Y)/\partial \beta \, \partial \beta')$ to

$$F_p(\beta) = -E\left(\frac{\partial^2 l_p(\beta|Y)}{\partial \beta \, \partial \beta'}\right) = F(\beta) + Q^{-1}.$$

For large n, $\hat{\beta}_p$ becomes approximately normal,

$$\hat{\beta}_p \stackrel{a}{\sim} N\big(\beta, F_p^{-1}(\hat{\beta}_p)\big),$$

under essentially the same conditions that assure asymptotic normality of the MLE. Then the posterior mode and the (expected) curvature $F_p^{-1}(\hat{\beta}_p)$ of $l_p(\beta)$, evaluated at the mode, are good approximations to the posterior mean (2.3.6) and covariance matrix (2.3.7).

Up to now, we have tacitly assumed that the prior is completely specified, that is, α and Q of the normal prior are known. Empirical Bayes analysis considers α and Q as unknown constants ("hyperparameters") that have to be estimated from the data. ML estimation by direct maximization of the marginal likelihood and indirect maximization by application of the exact EM algorithm again require numerical integration. This can be avoided if an *EM–type algorithm* is used instead, where posterior means and covariances appearing in the E–step are replaced by posterior modes and curvatures, see, e.g., Santner and Duffy (1989, p.249) and Duffy and Santner (1989). For simplicity, let us assume a normal prior with $\alpha = 0$ so that only Q has to be estimated. For an exact implementation of the EM algorithm, the next iterate Q_1 is obtained from the previous iterate Q_0 by maximizing the conditional expectation $E\left(l_p(\beta) \mid Y; Q_0\right)$ of the penalized likelihood. (See Appendix A.3 for a short discussion of the EM algorithm.) Since Q appears only in the log prior, this is equivalent to maximizing $E\left(\log p(\beta) \mid Y; Q_0\right)$. Assuming approximate posterior normality, maximization can be carried out analytically, leading to

$$Q_1 = \beta_0 \beta_0' + \operatorname{cov}(\beta|Y; Q_0),$$

where β_0 is obtained iteratively by application of Fisher scoring to (2.3.8) with $Q = Q_0$. Iteration of these steps leads to a joint estimation procedure for β and Q. We will use such EM–type algorithms in Chapters 7 and 8.

The Bayes models considered so far are based on the parameters β. Analytically more tractable expressions for the implementation of the estimation approach can be deduced, if the Bayes model is based on the means μ_i of the data densities (2.1.4). Albert (1988), for example, considers the means μ_i independent random variables with prior densities $p(\mu_i)$, where the prior means ν_i of μ_i are assumed to satisfy the GLM

$$E(\mu_i|x_i) = \nu_i = h(z_i'\beta). \qquad (2.3.10)$$

Model (2.3.10) allows for uncertainty concerning the specification of the means μ_i in the original GLM. The precision of the belief in the original GLM is reflected by the prior variance of μ_i. If the variance of the prior $p(\mu_i)$ approaches 0, the prior density becomes concentrated about the mean ν_i and the Bayesian model (2.3.10) in this limiting case is equivalent to the original GLM. The larger the prior variance the more uncertain is the original GLM. That means there are additional sources of variation that cannot be adequately explained by the original GLM.

Estimation of β and other unknown parameters in the prior $p(\mu_i)$ can be carried out by empirical or hierarchical Bayes procedures. For simplicity we restrict discussion to the case that only β is unknown. In the empirical Bayes context β is considered an unknown constant (hyperparameter). Estimation is based on the independent marginal densities

$$f(y_i|\beta) = \int f(y_i|\mu_i)\, p(\mu_i|\beta)\, \mathrm{d}\mu_i, \qquad (2.3.11)$$

where the data density $f(y_i|\mu_i)$ corresponds to the simple exponential family given in Appendix A1 and $p(\mu_i|\beta)$ denotes the prior density of μ_i, which depends on β via (2.3.10). Fortunately, the prior $p(\mu_i)$ can be chosen from the conjugate family so that analytic solutions of the integrations (2.3.11) are available. Densities that are conjugate to simple exponential families are described by Cox and Hinkley (1974), among others. For example, the conjugate prior for the Poisson density is the Gamma density, which yields a marginal density $f(y_i|\beta)$ of the negative binomial or Poisson–Gamma type. For such closed–form solutions of (2.3.11) estimation of β can be carried out by maximizing the marginal likelihood

$$L(\beta|Y) = \prod_{i=1}^{n} f(y_i|\beta) \tag{2.3.12}$$

with respect to β or applying the maximum quasi–likelihood principle (Section 2.3.1), which is based on the first two moments of $f(y_i|\beta)$. The latter is of advantage, if the marginal density $f(y_i|\beta)$ does not belong to the simple exponential family class. See, e.g., Williams (1982) for dichotomous data and Breslow (1984) and Lawless (1987) for count data.

Hierarchical Bayes estimation of β has been considered by Leonard and Novick (1986) and Albert (1988). In that context a two–stage prior is assigned to the means μ_i of the data density. In addition to the prior $p(\mu_i|\beta)$ a prior density $p(\beta)$ is assigned to β. This formulation is called hierarchical because there is a hierarchy of densities; one for the parameters in the data density and one for the parameters in the first–stage prior density. Estimation of β is based on the posterior density (2.3.5), where $L(\beta|Y)$ is given by (2.3.12), which represents the product of independent mixture densities of the form (2.3.11). However, as mentioned earlier, exact analytic solutions of the integrations in (2.3.5) are available only for some special cases.

2.3.3 Nonlinear and nonexponential family regression models*

In generalized linear (quasi–) models considered so far it was assumed that the predictors are linear in the parameters β. This assumption may be too restrictive and needs to be relaxed for some parts of the later chapters (e.g., in Section 3.5) by a nonlinear predictor, leading to *nonlinear exponential family regression models*. Then the mean structure is defined by

$$E(y|x) = h(x; \beta), \tag{2.3.13}$$

where the response function h has smoothness properties as in the original definition. Stressing dependence on β, we write $\mu(\beta) := h(x; \beta)$. Models with common response functions and nonlinear predictors $\eta(x; \beta)$,

$$\mu(\beta) = h\big(\eta(x; \beta)\big),$$

are of course covered by the general nonlinear model (2.3.13). Models with composite link functions (Thompson and Baker, 1981) and parametric link functions with linear predictors

$$\mu(\beta; \theta) = h(z'\beta; \theta)$$

(Pregibon, 1980; Scallan, Gilchrist and Green, 1984; Czado, 1992) are also within the general framework.

Given the data (y_i, x_i), $i = 1, 2, \ldots$, score functions and information matrices are now given by straightforward generalizations of (2.2.2) and (2.2.4):

$$s(\beta) = \sum_i M_i(\beta)\, \sigma_i^{-2}(\beta) \left[y_i - \mu_i(\beta) \right], \qquad (2.3.14)$$

$$F(\beta) = \sum_i M_i(\beta)\, \sigma_i^{-2}(\beta)\, M_i'(\beta), \qquad (2.3.15)$$

where $\mu_i(\beta) = h(x_i; \beta)$ and $M_i(\beta) = \partial\mu_i(\beta)/\partial\beta$.

Defining $M(\beta) = (M_1(\beta), \ldots, M_n(\beta))'$ and y, $\mu(\beta)$ and $\Sigma(\beta) = \text{diag}(\sigma_i^2(\beta))$ as in Subsection 2.2.1, we have

$$s(\beta) = M'(\beta)\, \Sigma^{-1}(\beta) \left[y - \mu(\beta) \right], \qquad F(\beta) = M'(\beta)\, \Sigma^{-1}(\beta)\, M(\beta)$$

in matrix notation. Fisher–scoring iterations for the MLE $\hat{\beta}$ can again be formulated as iterative weighted least squares (Green, 1984, 1989). While generalization of asymptotic theory is comparably straightforward under appropriate regularity conditions, questions of finite sample existence and uniqueness are difficult to deal with and no general results are available.

One may even go a step further and drop the exponential family assumption to obtain rather *general parametric regression models* as, e.g., in Green (1984, 1989) or Jorgenson (1983). Estimation for a wide class of (quasi–) likelihood models for independent observations is discussed by Gay and Welsch (1988). They consider objective functions of the form

$$l(\beta, \theta) = \sum_i l_i\big(\eta(x_i; \beta)\, ;\, \theta\big),$$

where the (quasi–) likelihood contribution l_i of observation i is a function of a nonlinear predictor $\eta(x_i; \beta)$ and a vector θ of nuisance parameters. This framework includes linear and nonlinear exponential family models, robust regression models as described in Huber (1981), Holland and Welsch (1977), and the (extended) quasi–likelihood models of Section 2.3.1. Green (1989) and Seeber (1989) show that Fisher scoring can again be written in the form of iteratively weighted least squares and give a geometric interpretation. Most of the extensions of GLM's in the following chapters, e.g., to multivariate, nonparametric or longitudinal data analysis, are written within the exponential family framework and, particularly, with a focus on discrete data. It should be noted, however, that much of the material can be extended to nonexponential family distributions without essential complications.

2.4 Further developments

There are other topics extending the basic framework of generalized linear models that are not described in this chapter or later. Generalized linear models with errors in variables have deserved considerable attention. For a survey see Carroll (1992) and the references given therein. Principal component and ridge regression are considered by Marx and Smith (1990) and Marx, Eilers and Smith (1992). For bootstrapping in generalized linear models, see, e.g., Moulton and Zeger (1989, 1991), Gigli (1992), Hinde (1992).

3
Models for multicategorical responses: multivariate extensions of generalized linear models

In this chapter the concept of generalized linear models is extended to the case of a vector–valued response variable. Consider Example 2.1, where we were interested in the effect of risk factors and antibiotics on infection following birth by Caesarean section. In this example the response was binary, only distinguishing between occurrence and nonoccurrence of infection, and thereby ignores that the data originally provided information on the type of infection (type I or II) as well. It is possible, however, to use this information by introducing a response variable with three categories (no infection / infection type I / infection type II). Naturally, these categories cannot be treated as a unidimensional response. We have to introduce a (dummy) variable for each category, thus obtaining a *multivariate* response variable. Therefore, link and response functions for the influence term will be vector-valued functions in this chapter. The focus is on multicategorial response variables and multinomial models. Variables of this type are often called polychotomous, the possible values are called categories. Extension to other multivariate exponential family densities is possible but not considered in this text.

After an introductory section we will first consider the case of a nominal response variable (Section 3.2). If the response categories are ordered, the use of this ordering yields more parsimoniouly parameterized models. In Section 3.3 several models of this type are discussed. Section 3.4 outlines statistical inference for the multicategorical case. In many applications, more than one response variable is observed, for example, when several measurements are made for each individual or unit or when measurements are observed repeatedly. Approaches for this type of multivariate responses with correlated components are outlined in Section 3.5.

3.1 Multicategorical response models

3.1.1 Multinomial distribution

For the categorical responses considered in this chapter the basic distribution is the multinomial distribution. Let the response variable Y have k possible values, which for simplicity are labelled $1, \ldots, k$. Sometimes consideration of $Y \in \{1, \ldots, k\}$ hides the fact that we actually have a *multivariate* response variable. This gets obvious by considering the response vector of the dummy variables $y = (y_1, \ldots, y_q)$, $q = k - 1$, with components

$$y_r = \begin{cases} 1, & \text{if } Y = r, \quad r = 1, \ldots, q. \\ 0, & \text{else.} \end{cases} \tag{3.1.1}$$

Then we have

$$Y = r \quad \Leftrightarrow \quad y = (0, \ldots, 1, \ldots, 0).$$

The probabilities are simply connected by

$$P(Y = r) = P(y_r = 1).$$

Given m independent repetitions $y_1, \ldots, y_m$ (or equivalently $Y_1, \ldots, Y_m$) it is useful to consider as a response variable the number of trials where we get outcome r. For the repetitions $(y_1, \ldots, y_m)$ we get the sum of vectors

$$y = \sum_{i=1}^{m} y_i.$$

Then the vector y is multinomially distributed with distribution function

$$P\left(y = (m_1, \ldots, m_q)\right) = \frac{m!}{m_1! \cdots m_q!(m - m_1 - \ldots - m_q)!} \cdot \pi_1^{m_1} \cdots \pi_q^{m_q} (1 - \pi_1 - \ldots - \pi_q)^{m - m_1 - \ldots - m_q} \tag{3.1.2}$$

where $\pi_r = P(Y_i = r)$, $i = 1, \ldots, m$. The multinomial distribution of y is abbreviated by

$$y \sim M(m, \pi) \quad \text{where} \quad \pi = (\pi_1, \ldots, \pi_q).$$

Sometimes it is useful to consider the scaled multinomial distribution $\bar{y} \sim M(m, \pi)/m$ where $\bar{y} = y/m$ instead of the multinomial $y \sim M(m, \pi)$. The mean $\bar{y}$ is an unbiased estimate of the underlying probability vector π with covariance matrix

$$\text{cov}(\bar{y}) = \frac{1}{m}(\text{diag}(\pi) - \pi\pi')$$

$$= \frac{1}{m} \begin{bmatrix} \pi_1(1-\pi_1) & -\pi_1\pi_2 \cdots & & -\pi_1\pi_q \\ -\pi_2\pi_1 & \ddots & & \\ \vdots & & \ddots & \\ -\pi_q\pi_1 & \cdots & \cdots & \pi_q(1-\pi_q) \end{bmatrix}.$$

3.1.2 Data

The data for regression type problems have the same form as in the case of univariate responses. Let

$$(y_i, x_i), \quad i = 1, \ldots, n$$

denote the observations of a cross section, where $x_i = (x_{i1}, \ldots, x_{im})$ is the vector of covariates and $y_i = (y_{i1}, \ldots, y_{iq})$ is the q–dimensional response vector, e.g., representing dummy variables for categories. Grouped and ungrouped data may be distinguished for the modelling of the conditional response y_i given x_i. In analogy to Section 2.1.1 the covariates of ungrouped data are given by a matrix X. However, observations of the dependent variable now form an $(n \times q)$-matrix.

	Response variable	Explanatory variables
Unit 1	$\begin{bmatrix} y_{11} & \cdots & y_{1q} \\ \vdots & & \vdots \\ y_{n1} & \cdots & y_{nq} \end{bmatrix}$	$\begin{bmatrix} x_{11} & \cdots & x_{1m} \\ \vdots & & \vdots \\ x_{n1} & \cdots & x_{nm} \end{bmatrix}$
Unit n		

If some of the covariates x_i are identical, the data may be grouped. Let $y_i^{(1)}, \ldots, y_i^{(n_i)}$ denote the n_i observed responses for fixed covariate x_i. Then the arithmetic mean

$$\bar{y}_i = \frac{1}{n_i} \sum_{j=1}^{n_i} y_i^{(j)}$$

or the sum of responses $n_i\bar{y}_i$ may be considered a response given fixed covariates x_i.

The grouped data may be condensed in the form

	Response variable	Explanatory variables
Group 1 (n_1 observations)	$\begin{bmatrix} y_{11} & \cdots & y_{1q} \\ \vdots & & \vdots \\ y_{g1} & \cdots & y_{gq} \end{bmatrix}$	$\begin{bmatrix} x_{11} & \cdots & x_{1m} \\ \vdots & & \vdots \\ x_{g1} & \cdots & x_{gm} \end{bmatrix}$
Group g (n_g observations)		

where now $x_i = (x_{i1}, \ldots, x_{im})$ stands for the covariates in the ith group and $y_i = (y_{i1}, \ldots, y_{iq})$ stands for the sum or the arithmetic mean of responses given fixed covariates x_i, $i = 1, \ldots, g$, $g \leq n$, where g is the number of covariates with different values in the data set.

3.1.3   The multivariate model

Univariate models with response y_i given x_i as considered in Chapter 2 have the form

$$\mu_i = h(z_i'\beta).$$

For a dichotomous response variable $y_i \in \{0, 1\}$, for example, the logistic model is given by

$$\pi_i = \frac{\exp(z_i'\beta)}{1 + \exp(z_i'\beta)}$$

where $\pi_i = P(y_i = 1|x_i)$. In the multinominal case $\pi_i = \mu_i = \mathrm{E}(y_i|x_i)$ is a $(q \times 1)$-vector $\pi_i = (\pi_{i1}, \ldots, \pi_{iq})$ rather than a scalar as earlier. Here the model has the form

$$\pi_i = h(Z_i\beta) \tag{3.1.3}$$

where h is a vector–valued response function, Z_i is a $(q \times p)$-design matrix composed from x_i, and β is a $(p \times 1)$-vector of unknown parameters. In analogy to the univariate case the influence term will be abbreviated $\eta_i = Z_i\beta$.

As an example consider the widely used multicategorical logit model that is treated more extensively in Section 3.2 and illustrated in Example 3.1. It is given by

$$P(Y_i = r) = \frac{\exp(\beta_{r0} + z_i'\beta_r)}{1 + \sum_{s=1}^{q} \exp(\beta_{s0} + z_i'\beta_s)}, \tag{3.1.4}$$

which may be written equivalently as

$$\log \frac{P(Y_i = r)}{P(Y_i = k)} = \beta_{r0} + z_i'\beta_r. \tag{3.1.5}$$

Here z_i is the vector of covariables determining the log odds for category r with respect to the reference category k. From the latter form of the model one immediately gets the response function $h = (h_1, \ldots, h_q)$ with

$$h_r(\eta_1, \ldots, \eta_q) = \frac{\exp(\eta_r)}{1 + \sum_{s=1}^{q} \exp(\eta_s)}, \quad r = 1, \ldots, q, \tag{3.1.6}$$

the design matrix

$$Z_i = \begin{bmatrix} 1 & z_i' & & \\ & 1 & z_i' & \\ & & \ddots & \\ & & & 1 & z_i' \end{bmatrix}$$

and the parameter vector $\beta = (\beta_{10}, \beta_1, \ldots, \beta_{q0}, \beta_q)$.

An alternative form of model (3.1.3) is given by using the link function g, which is the inverse of h, i.e., $g = h^{-1}$. Then the model has the form

$$g(\pi_i) = Z_i \beta.$$

As is immediately seen from (3.1.5) the link function of the logit model is given by $g = (g_1, \ldots, g_q)$, where

$$g_r(\pi_{i1}, \ldots, \pi_{iq}) = \log \frac{\pi_{ir}}{1 - (\pi_{i1} + \cdots + \pi_{iq})}.$$

This simple example shows that multivariate models (for multinomial responses) are characterized by two specifications:

- the response function h (or the link function $g = h^{-1}$),

- the design matrix that depends on the covariables and the model.

Example 3.1: Caesarian birth study

Consider the data on infection following Caesarian birth given in Table 1.1 of Chapter 1. The effect of risk factors and antibiotics on infection has already been examined in Example 2.1 of Chapter 2, though the analysis was simplified there by only distinguishing between occurrence and nonoccurrence of infection. In contrast to Chapter 2, we now want to distinguish between the two different types of infection. Therefore, our response variable Y has three possible outcomes (infection type I / infection type II / no infection), which are labeled with 1 to 3, thus having $Y \in \{1, 2, 3\}$. We introduce a multivariate response vector of dummy variables $y_i = (y_{i1}, y_{i2})$ to take into account the categorical character of Y. Let $y_{i1} = 1$ if the ith Caesarian was followed by infection I, and $y_{i2} = 1$ if it was followed by infection II. Assuming dummy coding, no infection leads to $y_i = (0, 0)$. There are three binary covariates, namely, NOPLAN, FACTOR, ANTIB, which we assume to be dummy coded with NOPLAN=1 for "the Caesarian has not been planned," FACTOR=1 for "there were risk factors present," ANTIB=1 for "there were antibiotics given as a prophylaxis." According to Subsection 3.1.2 the grouped data may be condensed in the form

		Response variable y_{i1}	y_{i2}	NOPLAN	ANTIB	FACTOR
Group 1	$n_1 = 40$	4	4	0	0	0
Group 2	$n_2 = 58$	11	17	0	0	1
Group 3	$n_3 = 2$	0	0	0	1	0
Group 4	$n_4 = 18$	0	1	0	1	1
Group 5	$n_5 = 9$	0	0	1	0	0
Group 6	$n_6 = 26$	10	13	1	0	1
Group 7	$n_7 = 98$	4	7	1	1	1

We fit a multicategorical logit model to the data taking 'no infection' as the reference category of Y. According to (3.1.5) the model can be written as

$$\log \frac{P(\text{infection type I})}{P(\text{no infection})} = \beta_{10} + z_i'\beta_1$$

$$\log \frac{P(\text{infection type II})}{P(\text{no infection})} = \beta_{20} + z_i'\beta_2.$$

Note that the link function g is now a *vector*-valued function of π_1 and π_2:

$$g(\pi_{i1}, \pi_{i2}) = \left(\log \frac{\pi_{i1}}{1 - \pi_{i1} - \pi_{i2}} \;,\; \log \frac{\pi_{i2}}{1 - \pi_{i1} - \pi_{i2}} \right).$$

The estimates of the multicategorical logit model are given as follows:

$\hat{\beta}_{10}$	−2.621	$\hat{\beta}_{20}$	−2.560
NOPLAN[1]	1.174	NOPLAN[2]	0.996
ANTIB[1]	−3.520	ANTIB[2]	−3.087
FACTOR[1]	1.829	FACTOR[2]	2.195

According to the given estimates, antibiotics given as a prophylaxis seem to decrease the relative risk of infection type I more than that of type II. On the contrary, the presence of risk factors seems to increase the relative risk of infection type II more than that of type I. In more detail, with risk factors present the relative risk of infection type I, P(infection type I)/P(no infection), is $\exp(-2.621 + 1.829) = 0.45$ times the relative risk with no risk factors present. For infection type II, however, the relative risk of infection with risk factors present is $\exp(-2.560 + 2.195) = 0.69$ times the relative risk with no risk factors present. We will examine the differences of parameter estimates for different types of infection in more detail in Section 3.4. □

3.1.4 Multivariate generalized linear models

Multinomial response models like (3.1.4) may be considered special cases of multivariate generalized linear models. In analogy to the univariate case (Section 2.1.2) multivariate generalized linear models are based on a distributional assumption and a structural assumption. However, the response variable y_i is now a q-dimensional vector with expectation
$\mu_i = E(y_i|x_i)$.

1. *Distributional assumption:*
   Given x_i, the y_i's are (conditionally) independent and have a distribution that belongs to a simple exponential family, which has the

form

$$f(y_i|\theta_i, \phi, \omega_i) = \exp \left\{ \frac{[y_i'\theta_i - b(\theta_i)]}{\phi} \omega_i + c(y_i, \phi, \omega_i) \right\}.$$

2. *Structural assumption:*
The expectation μ_i is determined by a linear predictor

$$\eta_i = Z_i \beta$$

in the form

$$\mu_i = h(\eta_i) = h(Z_i \beta)$$

where

– the response function $h : S \to M$ is defined on $S \subset \mathbb{R}^q$, taking values in the admissible set $M \subset \mathbb{R}^q$,
– Z_i is a $(q \times p)$-design matrix, and
– $\beta = (\beta_1, \ldots, \beta_p)$ is a vector of unknown parameters from the admissible set $B \subset \mathbb{R}^p$.

For the case of a multicategorial response one has to consider the multinomial distribution, which may be embedded into the framework of a simple (multivariate) exponential family. For $y_i = (y_{i1}, \ldots, y_{iq}) \sim M(n_i, \pi_i)$ the distribution of the arithmetic mean $\bar{y}_i = y_i/n_i$ has the form

$$f(\bar{y}_i|\theta_i, \phi, \omega_i) = \exp \left\{ \frac{\bar{y}_i'\theta_i - b(\theta_i)}{\phi} \omega_i + c(y_i, \phi, \omega_i) \right\} \qquad (3.1.7)$$

where the natural parameter θ_i is given by

$$\theta_i' = \left[\log \left(\frac{\pi_{i1}}{1 - \pi_{i1} - \ldots - \pi_{iq}} \right), \ldots, \log \left(\frac{\pi_{iq}}{1 - \pi_{i1} - \ldots - \pi_{iq}} \right) \right],$$

and

$$b(\theta_i) = \log(1 - \pi_{i1} - \ldots - \pi_{iq}),$$
$$c(y_i, \phi, \omega_i) = \log \left(\frac{n_i!}{y_{i1}! \cdots y_{iq}!(n - y_{i1} - \ldots - y_{iq})!} \right),$$
$$\omega_i = n_i.$$

The parameter ϕ may be treated as an additional dispersion parameter. In the following it is considered fixed with $\phi = 1$.

For the multinomial distribution the conditional expectation μ_i is the vector of probabilities $\pi(x_i) = (\pi_{i1}, \ldots, \pi_{iq})$, $\sum_{r=1}^{q} \pi_{ir} < 1$. Therefore the admissible set of expectations M is given by

$$M = \left\{ (z_1, \ldots, z_q) \, \Big| \, 0 < z_i < 1, \sum_i z_i < 1 \right\}.$$

For *metrical* response a multivariate generalized linear model being in wide use is the multivariate linear model assuming a normal distribution.

3.2   Models for nominal responses

Let the response variable Y have possible values $1,\ldots,k$ where the numbers are mere labels for the categories, i.e., neither ordering nor difference between category numbers is meaningful. As an example one may consider biometrical problems where the categories $1,\ldots,k$ stand for alternative types of infection (see Example 3.1). Nominal responses often occur in situations where an individual faces k choices. The categories refer then to the several alternatives. For example, in the choice of transportation mode the alternatives may be bus, train or automobile. In the following, models for nominal responses are motivated for the frequent situation where the response can be interpreted as resulting from an individual choice.

3.2.1   The principle of maximum random utility

Models for unordered categories may be motivated from the consideration of latent variables. In probabilistic choice theory it is often assumed that an unobserved utility U_r is associated with the rth response category. More general, let U_r be a latent variable associated with the rth category. For the choice of transportation mode the underlying variable may be interpreted as the consumers' utility connected to the transportation mode.

Let U_r be given by

$$U_r = u_r + \epsilon_r$$

where u_r is a fixed value associated with the rth response category and $\epsilon_1,\ldots,\epsilon_k$ are i.i.d. random variables with continuous distribution function F. Following the principle of maximum random utility the observable response Y is determined by

$$Y = r \quad \Leftrightarrow \quad U_r = \max_{j=1,\ldots,k} U_j. \tag{3.2.1}$$

This means that the response is category r if the latent variable U_r underlying this category is maximal. In choice situations the alternative is chosen which has maximal utility.

From (3.2.1) it follows that

$$
\begin{aligned}
P(Y = r) &= P(U_r - U_1 \geq 0, \ \ldots \ , U_r - U_k \geq 0) \\
&= P(\epsilon_1 \leq u_r - u_1 + \epsilon_r , \ldots, \epsilon_k \leq u_r - u_k + \epsilon_r) \\
&= \int_{-\infty}^{\infty} \prod_{s \neq r} F(u_r - u_s + \epsilon) \, f(\epsilon) \, d\epsilon \tag{3.2.2}
\end{aligned}
$$

where $f = F'$ is the density function of ϵ_r and F denotes the distribution function of ϵ_r.

Depending on the distributional assumption for the noise variables ϵ_r, equation (3.2.2) yields differing models. If the ϵ's are independently normally distributed, one gets the independent probit model. The more general multivariate probit model allows correlated noise variables. A simpler model is generated by assuming independent noise variables following the extreme value distribution

$$F(x) = \exp\left(-\exp(-x)\right). \tag{3.2.3}$$

Then by simple integration one gets the multinomial logit model

$$P(Y = r) = \frac{\exp(u_r)}{\sum\limits_{s=1}^{k} \exp(u_s)}. \tag{3.2.4}$$

Since only differences of utilities are identifiable it is useful to consider the alternative form

$$P(Y = r) = \frac{\exp(u_r - u_k)}{1 + \sum\limits_{s=1}^{q} \exp(u_s - u_k)} = \frac{\exp(\tilde{u}_r)}{1 + \sum\limits_{s=1}^{q} \exp(\tilde{u}_s)} \tag{3.2.5}$$

where $\tilde{u}_r = u_r - u_k$ is the difference between the rth utility and the reference utility u_k. We can see that the principle of maximum random utility in combination with a specific distribution function F do determine the link or response function of the model.

The response function $h : S \to M$, $S \subset \mathbb{R}^q$, determines how the expectation μ is related to the linear predictor $\eta = Z\beta$. For the multinomial logit model the response function $h = (h_1, \ldots, h_q)$ is given by (3.1.6).

The connection between the extreme value distribution and the logit model has been considered by Yellott (1977) and McFadden (1973). More general models based on stochastic utility maximization have been treated in the literature. McFadden (1981) considered the generalized extreme value distribution, Hausman and Wise (1978), Daganzo (1980), Lerman and Manski (1981), and McFadden (1984) considered probit models that do not assume independent utilities (see also Small, 1987, Börsch-Supan, 1990).

3.2.2 Modelling of explanatory variables: choice of design matrix

For multivariate models there are more possibilities of specifying the influence term than there are for the univariate models considered in Chapter 2. In particular there are several types of variables that might influence the response. Let us consider the situation where an individual faces k choices and a set of variables characterizes the individual. Let the ith individual be characterized by the vector $z_i = (z_{i1}, \ldots, z_{im})$ containing variables like age, sex and income. Since we now have to distinguish between individuals (observations) the index i is added. Consequently, u_{ir} will denote the utility of the rth category for individual i, Y_i will denote the categorical response variable. A simple linear model for the utility u_{ir} is given by

$$u_{ir} = \alpha_{r0} + z_i'\alpha_r$$

where $\alpha_r = (\alpha_{r1}, \ldots, \alpha_{rm})$ is a parameter vector. That means the preference of the rth alternative for the ith individual is determined by z_i and a parameter α_r that depends on the category. For example, in the choice of transportation mode, the individual income may be regarded as a covariate affecting the preference of different alternatives (automobile, bus, train). The according parameters α_r depend on the category, e.g. for increasing income an increasing preference of automobile but a decreasing preference of bus may be expected. In the following a parameter that depends on the category will be called *category–specific* . Since the explanatory variables z_i do not depend on the response categories the variables z_i are called *global.*

For the differences between utilities $\tilde{u}_{ir} = u_{ir} - u_{ik}$ one gets

$$\tilde{u}_{ir} = \beta_{r0} + z_i'\beta_r$$

where $\beta_{r0} = \alpha_{r0} - \alpha_{k0}$, $\beta_r = (\alpha_r - \alpha_k)$. Assuming only global variables z_i the multinomial logit model (3.2.5) has the form

$$P\left(Y_i = r|z_i\right) = \frac{\exp(\beta_{r0} + z_i'\beta_r)}{1 + \sum_{s=1}^{q} \exp(\beta_{s0} + z_i'\beta_s)}.$$

For the generalized linear model $\mu_i = h(Z_i\beta)$ with response function (3.1.6) one has the design matrix

$$Z_i = \begin{bmatrix} 1 & z_i' & & & 0 \\ & & 1 & z_i' & \\ & & & \ddots & \\ 0 & & & & 1 & z_i' \end{bmatrix}$$

and the parameter vector $\beta = (\beta_{10}, \beta_1, \ldots, \beta_{q0}, \beta_q)$, see Example 3.1.

In addition to global variables let the alternatives $1, \ldots, k$ themselves be characterized by variables. For example, when buying a new automobile the consumer faces choices characterized by prices, type, speed, etc. That means we have k sets of variables $w_1, \ldots, w_k$ that are connected to the k alternatives. Consequently variables of this type are called *alternative–specific* . Now let the utility be determined by

$$u_{ir} = \alpha_{r0} + z_i'\alpha_r + w_r'\gamma$$

where the additional term $w_r'\gamma$ with parameter γ accounts for the influence of the characteristics of alternatives (e.g., price, speed of transportation modes). Since γ does not depend on the category it is called a *global* parameter. The identifiable differences $\tilde{u}_{ir} = u_{ir} - u_{ik}$ are given by

$$\tilde{u}_{ir} = \beta_{r0} + z_i'\beta_r + (w_r - w_k)'\gamma.$$

The design matrix now has the form

$$Z_i = \begin{bmatrix} 1 & z_i' & & & & w_1' - w_k' \\ & & 1 & z_i' & & w_2' - w_k' \\ & & & & \ddots & \\ & & & & 1 & z_i' & w_q' - w_k' \end{bmatrix}$$

and the parameter vector is given by $\beta = (\beta_{10}, \beta_1, \ldots, \beta_{q0}, \beta_q, \gamma)$.

As is seen from the design matrices variables that are weighted by category–specific parameters induce a diagonal structure, whereas variables that are weighted by global parameters induce a column in the design matrix. This structure does not depend on whether a variable is global– (specific for the choice maker) or alternative–specific (specific for the alternatives at hand).

In an even more general case the alternative–specific variables may also vary across individuals. For the choice of transportation mode the prize often depends on the location of the residence. That means that price is a variable w_{ir} that depends on both the alternative $r = 1, \ldots, k$ *and* the individual $i = 1, \ldots, n$. We can include this type of alternative–specific variables w_{ir} into the design matrix Z_i by adding an extra column the same way we did with w_r. Note, however, that this column will contain values that differ between individuals, whereas the column containing variables w_r not depending on the individual was the same for all Z_i (see above).

Discrete choice modelling is treated extensively in Ben-Akiva and Lerman (1985), Maddala (1983), Pudney (1989) and Ronning (1991).

3.3   Models for ordinal responses

Response variables that have more than two categories often are ordinal. That means the events described by the category numbers $1, \ldots, k$ can be considered as ordered. However, one should keep in mind that only the ordering of the category numbers is meaningful. A formal theory of scale levels is developed in measurement theory (see, e.g., Roberts, 1979, or Krantz et al., 1971), but it is not necessary here. A formal characterization of a regression model as nominal or ordinal based on invariance principles is given in Tutz (1993). In the following we consider a more informal model ordinal if the ordering of response categories is taken into account. In particular for categorical data where the sample size is often critical, it is necessary

TABLE 3.1. Breathing results of Houston industrial workers

		Breathing Test Results		
Age	Smoking status	Normal	Borderline	Abnormal
< 40	Never smoked	577	27	7
	Former smoker	192	20	3
	Current smoker	682	46	11
40–59	Never smoked	164	4	0
	Former smoker	145	15	7
	Current smoker	245	47	27

to make use of all of the information available. Consequently the ordering of the response categories has to be taken into account allowing for simpler models.

Example 3.2: Breathing test results

Forthofer and Lehnen (1981, p.21) investigated the effect of age and smoking on breathing test results for workers in industrial plants in Texas (see also Agresti, 1984, p.96 ff). The test results have been classified in three categories, namely normal, borderline and abnormal. Thus, the response variable "breathing results" may be considered an ordinal variable. Table 3.1 gives the data from Forthofer and Lehnen (1981).   □

Example 3.3: Job expectation

In a study on the expectations of students at the University of Regensburg, psychology students have been asked if they expect to find an adequate employment within a reasonable time after getting their degree. The response categories were ordered with respect to their expectation. The categories 1 (don't expect adequate employment), 2 (not sure), and 3 (immediately after getting the degree) reflect this ordering. Table 3.2 shows the data for different ages of the students.   □

Ordinal variables may stem from quite different mechanisms. Anderson (1984) distinguishes between *"grouped continuous"* variables and *"assessed ordered"* categorical variables. The first type is a mere categorized version of a continuous variable, which in principle may be observed itself. For example, if the breathing test of Example 3.2 provides measurement on a physical scale, e.g., volume of breath, and the distinction normal, borderline and abnormal corresponds to intervals of the physical scale, then the variable "breath result" could be considered as a grouped continuous variable. The second type of ordered variable arises when an assessor processes an unknown amount of information leading to the judgment of the grade of

TABLE 3.2. Grouped data for job expectation of students of
psychology in Regensburg

Observation number	Age in years	Response categories 1	2	3	n_i
1	19	1	2	0	3
2	20	5	18	2	25
3	21	6	19	2	27
4	22	1	6	3	10
5	23	2	7	3	12
6	24	1	7	5	13
7	25	0	0	3	3
8	26	0	1	0	1
9	27	0	2	1	3
10	29	1	0	0	1
11	30	0	0	2	2
12	31	0	1	0	1
13	34	0	1	0	1

the ordered categorical scale. In the job expectation example the students
will weigh some information about their age, their grades at the university,
the level of unemployment in the country, etc., in making their judgment
on the category of their expectation. Consequently the response may be
considered an assessed ordered variable.

3.3.1   Cumulative models: the threshold approach

The most widely used model in ordinal regression is based on the so-called
category boundaries or threshold approach, which dates back at least to
Edwards and Thurstone (1952). It is assumed that the observable variable
Y is merely a categorized version of a latent continuous variable U. In the
case of a grouped continuous response variable, U may be considered the
unobserved underlying variable. In the case of an assessed ordered variable, U is the assessment on the underlying continuous scale. In both cases
the latent variable is primarily used for the construction of this type of
model. Although interpretation is simpler when the latent variable is taken
into account, the model may also be interpreted without reference to the
underlying continuous variable.

For a given vector x of explanatory variables the category boundary
approach postulates that the observable variable $Y \in \{1, \ldots, k\}$ and the
unobservable latent variable U are connected by

$$Y = r \quad \Leftrightarrow \quad \theta_{r-1} < U \leq \theta_r, \qquad r = 1, \ldots, k \qquad (3.3.1)$$

where $-\infty = \theta_0 < \theta_1 \ldots < \theta_k = \infty$. That means that Y is a coarser (categorized) version of U determined by the thresholds $\theta_1, \ldots, \theta_{k-1}$. Moreover, it is assumed that the latent variable U is determined by the explanatory variables by the linear form

$$U = -x'\gamma + \epsilon \tag{3.3.2}$$

where $\gamma = (\gamma_1, \ldots, \gamma_p)$ is a vector of coefficients and ϵ is a random variable with distribution function F.

From these assumptions it follows immediately that the observed variable Y is determined by the model

$$P(Y \leq r|x) = F(\theta_r + x'\gamma). \tag{3.3.3}$$

Since the left side of the equation is the sum of probabilities $P(Y = 1|x) + \ldots + P(Y = r|x)$, model (3.3.3) is called a cumulative model with distribution function F. Alternatively, the model is often called a threshold model. This name is due to the derivation based on the thresholds $\theta_1, \ldots, \theta_{k-1}$ of the latent variable.

Cumulative logistic model or proportional odds model

Specific choices of the distribution function lead to specific cumulative models. A common choice of the distribution function is the logistic distribution function $F(x) = 1/(1 + \exp(-x))$. Consequently, the cumulative logistic model has the form

$$P(Y \leq r|x) = \frac{\exp(\theta_r + x'\gamma)}{1 + \exp(\theta_r + x'\gamma)}, \tag{3.3.4}$$

$r = 1, \ldots, q = k - 1$. It is simple to show that equivalent forms are given by

$$\log\left\{\frac{P(Y \leq r|x)}{P(Y > r|x)}\right\} = \theta_r + x'\gamma \tag{3.3.5}$$

or

$$\frac{P(Y \leq r|x)}{P(Y > r|x)} = \exp(\theta_r + x'\gamma). \tag{3.3.6}$$

Another way of looking at model (3.3.3) is to consider the density $f = F'$ of the underlying continuous response U. The response mechanism (3.3.1) cuts the density of U into slices, whereby the cutoff points are determined by the thresholds. The explanatory term $-x'\beta$ in (3.3.2) determines the shift of the response U on the latent scale. Depending on the strength and direction of the shift the observable probabilities of response categories increase or decrease. In Figure 3.1 for three distribution functions the densities for two subpopulations corresponding to explanatory terms $-x_1'\beta$ and $-x_2'\beta$ are plotted.

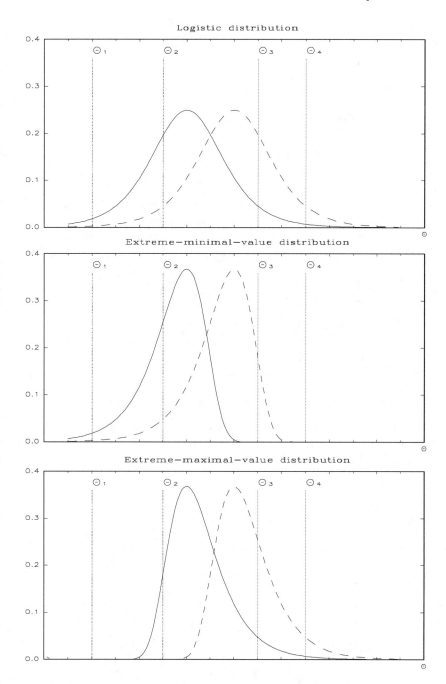

FIGURE 3.1. Densities of the latent response for two subpopulations with different values of x (logistic, extreme–minimal–value, extreme–maximal–value distributions).

On the left side of equation (3.3.6) we have the odds for the event $P(Y \leq r|x)$, since

$$\frac{P(Y \leq r|x)}{P(Y > r|x)} = \frac{P(Y \leq r|x)}{1 - P(Y \leq r|x)}.$$

According to (3.3.5) the logarithms of the cumulative odds

$$\log\left\{\frac{P(Y \leq r|x)}{P(Y > r|x)}\right\}, \quad r = 1, \ldots, q$$

are determined by a linear form of the explanatory variables. The logistic cumulative model has also been called the proportional odds model (Mc-Cullagh, 1980). This name is due to a special property of model (3.3.4). If two populations characterized by explanatory variables x_1 and x_2 are considered, the ratio of the cumulative odds for the two populations is given by

$$\frac{P(Y \leq r|x_1)/P(Y > r|x_1)}{P(Y \leq r|x_2)/P(Y > r|x_2)} = \exp\{(x_1 - x_2)'\gamma\}$$

and therefore does not depend on the category. That means the ratio of the cumulative odds for two populations is the same for all of the cumulative odds. This property yields a *strict stochastic ordering* of populations (see also Section 3.3.4).

Grouped Cox model or proportional hazards model

Another choice of the distribution function F is the extreme–minimal–value distribution $F(x) = 1 - \exp(-\exp(x))$. By inserting this distribution function in (3.3.3) one gets the model

$$P(Y \leq r|x) = 1 - \exp\{-\exp(\theta_r + x'\gamma)\}, \quad r = 1, \ldots, q, \quad (3.3.7)$$

or equivalently with complementary log-log links

$$\log[-\log P(Y > r|x)] = \theta_r + x'\gamma, \quad r = 1, \ldots, q.$$

The latter form shows that $\log(-\log P(Y > r|x))$ is determined by a linear influence term composed of the explanatory variables. The model is called *grouped Cox model* since it may be derived as a grouped version of the continuous Cox or proportional hazards model, which is well known in survival analysis (e.g., Kalbfleisch and Prentice, 1980).

As is seen in Figure 3.1 for small values the underlying density is rather similar to that of the cumulative logistic model. Consequently the models often yield very similar fits.

Extreme–maximal–value distribution model

Instead of the extreme-minimal-value distribution $F(x) = 1 - \exp(-\exp(x))$ the extreme-maximal-value distribution $F(x) = \exp(-\exp(-x))$ may be used. While the former is rather similar to the logistic distribution for small x, the latter is rather similar to the logistic distribution for large x. The model based on the extreme-maximal-value distribution is given by

$$P(Y \leq r|x) = \exp\left[-\exp\{-(\theta_r + x'\gamma)\}\right] \qquad (3.3.8)$$

or equivalently with log-log links

$$\log[-\log P(Y \leq r|x)] = -(\theta_r + x'\gamma). \qquad (3.3.9)$$

Although model (3.3.7) is not equivalent to model (3.3.8) it may be estimated using the latter one. To do so, one has to construct first a variable $\tilde{Y} = k + 1 - Y$ having support $1, ..., k$ but with inverse ordering of categories, and then estimate model (3.3.8) using $\tilde{Y}$ instead of Y. In a second step, the obtained estimates have to be multiplied by -1 to yield the estimates for model (3.3.7), and the order of the threshold parameters θ_r has to be reversed. More formally, let the ordering of categories be reversed by $\tilde{r} = k + 1 - Y$. Then the extreme-minimal-value model (3.3.7) for Y with parameters θ_r, β is equivalent to the extreme-maximal-value model (3.3.8) for $\tilde{Y}$ with parameters $\tilde{\theta}_r = -\theta_{k-\tilde{r}}, \tilde{\beta} = -\beta$.

Several families of link-functions have been proposed for the binary response model by Prentice (1976), Pregibon (1980), Aranda-Ordaz (1983), Morgan (1985), Stukel (1988), Friedl (1991) and Czado (1992). In principle those families may also be used for ordinal response variables. Genter and Farewell (1985) consider a family of link-functions within the cumulative model approach. Their family includes the probit model as well as extreme-minimal-value link and extreme-maximal-value link for the modelling of ordinal responses.

3.3.2   Extended versions of cumulative models

The simple cumulative model (3.3.3) is based on the assumption that the explanatory variable causes a shift on the latent scale but does not change the thresholds $\theta_1, \ldots, \theta_q$. In a more general model the threshold may also depend on explanatory variables $w = (w_1, \ldots, w_m)$ in the linear form

$$\theta_r = \beta_{r0} + w'\beta_r$$

where $\beta_r = (\beta_{r1}, \ldots, \beta_{rm})$ is a category–specific parameter vector (see Terza, 1985). The extended model follows directly from the mechanism (3.3.1) and the parameterization of the latent variable (3.3.2). It is given by

$$P(Y \leq r|x, w) = F(\beta_{r0} + w'\beta_r + x'\gamma). \qquad (3.3.10)$$

Model (3.3.10) still assumes the category boundaries mechanism, but now only for explanatory variables x. The variables w only determine the thresholds that lie on the latent score. As is immediately seen, the assumption $w_i = x_i$, $i = 1, \ldots, m$, $m = p$, makes γ an unidentifiable parameter. Thus one has to distinguish strictly between threshold variables w_i and shift variables x_i. Threshold variables are always weighted by a category–specific parameter vector β_r and shift variables are always weighted by a global parameter vector γ.

3.3.3   Link functions and design matrices for cumulative models

The embedding into the framework of generalized linear models is done by specifying the link (or response) function and the design matrix. The link function $g = (g_1, \ldots, g_q)$ is immediately given from (3.3.3) or (3.3.10) by

$$g_r(\pi_1, \ldots, \pi_q) = F^{-1}(\pi_1 + \ldots + \pi_r),$$

$r = 1, \ldots, q$. For the grouped Cox model one has

$$g_r(\pi_1, \ldots, \pi_q) = \log\{-\log(1 - \pi_1 - \ldots - \pi_r)\},$$

$r = 1, \ldots, q$. For the design matrix one has to distinguish between the simple model (3.3.3) and the general model (3.3.10). For the simple model the linear term $\eta_i = Z_i \beta$ is determined by

$$Z_i = \begin{bmatrix} 1 & & & x_i' \\ & 1 & & x_i' \\ & & \ddots & \vdots \\ & & & 1 & x_i' \end{bmatrix} \tag{3.3.11}$$

and the parameter vector $\beta = (\theta_1, \ldots, \theta_q, \gamma)$. For the more general model with threshold variables w_i and shift variables x_i the design matrix has the form

$$Z_i = \begin{bmatrix} 1 & w_i' & & & x_i' \\ & 1 & w_i' & & x_i' \\ & & & \ddots & \vdots \\ & & & 1 & w_i' & x_i' \end{bmatrix} \tag{3.3.12}$$

The parameter vector β is given by $\beta = (\beta_{1o}, \beta_1, \ldots, \beta_{qo}, \beta_q, \gamma)$. Here, neither the threshold variables nor the shift variables contain a constant. The constant itself corresponds to the thresholds β_{r0} that are category specific.

Sometimes, e.g., in random coefficient models (Chapter 7) or in dynamic modelling (Chapter 8), it is useful to consider an alternative form for link function and design matrix. For cumulative models the parameters may not vary freely. The parameters for the simple model (3.3.3) are restricted by $\theta_1 < \ldots < \theta_q$. For the general model, the restriction is determined by $\beta_{1o} + \beta_1'w < \ldots < \beta_{qo} + \beta_q'w$ for all possible values of the shift variables w. For link functions and design matrix as given earlier these restrictions are not taken into account. Since the restriction must hold for each w, the severeness of the restriction depends on the range of the covariates w. If the constraint is not explicitly used in estimation, the iterative estimation procedure may fail by fitting inadmissible parameters. There are no problems, if for the simple model (3.3.3) the thresholds are well separated. However, there might be numerical problems in the estimation procedure if some thresholds are very similar. For model (3.3.3) these problems may be simply avoided by using an alternative formulation. The model may be reparameterized by

$$\alpha_1 = \theta_1, \qquad \alpha_r = \log(\theta_r - \theta_{r-1}), \qquad r = 2, \ldots, q,$$

or respectively

$$\theta_1 = \alpha_1, \qquad \theta_r = \theta_1 + \sum_{i=2}^{r} \exp(\alpha_r), \qquad r = 2, \ldots, q.$$

Then the parameters $\alpha_1, .., \alpha_q$ are not restricted and we have $(\alpha_1, .., \alpha_q) \in \mathbb{R}^q$. The linear structure of the model becomes obvious in the form

$$F^{-1}\big(P(Y = 1|x)\big) = \alpha_1 + x'\gamma$$

$$\log\left[F^{-1}\{P(Y \le r|x)\} - F^{-1}\{P(Y \le r-1|x)\}\right] = \alpha_r, \qquad r = 2, \ldots, q.$$

Hereby the link function is determined.

For the special case of the logistic cumulative model one gets the link function

$$g_1(\pi_1, \ldots, \pi_q) = \log\left(\frac{\pi_1}{1 - \pi_1}\right)$$

$$g_r(\pi_1, \ldots, \pi_q) = \log\left[\log\left\{\frac{\pi_1 + \ldots + \pi_r}{1 - \pi_1 - \ldots - \pi_r}\right\}\right.$$

$$\left. - \log\left\{\frac{\pi_1 + \ldots + \pi_{r-1}}{1 - \pi_1 - \ldots - \pi_{r-1}}\right\}\right],$$

$r = 1, \ldots, q$. Of course, when using this alternative link function for the logit–type model the design matrix has to be adapted. The design matrix for observation (y_i, x_i) now has the form

$$Z_i = \begin{bmatrix} 1 & & & x_i' \\ & 1 & & 0 \\ & & \ddots & \vdots \\ & & & 1 & 0 \end{bmatrix} \qquad (3.3.13)$$

TABLE 3.3. Estimates of cumulative models for breathing test data (p-values in each second column)

	Cumulative logistic		Proportional hazards		Extreme–maximal–value	
Threshold 1	2.370	0.0	0.872	0.0	2.429	0.0
Threshold 2	3.844	0.0	1.377	0.0	3.843	0.0
AGE[1]	0.114	0.29	0.068	0.04	0.095	0.37
SMOKE[1]	0.905	0.0	0.318	0.0	0.866	0.19
SMOKE[2]	−0.364	0.01	−0.110	0.02	−0.359	0.14
AGE[1]*SMOKE[1]	−0.557	0.0	−0.211	0.00	−0.529	0.19
AGE[1]*SMOKE[2]	0.015	0.91	0.004	0.92	0.021	0.14
Deviance	8.146		3.127		9.514	

and the parameter vector is given by $\beta = (\alpha_1, \ldots, \alpha_q, \gamma)$.

Example 3.4: Breathing test results (Example 3.2 continued)
The data given in Table 3.1 are used to investigate the effect of age and smoking upon breathing test results. The analysis is based on the cumulative model in three variants, namely, the cumulative logistic model, the proportional hazards model and the extreme-maximal-value model. Taking the deviance (see Section 3.4) as a measure of distance between the data and the fitted values, the proportional hazards model shows the best fit. The extreme-maximal-value model shows the worst fit. Thus it is neglected in the following.

Table 3.3 gives the estimates for the three models of the simple type (3.3.3). All variables are given in effect coding, with −1 for the last category. Positive sign of the coefficient signals a shifting on the latent scale to the left end yielding higher probabilities for normal and borderline categories (i.e., categories 1 and 2). As is to be expected, low age and low smoking categories produce better breathing test results.

Parameter estimates must be different for the models because of the differing variance of the logistic and the extreme-value distributions. As is seen from Figure 2.2 (Chapter 2) the extreme–minimal–value distribution function, underlying the proportional hazards model, is steeper than the logistic distribution function. Therefore, to achieve the same amount of shifting on the latent scale a larger effect (measured by the parameter) is necessary for the latter distribution function. Consequently the parameters for the logistic model are larger for all of the variables. However, the tendency of the parameters is about the same.

In this data set the strong interaction effect of age and smoking is interesting. It is much more impressive than the main effect of age. The interaction AGE[1]*SMOKE[1] $= -0.211$ (proportional hazards model) shows that the positive tendency given by the strong influence SMOKE[1] $= 0.318$ is not so impressive when the person is still young. Since effect coding was used, the interaction effects not given in Table 3.3 are easily computed via the restriction so that they sum zero. This leads to the following table of interaction effects:

	SMOKE[1]	SMOKE[2]	SMOKE[3]
AGE[1]	-0.211	0.004	0.207
AGE[2]	0.211	-0.004	-0.207

From this table of interactions it can be seen very easily that the smoking history becomes really influential for higher age. Note that the interaction AGE[2]*SMOKE[3]$=-0.207$ has to be added to the negative effects AGE[2] $= -0.068$ and SMOKE[3]$=-0.208$. The same conclusions are drawn from the logistic model, although the fit is inferior in comparison to the proportional hazards model.   □

Example 3.5: Job expectation (Example 3.3 continued)
For the job expectation data given in Table 3.2 log AGE is considered in the influence term. Two models have been fitted:

the simple cumulative logistic model

$$P(Y \leq r | AGE) = F(\theta_r + \gamma \log AGE)$$

and the extended version

$$P(Y \leq r | AGE) = F(\beta_{r0} + \beta_r \log AGE)$$

where log AGE is a threshold variable.

Estimates are given in Table 3.4. In particular the simple model where log AGE is a shift variable shows a rather bad fit in Pearson's χ^2. The fit of the extended version is not overwhelming but not so bad. The strong difference between deviance and Pearson's χ^2 may be a hint that the assumptions for asymptotics may be violated (see increasing cells asymptotic in Section 3.4.3). The extended version yields estimates for the parameters and an obvious change in the threshold values. The negative values of $\hat{\beta}_r$

TABLE 3.4. Cumulative model for job expectation data

	Log AGE as global shift variable		Log AGE as threshold variable	
Threshold 1	14.987	(0.010)	9.467	(0.304)
Threshold 2	18.149	(0.002)	20.385	(0.002)
Slope γ	-5.402	(0.004)	—	
Category–specific slope β_1	—		-3.597	(0.230)
Category-specific slope β_2	—		-6.113	(0.004)
Pearson's χ^2	42.696	(0.007)	33.503	(0.055)
Deviance	26.733	(0.267)	26.063	(0.248)

(and $\hat{\gamma}$) signal that increasing age yields lower probabilities for low categories. Students seem to get more optimistic when getting closer to the examinations. However, the effect on the cumulative odds

$$\frac{P(Y \leq r|x_1)/P(Y > r|x_1)}{P(Y \leq r|x_2)/P(Y > r|x_2)} = \exp(\beta_{r0} + \beta_r(x_1 - x_2))$$

depends on the category r. Consider age groups $x_1 > x_2$. Then the cumulative odds for $r = 1$ measures the tendency toward strongly negative expectation (category 1 : don't expect adequate employment); for $r = 2$ the cumulative odds measure the tendency towards strongly negative *or* uncertain expectation (category 2 : not sure) in comparison to the positive statement of category 3. Since $\hat{\beta}_2 < \hat{\beta}_1$ the effect of age on the latter cumulative odds is stronger than for the former cumulative odds. Looking at the p–value 0.230 of β_1 shows that the real negative expectation (category 1 versus categories 2 and 3) may even be influenced by age. However, even the extended model shows a very high value of Pearsons's χ^2, which is a goodness–of–fit measure given in Section 3.4.2. The bad fit suggests further investigation of the residuals, which is given in Chapter 4.   □

3.3.4   Sequential models

In many applications the ordering of the response categories is due to a sequential mechanism. The categories are ordered since they can be reached only successively. A response variable of this type is tonsil size, as considered in the following example.

Example 3.6: Tonsil size

A data set that has been considered by several authors (e.g., Holmes and Williams, 1954, McCullagh, 1980) focuses on the tonsil size of children. Children have been classified according to their relative tonsil size and whether or not they are carriers of Streptococcus pyogenes. The data are given in Table 3.5.

TABLE 3.5. Tonsil size and Streptococuss pyogenes
(Holmes and Williams, 1954)

	Present but not enlarged	Enlarged	Greatly enlarged
Carriers	19	29	24
Noncarriers	497	560	269

It may be assumed that tonsil size always starts in the normal state "present but not enlarged" (category 1). If the tonsils grow abnormally, they may become "enlarged" (category 2), if the process does not stop, tonsils may become "greatly enlarged" (category 3). But in order to get greatly enlarged tonsils, they first have to be enlarged whatever the duration of the intermediate state "enlarged" is.   □

For data of this type models based on a sequential mechanism will often be more appropriate. In a similar way as for the cumulative model, sequential models may be motivated from latent variables.

Let latent variables U_r, $r = 1, \ldots, k - 1$, have the linear form $U_r = -x'\gamma + \epsilon_r$ where ϵ_r is a random variable with distribution function F. The response mechanism starts in category 1 and the first step is determined by

$$Y = 1 \quad \Leftrightarrow \quad U_1 \leq \theta_1$$

where θ_1 is a threshold parameter. If $U_1 \leq \theta_1$ the process stops. For the tonsil size data U_1 may represent the latent tendency of growth in the initial state of normal tonsil size. If U_1 is below threshold θ_1 the tonsil size remains normal ($Y = 1$), if not, at least enlarged tonsils result ($Y \geq 2$). That means that, if $U_1 > \theta_1$, the process is continuing in the form

$$Y = 2 \quad \text{given} \quad Y \geq 2 \quad \Leftrightarrow \quad U_2 \leq \theta_2$$

and so on. The latent variable U_2 may represent the unobservable tendency of growth when the tonsils are already enlarged. Generally, the complete mechanism is specified by

$$Y = r \quad \text{given} \quad Y \geq r \quad \Leftrightarrow \quad U_r \leq \theta_r$$

or equivalently

$$Y > r \quad \text{given} \quad Y \geq r \quad \Leftrightarrow \quad U_r > \theta_r \qquad (3.3.14)$$

$r = 1, \ldots, k - 1$. The sequential mechanism (3.3.14) models the transition from category r to category $r + 1$ given category r is reached. A transition takes place only if the latent variable determining the transition is above a threshold that is characteristic for the category under consideration.

The main difference to the category boundaries approach is the conditional modelling of transitions. The sequential mechanism assumes a binary decision in each step. Given category r is reached it must be decided whether the process stops (thus getting r as the final category) or whether it continues with a resulting higher category. Only the finally resulting category is observable.

The sequential response mechanism (3.3.14) combined with the linear form of the latent variables $U_r = -x'\gamma + \epsilon_r$ immediately leads to the sequential model with distribution function F

$$P\big(Y = r | Y \geq r, x\big) \ = \ F(\theta_r + x'\gamma), \qquad (3.3.15)$$

$r = 1, \ldots, k$, where $\theta_k = \infty$. The probabilities of the model are given by

$$P\big(Y = r | x\big) \ = \ F(\theta_r + x'\gamma) \prod_{i=1}^{r-1} \{1 - F(\theta_i + x'\gamma)\},$$

$r = 1, \ldots, k$, where $\prod_{i=1}^{0} \{\cdot\} = 1$. Model (3.3.15) is also called continuation ratio model (e.g., Agresti 1984). Note that for the parameters $\theta_1, \ldots, \theta_q$ no ordering restriction is needed as was the case for the cumulative–type model.

So far we have only considered the general form of the sequential model. There are several sequential models depending on the choice of the distribution function F. If F is chosen as the logistic distribution $F(x) = 1/(1 + \exp(-x))$ we get the sequential logit model

$$P\big(Y = r | Y \geq r, x\big) \ = \ \frac{\exp(\theta_r + x'\gamma)}{1 + \exp(\theta_r + x'\gamma)}$$

or equivalently

$$\log\left\{ \frac{P(Y = r | x)}{P(Y > r | x)} \right\} \ = \ \theta_r + x'\gamma.$$

For the extreme–value–distribution $F(x) = 1 - \exp(-\exp(x))$ the model has the form

$$P\big(Y = r | Y \geq r, x\big) \ = \ 1 - \exp\big(-\exp(\theta_r + x'\gamma)\big) \qquad (3.3.16)$$

or equivalently

$$\log\left[-\log\left\{ \frac{P(Y > r | x)}{P(Y \geq r | x)} \right\} \right] \ = \ \theta_r + x'\gamma. \qquad (3.3.17)$$

It is noteworthy that this sequential model is equivalent to the *cumulative* model with distribution function $F(x) = 1 - \exp(-\exp(x))$. That means model (3.3.17) is a special parametric form of the grouped Cox model. The equivalence is easily seen by the reparameterization

$$\theta_r = \log\left\{ \exp(\tilde{\theta}_r) - \exp(\tilde{\theta}_{r-1}) \right\}, \qquad r = 1, \ldots, k-1$$

or

$$\tilde{\theta}_r = \log\left(\sum_{i=1}^{r} \exp(\theta_i) \right).$$

If θ_r is inserted in (3.3.17) one gets the grouped Cox model

$$P(Y \leq r|x) \;=\; 1 - \exp\left(-\exp(\tilde{\theta}_r + x'\gamma)\right)$$

(see also Läärä and Matthews 1985, Tutz, 1991a).

Another member of the sequential family is the exponential sequential model, which is based on the exponential distribution $F(x) = 1 - \exp(-x)$. It is given by

$$P(Y = r|Y \geq r, x) \;=\; 1 - \exp\left(-(\theta_r + x'\gamma)\right)$$

or equivalently

$$-\log\left(\frac{P(Y > r|x)}{P(Y \geq r|x)} \right) \;=\; \theta_r + x'\gamma.$$

Generalized sequential models

In the same way as for the cumulative model the thresholds θ_r of the sequential model may be determined by covariates $z = (z_1, \ldots, z_m)$. From the linear form

$$\theta_r = \delta_{ro} + z'\delta_r$$

where $\delta_r = (\delta_{r1}, \ldots, \delta_{rm})$ is a category–specific parameter vector one gets the generalized sequential model

$$P(Y = r|Y \geq r, x) \;=\; F(\delta_{r0} + z'\delta_r + x'\gamma). \qquad (3.3.18)$$

Alternatively model (3.3.18) can be derived directly from the sequential mechanism. One has to assume that the latent variables U_r have the form $U_r = -x'\gamma - z'\delta_r + \epsilon_r$ and that the response mechanism

$$Y > r \quad \text{given} \quad Y \geq r \quad \text{if} \quad U_r > \delta_{r0}$$

is given for the fixed thresholds δ_{r0}.

Implicitly it is assumed that the influence of variables z on the transition from category r to $r+1$ depends on the category. The effect of the variables z is nonhomogeneous over the categories whereas the effect of variables x

TABLE 3.6. Fits for tonsil size data

	Cumulative logit	Sequential logit
θ_1	−0.809 (0.013)	−0.775 (0.011)
θ_2	1.061 (0.014)	0.468 (0.012)
γ	−0.301 (0.013)	−0.264 (0.010)
Pearson	0.301	0.005
Deviance	0.302	0.006
DF	1	1

The variable is given in effect coding, p–values are given in parentheses.

is homogeneous. Given category r is reached, the transition to a higher category is always determined by $x'\gamma$. Since the shift of the underlying score is determined by $x'\gamma$ and is constant over categories, in analogy to the general cumulative model, variables x are called shift variables. Although the assumption of linearly determined thresholds $\theta_r = \delta_{r0} + z'\delta_r$ is optional, variables z with a category–specific weight δ_r are called threshold variables. Thus the distinction between shift and threshold variables corresponding to global or category–specific weighting is the same as for cumulative models.

Example 3.7: Tonsil size (Example 3.6 continued)
Table 3.6 shows the estimates for the tonsil size data for the cumulative logit model and for the sequential logit model. Pearson statistic and deviance suggest a better fit of the sequential model. However, the extremely small values of the statistics for the sequential model may be a hint for underdispersion (see Section 3.4). Parameters have to be interpreted with reference to the type of model used. The estimated parameter $\gamma = -0.301$ for the cumulative model means that the odds $P(Y \le r|x)/P(Y > r|x)$ of having normal–sized tonsils ($r = 1$) as well as the odds of having normal–sized or enlarged tonsils ($r = 2$) is $\exp((x_1 - x_2)'\gamma) = \exp(-0.301(-1 - 1)) \approx 1.8$ times as large for noncarriers ($x = -1$) as for carriers ($x = 1$; x was effect–coded). Within the sequential model the parameter γ gives

TABLE 3.7. Sequential logit models for the breathing test data (p–values in brackets)

	Simple sequential logit model		Sequential logit model with threshold variables	
Threshold 1	2.379	(0.0)	2.379	(0.0)
Threshold 2	1.516	(0.0)	1.510	(0.0)
AGE	0.094	(0.368)	0.092	(0.385)
SMOKE[1]	0.882	(0.0)	0.915	(0.0)
			0.675	(0.108)
SMOKE[2]	−0.356	(0.008)	−0.375	(0.008)
			−0.163	(0.609)
AGE*SMOKE[1]	−0.601	(0.001)	−0.561	(0.003)
			−0.894	(0.047)
AGE*SMOKE[2]	0.092	(0.492)	0.015	(0.912)
			0.532	(0.161)

the strength with which the transition from category 1 to 2, and from 2 to 3, is determined. The estimated value $\gamma = -0.264$ means that the odds $P(Y = r|x)/P(Y > r|x)$ of having normal–sized tonsils ($r = 1$) is $\exp((x_1 - x_2)'\gamma) = \exp(-0.264(-1 - 1)) \approx 1.7$ times as large for non-carriers as for carriers. The same proportion holds for the odds $P(Y = 2|x)/P(Y > 2|x)$ of having merely enlarged tonsils given the tonsils are not normal. The sequential model as a step-wise model assumes that the process of enlargement as far as the comparison of carriers and noncarriers is concerned does not stop in the "enlarged" category but is going on and leads to greatly enlarged tonsils where $(x_1 - x_2)'\gamma = 2\gamma$ determines the comparison between populations.   □

Example 3.8: Breathing test results (Example 3.2 and 3.4 continued) The effect of age and smoking history on breathing test results has already been investigated by use of the cumulative model. However, the categories "normal," "borderline" and "abnormal" may be seen as arising from a sequential mechanism starting with "normal test results." Consequently the sequential model may be used for this data set. Table 3.7 gives the estimates for two sequential logit models. The first is of the simple type (3.3.15) with AGE, SMOKE and the interaction AGE*SMOKE. The second one is of the generalized type (3.3.18) where AGE is a shift variable (with global weight) and SMOKE and the interaction AGE*SMOKE are threshold variables (with category–specific weight). The deviance for the first model is 4.310 on 5 df, which shows that the sequential model fits the data better than the cumulative model (see Table 3.3). The estimates of the generalized–type model are given for comparison. It is seen that for variable SMOKE the category–specific effects are not so different for the two thresholds. The first has significant effects, for the second threshold the p–values are rather high, an effect due to the low number of observations in category 3. The thresholds for the interaction AGE*SMOKE[2] have quite different effects but both of them turn out to be not significant.   □

Link functions of sequential models

Response and link function may be derived directly from model (3.3.15). The link function $g = (g_1, \ldots, g_q)'$ is given by

$$g_r(\pi_1, \ldots, \pi_q) = F^{-1}\big(\pi_r/(1 - \pi_1 - \ldots - \pi_{r-1})\big), \qquad r = 1, \ldots, q,$$

and the response function $h = (h_1, \ldots, h_q)'$ has the form

$$h_r(\eta_1, \ldots, \eta_q) = F(\eta_r) \prod_{i=1}^{r-1} \Big(1 - F(\eta_i)\Big), \qquad r = 1, \ldots, q.$$

For the sequential logit model we have

$$g_r(\pi_1, \ldots, \pi_q) = \log\left(\frac{\pi_r}{1 - \pi_1 - \ldots - \pi_r}\right), \qquad r = 1, \ldots, q,$$

and

$$h_r(\eta_1, \ldots, \eta_q) = \exp(\eta_r) \prod_{i=1}^{r} \Big(1 + \exp(\eta_i)\Big)^{-1}, \qquad r = 1, \ldots, q.$$

For the other models the functions may be easily derived.

Naturally, the design matrices depend on the specification of shift and threshold variables. But as far as design matrices are concerned there is no difference between sequential and cumulative models. Thus the design matrices from Section 3.3.3 apply.

3.3.5   Strict stochastic ordering*

In Section 3.3.1 it is shown that the ratio of the cumulative odds for the subpopulations does not depend on the category. This property, called strict stochastic ordering (McCullagh, 1980), is shared by all simple cumulative models of the type (3.3.3). Let two groups of subpopulations be represented by covariates x_1, x_2. Then for the cumulative model (3.3.3) the difference

$$\Delta_c(x_1, x_2) = F^{-1}\Big\{P(Y \leq r|x_1)\Big\} - F^{-1}\Big\{P(Y \leq r|x_2)\Big\}$$

is given by $\gamma'(x_1 - x_2)$. That means that, e.g., for the logistic cumulative model, where $F^{-1}\{P(Y \leq r|x)\} = \log\{P(Y \leq r|x)/P(Y > r|x)\} = l_r(x)$, the difference between "cumulative" log-odds

$$\Delta_c(x_1, x_2) = l_r(x_1) - l_r(x_2) = \gamma'(x_1 - x_2)$$

does not depend on the category. Thus if one of the cumulative log-odds l_r is for population x_1 larger than for population x_2, then this holds for the cumulative log-odd of any category. For the grouped Cox model one gets

$$F^{-1}\{P(Y \leq r|x)\} = \log\left(-\log P(Y > r|x)\right).$$

Thus the difference of log-log transformed cumulative probabilities does not depend on r.

It should be noted that for the *general* cumulative model (3.3.10) the strict stochastic ordering of subpopulations no longer holds. Omitting the term $x'\gamma$ and using $w = x$ yields a special generalization of the simple model (3.3.3). For this model a test of $\beta_1 = \ldots = \beta_q$ may be considered a test of strict stochastic ordering. Armstrong and Sloan (1989) used this approach and considered relative efficiency of using dichotomized outcomes compared with ordinal models. For tests of the type $\beta_1 = \ldots = \beta_q$ see Section 3.4.2.

Strict stochastic ordering is not restricted to the cumulative models. As is seen from the sequential model (3.3.15), for two subpopulations characterized by covariates x_1, x_2 we get

$$
\begin{aligned}
\Delta_s(x_1, x_2) &= F^{-1}\{P(Y = r|Y \geq r, x_1)\} - F^{-1}\{P(Y = r|Y \geq r, x_2)\} \\
&= \gamma'(x_1 - x_2).
\end{aligned}
$$

That means the difference $\Delta_s(x_1, x_2)$ does not depend on the category. If the hazard $P(Y = r|Y \geq r)$ for subpopulation x_2 is larger than the hazard for subpopulation x_1 this relation holds for any category r (see also Tutz, 1991a).

3.3.6   Two–step models

Both types of models, cumulative and sequential, only make use of the ordering of the response categories $1, \ldots, k$. However, often the response categories quite naturally may be divided into sets of categories with very homogeneous responses where the sets are heterogeneous.

Example 3.9: Rheumatoid arthritis
Mehta, Patel and Tsiatis (1984) analysed data of patients with acute rheumatoid arthritis. A new agent was compared with an active control, and each patient was evaluated on a five-point assessment scale. The data are given in Table 3.8.
The global assessment in this example may be subdivided in the coarse response "improvement," "no change" and "worse." On a higher level improvement is split up into "much improved" and "improved" and the

TABLE 3.8. Clinical trial of a new agent and an active control

Drug	Global assessment				
	Much improved	Improved	No change	Worse	Much worse
New agent	24	37	21	19	6
Active control	11	51	22	21	7

"worse" category is split up into "worse" and "much worse." Thus there is a split on the response scale after category 2 and a split after category 3, yielding three sets of homogeneous responses.   □

For data of this type it is useful to model in a first step the coarse response and in a second step the response within the sets of homogeneous categories. More general, let the categories $1, \ldots, k$ be subdivided into t basic sets $S_1, \ldots, S_t$ where $S_j = \{m_{j-1} + 1, \ldots, m_j\}$, $m_o = 0$, $m_t = k$.

Let the response in the *first* step in one of the sets be determined by a cumulative model, based on an underlying latent variable $U_o = -x'\gamma_o + \epsilon$, where the random variable ϵ has distribution function F. The response mechanism is given by

$$Y \in S_j \quad \Leftrightarrow \quad \theta_{j-1} < U_o \leq \theta_j.$$

In the *second* step let the conditional mechanism conditioned on S_j also be determined by a cumulative model based on a latent variable $U_j = -x'\gamma_j + \epsilon_j$, with ϵ_j also having distribution function F. We assume

$$Y = r | Y \in S_j \quad \Leftrightarrow \quad \theta_{j,r-1} < U_j \leq \theta_{jr}.$$

With the assumption of independent noise variables ϵ_j, the resulting model is given by

$$
\begin{aligned}
P(Y \in T_j | x) &= F(\theta_j + x'\gamma_o) \\
P(Y \leq r | Y \in S_j, x) &= F(\theta_{jr} + x'\gamma_j)
\end{aligned}
\tag{3.3.19}
$$

where $T_j = S_1 \cup \ldots \cup S_j$, $\theta_1 < \ldots < \theta_{t-1}$, $\theta_t = \infty$,

$$\theta_{j,m_{j-1}+1} < \ldots < \theta_{j,m_j-1}, \quad \theta_{j,m_j} = \infty, \quad j = 1, \ldots, t.$$

The underlying process in both steps is based on cumulative mechanisms. Thus, the model is called a *two-step cumulative model* or a *compound cumulative (-cumulative) model*.

An advantage of the two–step model is that different parameters are involved in different steps. The choice between the basic sets is determined by the parameter γ_o, the choice on the higher level is determined by the parameters γ_i. Thus, in the model the influence of the explanatory variables on the dependent variable may vary. The choice between the basic sets, e.g., the alternatives improvement or no improvement, may be influenced by different strength or even by different variables than the choice within the "improvement" set and the "no improvement" set.

Example 3.10: Rheumatoid arthritis (Example 3.9 continued)
The arthritis data may serve as an illustration example. Here the basic sets are given by $S_1 = \{1, 2\}$, $S_2 = \{3\}$, $S_3 = \{4, 5\}$. Table 3.9 gives the results for the fitting of the compound logit model. The standardized estimates $\hat{\beta}/\sqrt{\text{var}(\hat{\beta})}$ are given in brackets. The simple cumulative model yields deviance 5.86 with 3 d.f. Thus it will not be rejected but it does not fit very well. The cumulative model assumes that the distribution of the underlying variable for the active control treatment group is shifted with respect to the new agent treatment group. Thus the response probabilities of all categories are changed simultaneously with the consequence of increasing probability for low response categories and decreasing probability for high

TABLE 3.9. Analysis of clinical trial data on rheumatoid arthritis

	$\hat{\gamma}_0$	$\hat{\gamma}_1$	$\hat{\gamma}_2$	Log–likelihood	Deviance	DF
Two–step cumulative model	0.04 (0.28)	0.55 (2.60)	-0.03 (-0.08)	-315.502	0.008	1
Two–step cumulative model with $\gamma_0 = 0$	—	0.55	-0.02	-315.541	0.087	2
Two–step cumulative model with $\gamma_0 = \gamma_2 = 0$	—	0.55	—	-315.545	0.094	3
Two–step cumulative model with $\gamma_0 = \gamma_1 = \gamma_2 = 0$				-319.132	7.269	4

response categories.

The compound cumulative model in the first step models whether according to the assessment health is getting better or worse. With $\hat{\gamma}_o = 0.04$ the effect is negligible. On a higher level it is investigated whether there is a "conditional shift" between treatment groups within the basic sets. The effect within the basic sets seems not to be negligible for the improvement categories. From the results in Table 3.9 it follows that for the compound model γ_o and γ_2 may be omitted. If, in addition we consider the model with $\gamma_1 = 0$ the deviance is rather large. Thus the only effect that may not be neglected is γ_1. The new drug seems to have an effect only if there is an improvement. The cumulative compound model with $\gamma_o = \gamma_2 = 0$ has the same number of parameters as the simple cumulative model but with deviance 0.094 a much better fit. This is because the cumulative model is unable to detect that the only essential transition is from category 1 to category 2.

Model (3.3.19) is only one representative of two–step models. We get alternative models, e.g., the cumulative–sequential model if the first step is based on a cumulative model and the second step is based on a sequential model. A model of this type is appropriate if we have three basic sets where S_2 is a sort of starting point and S_1, S_3 are stages of change in differing directions. The arthritis example is of this type. However, for this example S_1, S_3 have only two categories and therefore there is no difference between a cumulative and a sequential model in the second step. For alternative models and examples see Tutz (1989b), Morawitz and Tutz (1990).   □

Link function and design matrix for two–step models

Two–step models may be written in the form of a generalized linear model. For the cumulative (-cumulative) model the link function is given by

$$
\begin{aligned}
g_j(\pi_1,\ldots,\pi_q) &= F^{-1}(\pi_1 + \pi_2 + \ldots + \pi_{m_j}), \qquad j = 1,\ldots,t, \\
g_{jr}(\pi_1,\ldots,\pi_q) &= F^{-1}\left\{ \frac{\pi_{m_{j-1}+1} + \ldots + \pi_{m_{j-1}+1+r}}{\pi_{m_{j-1}+1} + \ldots + \pi_{m_j}} \right\} \\
j = 1,\ldots,t, &\qquad r = 1,\ldots,m_j - 1.
\end{aligned}
$$

The design matrix is somewhat more difficult than for the simple cumulative and sequential model. It has the form

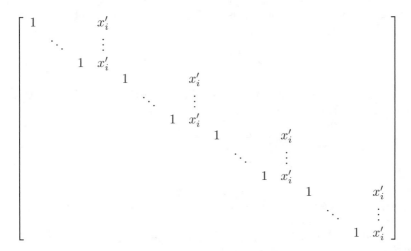

with parameter vector

$$\beta' = (\ \theta_1, .., \theta_t, \gamma'_o, \quad \theta_{11}, .., \theta_{1,m_1-1}, \gamma'_1, \quad ...., \quad \theta_{t1}, .., \theta_{t,m_t-1}, \gamma'_t\ ).$$

3.3.7 Alternative approaches*

Alternative approaches in ordinal regression have been considered and are shortly sketched in the following. Anderson (1984) introduced the so–called stereotype regression model, which in the simple one–dimensional form is given by

$$P(Y = r|x) = \frac{\exp(\beta_{r0} - \phi_r \beta' x)}{1 + \sum_{i=1}^{q} \exp(\beta_{i0} - \phi_i \beta' x)},$$

$r = 1, ...q$. In order to get an ordered regression model the parameters $\phi_1, ..., \phi_k$ must fulfil the constraints

$$1 = \phi_1 > ... > \phi_k = 0.$$

Most often the model is estimated without imposing the constraints a priori. However, if the estimated values $\hat{\phi}_i$ are allowed to yield the ordering of categories the order is a result of the model and not a trait of the variable considered. Then the model is no ordinal regression model because it makes no use of the information provided by the ordering. Anderson (1984) also considered the concept of indistinguishability, meaning that response categories are indistinguishable if x is not predictive between these categories. For a similar concept in cumulative and sequential models see Tutz (1991a). A comparison of the proportional odds model and the stereotype model is given by Holtbrügge and Schuhmacher (1991).

Another type of model assumes given scores for the categories of the response Y. Williams and Grizzle (1972) consider the model

$$\sum_{r=1}^{k} s_r P(Y = r|x) = x'\beta,$$

where $s_1, ..., s_k$ are given scores. Instead of using the support $\{1, ..., k\}$ one may consider $Y \in \{s_1, ..., s_k\}$ and write the model by

$$\sum_{r=1}^{k} s_r P(Y = s_r|x) = x'\beta.$$

Obviously models of this type are not suited for responses that are measured on ordinal scale level. By introducing scores, a higher–scale level is assumed for the discrete response.

A third type of model that are only mentioned here are adjacent categories logits (e.g., Agresti, 1984). The model

$$\log\left[P(Y = r|x)/P(Y = r - 1|x)\right] = x'\beta_r$$

is based on the consideration of the adjacent categories $\{r - 1, r\}$. Logits are built locally for these adjacent categories. Another form of the model is

$$P(Y = r|Y \in \{r, r + 1\}, x) = F(x'\beta_r),$$

where F is the logistic distribution function. The latter form shows that the logistic distribution function may be substituted for any strictly monotone increasing distribution function. Moreover, it shows that it may be considered as a dichotomous response model given $Y \in \{r, r + 1\}$. Very similar models are used in item response theory (Masters, 1982) and are often misinterpreted as sequential process models (see Tutz, 1990c). The model may also be considered the corresponding regression model arising from the row-column (RC-) association model considered by Goodman (1979, 1981a,b).

3.4   Statistical inference

Since maximum likelihood estimation plays a central role it is considered separately in Section 3.4.1. The testing of linear hypotheses, which has been considered already in the univariate case, is only mentioned in the following section. In Section 3.4.2 we consider more general goodness–of–fit statistics than for the univariate case. The family of power-divergence statistics due to Cressie and Read (1984) and the associated minimum power-divergence estimation principle together with asymptotics for the classical and the "increasing-cells" case are given in Section 3.4.3.

3.4.1   Maximum likelihood estimation

In the following, estimation is described in the general form of multivariate exponential families (compare Appendix A1). The special case of the multinomial distribution is given by (3.1.7). Based on the exponential family

$$f\left(y_i|\theta_i, \phi, \omega_i\right) \;=\; \exp\left\{\frac{y_i'\theta_i - b(\theta_i)}{\phi}\omega_i + c(y_i, \phi, \omega_i)\right\}$$

for observation vectors $y_1, \ldots, y_n$, maximum likelihood estimation may be derived in analogy to the one-dimensional case outlined in Section 2.2. The log–likelihood kernel for observation y_i is given by

$$l_i(\mu_i) \;=\; \frac{y_i'\theta_i - b(\theta_i)}{\phi}\omega_i, \qquad \theta_i = \theta(\mu_i), \tag{3.4.1}$$

and the log–likelihood for the sample has the form

$$l(\beta) = \sum_{i=1}^{n} l_i(\mu_i).$$

Note that in the grouped case n stands for the number of groups, whereas in the ungrouped case n stands for the number of units. Using the link $\mu_i = h(Z_i\beta)$, the score function $s(\beta) = \partial l/\partial \beta = \sum_{i=1}^{n} s_i(\beta)$ has components

$$s_i(\beta) = Z_i'\, D_i(\beta)\, \Sigma_i^{-1}(\beta) \left[y_i - \mu_i(\beta)\right] \tag{3.4.2}$$

where

$$D_i(\beta) = \frac{\partial h(\eta_i)}{\partial \eta}$$

is the derivative of $h(\eta)$ evaluated at $\eta_i = Z_i\beta$ and

$$\Sigma_i(\beta) = \operatorname{cov}(y_i)$$

denotes the covariance matrix of observation y_i given parameter vector β. Equation (3.4.2) is a direct generalization of (2.2.2), in Chapter 2. The alternative form

$$s_i(\beta) = Z_i'\, W_i(\beta)\, \frac{\partial g(\mu_i)}{\partial \mu'} \left[y_i - \mu_i(\beta)\right]$$

makes use of the weight matrix

$$W_i(\beta) = D_i(\beta)\, \Sigma_i^{-1}(\beta)\, D_i'(\beta) \;=\; \left\{\frac{\partial g(\mu_i)}{\partial \mu'}\, \Sigma_i(\beta)\, \frac{\partial g(\mu_i)}{\partial \mu}\right\}^{-1} \tag{3.4.3}$$

which may be considered an approximation of the inverse of the covariance matrix of the "transformed" observation $g(y_i)$ in cases where $g(y_i)$ exists. The expected Fisher information is given by

$$F(\beta) = \operatorname{cov}\big(s(\beta)\big) = \sum_{i=1}^{n} Z_i'\, W_i(\beta)\, Z_i$$

which is a direct generalization of the form in Chapter 2.

In matrix notation score function and Fisher matrix have the same form as in Chapter 2, namely,

$$s(\beta) = Z' \, D(\beta) \, \Sigma^{-1}(\beta) \left[y - \mu(\beta) \right], \qquad F(\beta) = Z' \, W(\beta) \, Z$$

where y and $\mu(\beta)$ are given by

$$y' = (y_1', \ldots, y_n'), \qquad \mu(\beta)' = (\mu_1(\beta)', \ldots, \mu_n(\beta)').$$

The matrices have block diagonal form

$$\Sigma(\beta) = \mathrm{diag}\big(\Sigma_i(\beta)\big), \qquad W(\beta) = \mathrm{diag}\big(W_i(\beta)\big), \qquad D(\beta) = \mathrm{diag}\big(D_i(\beta)\big),$$

and the total design matrix is given by

$$Z = \begin{bmatrix} Z_1 \\ Z_2 \\ \vdots \\ Z_n \end{bmatrix}.$$

These formulas are given for individual observations $y_1, \ldots, y_n$. For grouped observations, which are more convenient for computational purposes, the formulas are the same. The only difference is that the summation is over the grouped observations $y_1, \ldots, y_g$ where y_i is the mean over n_i observations, and $\Sigma_i(\beta)$ is replaced by $\Sigma_i(\beta)/n_i$.

Under regularity assumptions (comparable to the assumptions in Chapter 2) one gets asymptotic normality of the estimate:

$$\hat{\beta} \overset{a}{\sim} N(\beta, F^{-1}(\hat{\beta})).$$

That means $\hat{\beta}$ is approximately normal with covariance matrix $\mathrm{cov}(\hat{\beta}) = F^{-1}(\hat{\beta})$.

Numerical computation

Numerical computation of maximum likelihood estimates has already been given for univariate models in Chapter 2. In the multivariate case one merely has to substitute vectors and matrices by the multivariate versions. The working or pseudo observation vector now is given by $\tilde{y}(\beta) = (\tilde{y}_1(\beta), \ldots, \tilde{y}_n(\beta))$ where

$$\tilde{y}_i(\beta) = Z_i\beta + \left(D_i^{-1}(\beta) \right)' \left[y_i - \mu_i(\beta) \right]$$

is an approximation for $g(\mu_i(\beta))$. Given the kth estimate $\hat{\beta}^{(k)}$ the weighted least–squares estimate

$$\hat{\beta}^{(k+1)} = \left(Z' \, W(\hat{\beta}^{(k)}) \, Z \right)^{-1} Z' \, W(\hat{\beta}^{(k)}) \, \tilde{y}(\hat{\beta}^{(k)})$$

is equivalent to the Fisher scoring iteration

$$\hat{\beta}^{(k+1)} = \hat{\beta}^{(k)} + \left(Z' \, W\big(\hat{\beta}^{(k)}\big) \, Z \right)^{-1} s\big(\hat{\beta}^{(k)}\big).$$

Thus the Fisher scoring can be viewed as an iteratively weighted least–squares estimate.

3.4.2   Testing and goodness–of–fit

Testing of linear hypotheses

The testing of linear hypotheses of the form $H_0 : C\beta = \xi$ against $H_1 : C\beta \neq \xi$ has already been considered in Section 2.2.2. The statistics given there may be used in the multivariate case, too. One only has to substitute score functions and Fisher matrices by their multivariate versions.

Goodness–of–fit statistics

Goodness–of–fit of the models may again be checked by the Pearson statistic and the deviance. Since we are considering multinomial data the expectation μ_i is equivalent to the probability vector $\pi_i = (\pi_{i1}, \ldots, \pi_{iq})$, $\pi_{ik} = 1 - \pi_{i1} - \ldots - \pi_{iq}$, where $\pi_{ir} = P(Y = r | x_i)$. The estimate of π_i based on the model is denoted by $\hat{\mu}_i = \hat{\pi}_i = (\hat{\pi}_{i1}, \ldots, \hat{\pi}_{iq})$. For the following data should be grouped as far as possible so that the observation vector $y_i = (y_{i1}, \ldots, y_{iq})$ consists of relative frequencies.

The *Pearson statistic* in general is given by

$$\chi^2 = \sum_{i=1}^{g} (y_i - \hat{\mu}_i)' \, \Sigma_i^{-1}(\hat{\beta}) \, (y_i - \hat{\mu}_i).$$

In the case of a multicategorical response variable with multinomial distribution $n_i y_i \sim M(n_i, \pi_i)$ χ^2 may be written in the more familiar form

$$\chi^2 = \sum_{i=1}^{g} \chi_P^2(y_i, \hat{\pi}_i)$$

where

$$\chi_P^2(y_i, \hat{\pi}_i) = n_i \sum_{j=1}^{k} \frac{(y_{ij} - \hat{\pi}_{ij})^2}{\hat{\pi}_{ij}}$$

is the Pearson residual for the ith (grouped) observation with $y_{ik} = 1 - y_{i1} - \ldots - y_{iq}$, $\hat{\pi}_{ik} = 1 - \hat{\pi}_{i1} - \ldots - \hat{\pi}_{iq}$.

The *deviance* or likelihood ratio statistic is given by

$$D = -2 \sum_{i=1}^{g} \left\{ l_i(\hat{\pi}_i) - l_i(y_i) \right\}.$$

For multinomial data the more familiar form is given by

$$D = 2 \sum_{i=1}^{g} \chi_D^2(y_i, \hat{\pi}_i)$$

where

$$\chi_D^2(y_i, \hat{\pi}_i) = n_i \sum_{j=1}^{k} y_{ij} \log \left(\frac{y_{ij}}{\hat{\pi}_{ij}} \right)$$

is the deviance residual. If $y_{ij} = 0$ the term $y_{ij} \log(y_{ij}/\hat{\pi}_{ij})$ is set to zero.

Under "regularity conditions" including in particular increasing sample sizes $n_i \to \infty$, $i = 1, \ldots, g$ such that $n_i/n \to \lambda_i > 0$ one gets approximately χ^2–distributed goodness–of–fit statistics

$$\chi^2, \; D \overset{a}{\sim} \chi^2(g(k-1) - p)$$

where g denotes the number of groups, k is the number of response categories and p is the number of estimated parameters. For sparse data with small n_i alternative asymptotics is appropriate (see end of next section).

Example 3.11: Caesarean birth study (Example 3.1 continued)
In Example 3.1 a multicategorical logit model was fitted to the data on infection following birth by Caesarean section. The response Y had three levels (infection type I (1) / infection type II (2) / no infection (3)), and there were three binary, dummy–coded covariates included in the model, namely, NOPLAN (= 1 for the Caesarean was not planned), FACTOR (= 1 for risk factors were present) and ANTIB (= 1 for antibiotics were given as a prophylaxis). Before testing differences between parameters, we recall the model fitted:

$$\log \frac{P(Y=1)}{P(Y=3)} = \beta_{10} + \beta_{1N} \cdot \text{NOPLAN} + \beta_{1F} \cdot \text{FACTOR}$$
$$+ \beta_{1A} \cdot \text{ANTIB}$$
$$\log \frac{P(Y=2)}{P(Y=3)} = \beta_{20} + \beta_{2N} \cdot \text{NOPLAN} + \beta_{2F} \cdot \text{FACTOR}$$
$$+ \beta_{2A} \cdot \text{ANTIB}.$$

The parameter estimates from Example 3.1 were as follows:

$\hat{\beta}_{10}$	−2.621	$\hat{\beta}_{20}$	−2.560
$\hat{\beta}_{1N}$	1.174	$\hat{\beta}_{2N}$	0.996
$\hat{\beta}_{1F}$	1.829	$\hat{\beta}_{2F}$	2.195
$\hat{\beta}_{1A}$	−3.520	$\hat{\beta}_{2A}$	−3.087

Looking at the parameter estimates we might assume that ANTIB has different effects on the relative risks of infection types I and II. On the contrary, the effects of NOPLAN and FACTOR seem to be about the same for both types of infection. We want to test the latter assumption, i.e., $H_0 : (\beta_{1N} = \beta_{2N}$ and $\beta_{1F} = \beta_{2F})$ against $H_1 : (\beta_{1N} \neq \beta_{2N}$ or $\beta_{1F} \neq \beta_{2F})$. The deviance for the unrestricted model under H_1 is 11.830 with $14 - 8 = 6$ degrees of freedom. The restricted model under H_0 has two parameters less and thus 8 degrees of freedom; the deviance is 12.162. We can calculate the likelihood ratio statistic λ (see Section 2.2.2) as the difference of the deviances $\lambda = 12.162 - 11.830 = 0.8467$. Assuming λ to be χ^2-distributed with 2 df, we obtain a p-value of 0.847 for the test, which is far from being significant.

Starting off with a new model where $\beta_{1N} = \beta_{2N}$ and $\beta_{1F} = \beta_{2F}$ we might now test $H_0 : \beta_{1A} = \beta_{2A}$ against $H_1 : \beta_{1A} \neq \beta_{2A}$. The deviance of the former model has been 12.162 at 8 degrees of freedom. The deviance of the H_0-model is 12.506 at 9 degrees of freedom. Thus, $\lambda = 0.3437$ and is far from being significant (the p-value resulting from the $\chi^2(1)$-distribution is 0.557). $\square$

3.4.3   Power–divergence family*

For categorical data a more general single-parameter family of goodness–of–fit statistics is the power-divergence family, which was introduced by Cressie and Read (1984). The power-divergence statistic with parameter $\lambda \in \mathbb{R}$ is given by

$$S_\lambda = \sum_{i=1}^{g} SD_\lambda(y_i, \hat{\pi}_i)$$

where the sum of deviations over observations at fixed point y_i is given by

$$SD_\lambda(y_i, \hat{\pi}_i) = \frac{2n_i}{\lambda(\lambda + 1)} \sum_{j=1}^{k} y_{ij} \left[\left(\frac{y_{ij}}{\hat{\pi}_{ij}} \right)^\lambda - 1 \right] \tag{3.4.4}$$

where $-\infty < \lambda < \infty$ is a fixed value. The term power divergence is derived from the fact that the divergence of the empirical distribution y_i (the vector of relative frequencies) from the hypothesized distribution $\hat{\pi}_i$ is measured through a sum of powers of the term $y_{ij}/\hat{\pi}_{ij}$. The cases $\lambda = -1$ and $\lambda = 0$ are defined by the continuous limits $\lambda \to -1$, $\lambda \to 0$. Many statistics that are in common use in discrete data analysis turn out to be special cases of the power–divergence family. In particular for $\lambda = 1$ one obtains the Pearson statistic

$$S_1 = \sum_{i=1}^{g} n_i \sum_{j=1}^{k} \frac{(y_{ij} - \hat{\pi}_{ij})^2}{\hat{\pi}_{ij}}$$

and for $\lambda \to 0$ one obtains the likelihood ratio statistic

$$S_0 = 2 \sum_{i=1}^{g} n_i \sum_{j=1}^{k} y_{ij} \log \left(\frac{y_{ij}}{\hat{\pi}_{ij}} \right).$$

Further interesting cases are Kullback's minimum discrimination information statistic

$$S_{-1} = 2 \sum_{i=1}^{g} n_i \sum_{j=1}^{k} \hat{\pi}_{ij} \log \left(\frac{\hat{\pi}_{ij}}{y_{ij}} \right)$$

and Neyman's minimum modified χ^2 statistic

$$S_{-2} = \sum_{i=1}^{g} n_i \sum_{j=1}^{k} \frac{(y_{ij} - \hat{\pi}_{ij})^2}{y_{ij}}$$

where in comparison to the Pearson statistic only the denominator $\hat{\pi}_{ij}$ is substituted for y_{ij}. Moreover, the Freeman–Tukey statistic

$$S_{-1/2} = 4 \sum_{i=1}^{g} \sum_{j=1}^{k} \left(\sqrt{n_i y_{ij}} - \sqrt{n \hat{\pi}_{ij}} \right)^2$$

is a special case (see Read and Cressie, 1988, Bhapkar, 1980).

The power-divergence family provides a number of goodness–of–fit statistics that are determined by the single parameter λ. Read and Cressie (1988) recommend for λ values from the interval $[-1, 2]$. For λ outside this interval the values of the statistics tend to be too small and the statistic becomes too sensitive to single–cell departures.

The maximum likelihood estimator of π_i is given as the probability vector $\hat{\pi}_i$, which maximizes the likelihood or equivalently minimizes the deviance given the model is true, i.e., $g(\pi_i) = Z_i \beta$ holds. Thus maximum likelihood estimation may be considered a minimum distance estimation procedure.

Alternative minimum distance estimators like the minimum discrimination information estimator (Kullback, 1959, 1985) or Neyman's minimum χ^2 estimator (Neyman, 1949) have been considered in the literature. These estimation procedures follow quite naturally as special cases of minimum power–divergence estimation. Let the estimates $\hat{\pi}_i = h(Z_i \hat{\beta})$ be determined by minimization of the power–divergence statistic for fixed parameter

$$S_\lambda = \sum_{i=1}^{g} SD_\lambda(y_i, \hat{\pi}_i) \quad \to \quad \text{min.}$$

Then special cases of minimum power–divergence estimates are the maximum likelihood estimate ($\lambda = 0$), the minimum χ^2 estimate ($\lambda = 1$), Kullback's minimum discrimination information estimate ($\lambda = -1$), and Neyman's minimum modified estimate ($\lambda = -2$). For further distance measures see Read and Cressie (1988) and Parr (1981).

Asymptotic properties under classical "fixed cells" assumptions

Classical assumptions in asymptotic theory for grouped data imply

- a fixed number of groups

- increasing sample sizes $n_i \to \infty$, $i = 1, \ldots, g$, such that $n_i/n \to \lambda_i$ for fixed proportions $\lambda_i > 0$, $i = 1, \ldots, n$

- a fixed "number of cells" k in each group

- a fixed number of parameters.

If Birch's (1963) regularity conditions hold, minimum power–divergence estimates will be best asymptotically normally (BAN -) distributed estimates. If BAN estimates $\hat{\pi}_i$ are used under the assumption that the model holds, the power–divergence statistic for any λ is asymptotically χ^2-distributed with

$$S_\lambda \sim \chi^2\big(g(k-1) - p\big)$$

where g is the number of groups, k denotes the number of possible outcomes and p the number of estimated parameters (see Read and Cressie, 1988, Appendix A6). Therefore, in this setting the statistics are asymptotically equivalent for all $\lambda \in \mathbb{R}$. Moreover, the asymptotic equivalence still holds under local alternatives where the limit distribution is a noncentral χ^2 distribution.

Sparseness and "increasing–cells" asymptotics

If several explanatory variables are considered, the number of observations for fixed explanatory variable x is often small and the usual asymptotic machinery will fail. Alternatively under such sparseness conditions it may be assumed that with increasing sample size $n \to \infty$ the number of groups (values of the explanatory variables) is also increasing with $g \to \infty$. Read and Cressie (1988, Sections 4.3 and 8.1) give a review of "increasing cells" asymptotics for parametric models. For the product–multinomial sampling scheme, Dale (1986) investigated the asymptotic distribution of Pearson's χ^2 and the deviance. Extensions for the power–divergence family have been given by Rojek (1989), and Osius and Rojek (1992).

   The main result in "increasing–cells" asymptotics is that S_λ is no longer χ^2 distributed but has an asymptotic normal distribution under the null–hypothesis that the model holds. Essential conditions beside smoothness and differentiability as considered by Osius and Rojek (1992) are the increase of information, i.e., there is a positive–definite limit of $F(\beta)/n$ and that the probabilities $P(Y = r|x)$ fullfill side conditions, e.g., are all bounded away from 0 and 1. Moreover, one needs consistency of the estimator $\hat{\beta}$. Further conditions assume that $g/\text{var}(S_\lambda)$ is bounded and that group sizes are increasing such that $\sum_i n_i/g \to \infty$ or $(\sum_i n_i^2/ng)^{1/2} \to \infty$. Then for the power–divergence statistic we have asymptotically

$$S_\lambda \sim N\!\left(\mu_\lambda, \sigma_\lambda^2\right) \qquad \text{resp.} \qquad T_\lambda = \frac{(S_\lambda - \mu_\lambda)}{\sigma_\lambda} \quad \to \quad N(0,1)$$

where $\lambda > 1$ is assumed. As the notation already suggests, the parameters of the limiting distribution of S_λ depend on λ. In contrast to the classical asymptotics, the asymptotic distributions for different λ's are not equivalent under sparseness assumptions. Although μ_λ and σ_λ^2 may be computed the computational effort becomes high for $k > 2$. Simple expressions are available for Pearsons' statistic $\lambda = 1$ and several special cases of increasing sample sizes (see Osius and Rojek, 1992).

3.5 Multivariate models for correlated responses

Up to now models with only *one*, though possibly multicategorical, response variable have been considered. In many applications, however, one is confronted with the truly multivariate case: A vector of correlated or clustered response variables is observed, together with covariates, for each unit in the sample. The response vector may include repeated measurements of units on the *same* variable, as in longitudinal studies or in subsampling primary units. Examples for the latter situation are common in genetic studies where the family is the cluster but responses are given by the members of the family, or in ophthalmology where two eyes form a cluster with observations taken for each eye. In other situations the response vector consists of *different* variables, e.g., different questions in an interview. The important case of repeated measurements or, in other words, longitudinal data with many short time series is treated in Section 6.2.

For approximately Gaussian variables multivariate linear models have been extensively studied and applied for a long time. Due to the lack of analytically and computationally convenient multivariate distributions for discrete or mixed discrete/continuous variables, multivariate regression analysis for non–Gaussian data becomes more difficult, and research is more recent and still in progress. One can distinguish three main approaches: *Conditional models* (Section 3.5.1), often called data–driven models, specify the conditional distribution of each component of the response vector given covariates and the remaining components. If primary interest is in effects of covariates on the responses, *marginal models* (Section 3.5.2), which specify marginal distributions, are more appropriate. In contrast to this population–averaged approach, *random effects models* allow for cluster– or subject–specific effects. The distinction of these types of modelling is studied, e.g., by Neuhaus, Hauck and Kalbfleisch (1991), and Agresti (1993b).

Cluster–specific approaches are considered in Chapter 7 in the context of random effects models.

This section provides only a short survey on work in this area. For a bibliography of methods for correlated categorical data, see Ashby et al. (1992). The focus is on binary and categorical responses, however, the basic ideas extend to other types of response. In the following we do not consider models based on a multivariate structure of latent variables to which the interpretation necessarily refers. Thus we will not consider factor analysis for categorical variables (Bartholomew, 1980) or structural equation models with explanatory variables (e.g., Muthén, 1984, Arminger and Küsters, 1985, Küsters, 1987, Arminger and Sobel, 1990). A further important class of alternative models for the investigation of the association structure in a given set of variables is graphical chain models (see Lauritzen and Wermuth, 1989, Wermuth and Lauritzen, 1990, Whittaker, 1990, and the references therein). These models, which also allow for sets of both discrete and continuous variables, are beyond the scope of this book.

3.5.1   Conditional models

Asymmetric models

In many applications the components of the response vector are ordered in a way that some components are prior to other components, e.g., if they refer to events that take place earlier. Let us consider the simplest case of a response vector $Y = (Y_1, Y_2)$ with categorical components $Y_1 \in \{1, \ldots, k_1\}$, $Y_2 \in \{1, \ldots, k_2\}$. Let Y_2 refer to events that may be considered conditional on Y_1. It is quite natural to model first the dependence of Y_1 on the explanatory variables x and then the conditional response $Y_2|Y_1$. In both steps the models of Sections 3.2 and 3.3 may be used according to the scale level of Y_1 and Y_2. More generally, consider m categorical responses $Y_1, \ldots, Y_m$ where Y_j depends on $Y_1, \ldots, Y_{j-1}$ but not on $Y_{j+1}, \ldots, Y_m$. A simple case when this assumption is appropriate is repeated measurements on the same variable. However, the components of $Y = (Y_1, \ldots, Y_m)$ may also stand for different variables when the causal structure is known, e.g., when Y_j stands for events that are previous to the events coded by Y_{j+1}. In Example 3.12 we will consider the responses years in school (Y_1) and reported happiness (Y_2). Since the years in school are previous to the statement about happiness, Y_2 may be influenced by Y_1 but not vice versa. In the following categorical variables with $Y_j \in \{1, \ldots, k_j\}$ are considered. For the special case of binary responses $Y_j \in \{1, 2\}$, we use 0–1 dummies $y_j = 2 - Y_j$.

Models that make use of this dependence structure are based on the decomposition

$$P(Y_1, \ldots, Y_m | x) = P(Y_1 | x) \cdot P(Y_2 | Y_1, x) \cdots P(Y_m | Y_1, \ldots, Y_{m-1}, x). \quad (3.5.1)$$

Simple models arise if each component of the decomposition is specified by a generalized linear model

$$P(Y_j = r | Y_1, \ldots, Y_{j-1}, x) = h_j(Z_j \beta) \qquad (3.5.2)$$

where $Z_j = Z(Y_1, \ldots, Y_{j-1}, x)$ is a function of previous outcomes $Y_1, \ldots$ $\ldots, Y_{j-1}$ and the vector of explanatory variables x. Conditional models of this type sometimes are called *data–driven*, since the response is determined by previous outcomes.

Markov–type transition models follow from the additional assumption $P(Y_j = r | Y_1, \ldots, Y_{j-1}, x) = P(Y_j | Y_{j-1}, x)$. For binary outcomes a simple model assumes

$$\log\left(P(y_1 = 1 | x) / P(y_1 = 0 | x)\right) = \beta_{01} + z_1' \beta_1$$

$$\log\left(\frac{P(y_j = 1 | y_1, \ldots, y_{j-1}, x)}{P(y_j = 0 | y_1, \ldots, y_{j-1}, x)}\right) = \beta_{0j} + z_j' \beta_j + y_{j-1} \gamma_j.$$

In the multicategorical case the binary model may be substituted by the nominal logit model. Repeated measurements Markov–type models (of first or higher order) may be interpreted as transition models since the transition from previous states to the actual state is modelled. Models of this type are further discussed in the framework of categorical time series and longitudinal data (Sections 6.1.1, 6.2.1). Similar models are also considered in discrete survival analysis (Section 9.2).

Regressive logistic models as considered by Bonney (1987) have the form

$$\log\left(\frac{P(y_j = 1 | y_1, \ldots, y_{j-1}, x)}{P(y_j = 0 | y_1, \ldots, y_{j-1}, x)}\right) = \beta_0 + \gamma_1 y_1 + \ldots + \gamma_{j-1} y_{j-1} + z_j \beta.$$

In this model the number of included previous outcomes depends on the component y_j. Thus no Markov–type assumption is implied. If variables are multicategorical (with possibly varying numbers of categories) regressive models may be based on the modelling approaches from previous sections as building blocks.

As an example, let us consider the case of only two ordinal components $Y_1 \in \{1, \ldots, k_1\}$, $Y_2 \in \{1, \ldots, k_2\}$. A simple regressive cumulative model is given by

$$P(Y_1 \leq r | x) = F\left(\theta_r + x' \beta_r^{(1)}\right)$$

$$P(Y_2 \leq s | Y_1 = r, x) = F\left(\theta_{rs} + x' \beta_s^{(2)}\right). \qquad (3.5.3)$$

Here the marginal distribution of Y_1 is given by a cumulative model. The conditional distribution is again a cumulative model where the previous outcome influences the thresholds θ_{rs} of the conditional responses. Of course the marginal distribution of Y_2 does not in general follow a cumulative model.

TABLE 3.10. Cross classification of sex, reported happiness and years of schooling

Sex	Reported happiness	Years of School Completed			
		<12	12	13–16	≥17
Males	Not too happy	40	21	14	3
	Pretty happy	131	116	112	27
	Very happy	82	61	55	27
Females	Not too happy	62	26	12	3
	Pretty happy	155	156	95	15
	Very happy	87	127	76	15

Example 3.12: Reported happiness

Clogg (1982) investigated the association between sex (x), years in school (Y_1) and reported happiness (Y_2). The data are given in Table 3.10. Since sex and years in school are prior to the statement about happiness the latter variable is modelled conditionally on Y_1. Since the dependent variables are ordinal the analysis is based on the regressive cumulative model (3.5.3). The deviance and Pearson's χ^2 are 1.55 on 4 degrees of freedom. If x is considered a shift variable for the marginal distribution of Y_1, i.e., $\beta_1^{(1)} = \beta_2^{(1)} = \beta^{(1)}$, the fit is very bad with deviance 19.54 on 6 degrees of freedom. That means within a logit model sex may not be seen as producing a simple shift on the response years in school. However the conditional modelling of $Y_2|Y_1$ may be simplified. The model where $\beta_s^{(2)} = 0$ yields deviance 13.27 and Pearson's χ^2 is 12.20 on 8 degrees of freedom, thus the model shows a satisfying fit. That means given the years of school (Y_1), sex does not influence the reported happiness within this modelling approach. The estimates of the latter model are given in Table 3.11. □

Models of the type (3.5.2) may be embedded into the framework of generalized linear models. Link function and design matrix follow from (3.5.1) and (3.5.2). In the general case link function and design matrix might be very complex and are not available in standard programs. However, if $k_1 = \ldots = k_m$ the local models (3.5.2) may often be chosen to have a link function h that does not depend on j. By choosing an appropriate design matrix (filled up with zeros) the decomposition (3.5.1) allows to use standard software, where $Y_1, Y_2|Y_1, \ldots, Y_m|Y_1, \ldots, Y_{m-1}$ (always given x) are treated as separate observations (see, e.g., Bonney, 1987).

TABLE 3.11. Estimates for the cross classification of sex, reported happiness and years of schooling

	Estimate	Standard deviation	p–value
θ_1	-0.545	0.053	0.000
θ_2	0.841	0.056	0.000
θ_3	2.794	0.112	0.000
$\beta_1^{(1)}$	0.001	0.053	0.984
$\beta_2^{(1)}$	-0.201	0.056	0.000
$\beta_3^{(1)}$	-0.388	0.112	0.000
θ_{11}	-1.495	0.109	0.000
θ_{12}	0.831	0.092	0.000
θ_{21}	-2.281	0.153	0.000
θ_{22}	0.528	0.091	0.000
θ_{31}	-2.564	0.203	0.000
θ_{32}	0.575	0.109	0.000
θ_{41}	-2.639	0.422	0.000
θ_{42}	0.133	0.211	0.527

The estimates in Table 3.11 refer to model (3.5.3) where $\beta_j^{(2)}=0$.

Symmetric models

If there is no natural ordering of the components of the response vector, or if one does not want to use this ordering, models that treat response components in a symmetric way are more sensible. An example from ophthalmology is visual impairment data from the Baltimore Eye Survey (Tielsch et al.,1989) analysed in Liang, Zeger and Qaqish (1992), see Example 1.6 in Chapter 1 and Example 3.13 at the end of Section 3.5.2. Binary observations of visual impairment for both (m=2) eyes are taken for more than 5000 persons, together with demographic covariates such as age, race, sex, etc. Conditional models are useful if conditional distributions or moments of one response component given the others are of interest, e.g., for the purpose of prediction. If the main scientific objective is to analyze effects of covariates on responses, marginal models (Section 3.5.2) rather than conditional ones are useful.

For the following, attention is restricted to binary response components $y_1, \ldots, y_m$, but ideas generalize to multicategorical and other types of variables. Symmetric *conditional* models can be developed by specification of the conditional distributions

$$P(y_j = 1|y_k, \ k \neq j; \ x_j) \qquad (3.5.4)$$

of one response given the others. For binary responses such conditional distributions uniquely determine the joint distribution. Logistic regression models are a natural choice for the conditional distributions in (3.5.4): Starting from a log–linear model including all interactions between $y_1, \ldots,$ $\ldots, y_m$, one arrives at logit models where the linear predictor includes covariates x_j, main effects y_k, $k \neq j$ and second– (and higher–) order interaction effects such as $y_k y_l$, $k, l \neq j$ in additive form, see Zeger and Liang(1989). A parsimonious class of conditional logistic models, treating the individual binary variables symmetrically, is considered by Qu et al. (1987):

$$\pi_j = P(y_j = 1 | y_k, \ k \neq j; \ x_j) = h\big(\alpha(w_j; \theta) + x_j' \beta_j\big) \qquad (3.5.5)$$

where h is the logistic function and α is an arbitrary function of a parameter θ and the sum $w_j = \sum_{k \neq j} y_k$ of the conditioning $k \neq j$. In model (3.5.5), the "location parameter" α depends on the conditioning y's, whereas covariate effects are kept constant. For the case of two components (y_1, y_2), the sums w_1, w_2 reduce to y_2, y_1, respectively. Then the simplest choice is a logistic model including the conditioning response as a further covariate:

$$\pi_j = P(y_j = 1 | y_k, \ k \neq j; \ x_j) = h(\theta_0 + \theta_1 y_k + x_j' \beta_j), \qquad j, k = 1, 2. \qquad (3.5.6)$$

Other choices for $\alpha(w; \theta)$ are discussed in Qu et al. (1987) and Conolly and Liang (1988). The joint density $P(y_1, \ldots, y_m; x_1, \ldots, x_m)$ can be derived from (3.5.5), however, it involves a normalizing constant, which is a complicated function of the unknown parameters θ and β, see Prentice (1988), and Rosner (1984). Full likelihood estimation may therefore become computationally cumbersome. Conolly and Liang (1988) propose a quasi–likelihood approach, with an "independence working" quasi–likelihood and quasi–score function for each cluster $i = 1, \ldots, n$:

$$L_i(\beta, \theta) = \prod_{j=1}^{m} \pi_{ij}^{y_{ij}} (1 - \pi_{ij})^{1 - y_{ij}},$$

$$s_i(\beta, \theta) = \sum_{j=1}^{m} \frac{\partial \pi_{ij}}{\partial(\beta, \theta)} \sigma_{ij}^{-2} \big(y_{ij} - \pi_{ij}(\beta, \theta)\big),$$

where $y_i = (y_{i1}, \ldots, y_{ij}, \ldots, y_{im})$ are the responses in cluster i, $\pi_{ij}(\beta, \theta) = P(y_{ij} = 1 | \cdot)$ is defined by (3.5.5), and $\sigma_{ij}^2 = \pi_{ij}(\beta, \theta)\big(1 - \pi_{ij}(\beta, \theta)\big)$. If α is a linear function of θ, as, e.g., in (3.5.6), this is the common form of the score function for m independent binary responses. Setting

$$M_i = \left(\frac{\partial \pi_{i1}}{\partial(\beta, \theta)}, \ldots, \frac{\partial \pi_{im}}{\partial(\beta, \theta)} \right), \quad \Sigma_i = \mathrm{diag}(\sigma_{i1}^2, \ldots, \sigma_{im}^2),$$

$$\pi_i = (\pi_{i1}, \ldots, \pi_{im})'$$

we have
$$s_i(\beta, \theta) = M_i \, \Sigma_i^{-1} \, (y_i - \pi_i),$$
which is a multivariate extension of the (quasi–) score function in Section 2.3.1. The roots $(\hat{\beta}, \hat{\theta})$ of the resulting generalized estimation equation
$$s(\beta, \theta) = \sum_{i=1}^{n} s_i(\beta, \theta) = 0$$
are, under regularity assumptions, consistent and asymptotically normal:
$$(\hat{\beta}, \hat{\theta}) \overset{a}{\sim} N\big((\beta, \theta) \, , \, \hat{F}^{-1} \hat{V} \hat{F}^{-1}\big),$$
with
$$\hat{F} = \sum_{i=1}^{n} \hat{M}_i \, \hat{\Sigma}_i^{-1} \, \hat{M}_i, \qquad \hat{V} = \sum_{i=1}^{n} \hat{s}_i \hat{s}_i',$$
where "ˆ" means evaluation at (β, θ).

Two drawbacks of conditional models should be kept in mind: First, by construction, they measure the effect of covariates on a binary component y_i having already accounted for the effect of other responses y_k, $k \neq j$. If regression of covariates on the response is the scientific focus, this will often condition away covariate effects. A second drawback is that interpretation of effects depends on the dimension of y: Excluding components of y leads to different kinds of "marginal" distributions for the remaining components; the models are not "reproducible." Therefore, conditional models are only meaningful if the dimension of y is the same for all observations. Both disadvantages can be avoided by the following marginal modelling approach.

3.5.2 Marginal models

In many situations the primary scientific objective is to analyze the *marginal mean* of the responses given the covariates. The association between responses is often of secondary interest. In the Baltimore Eye Survey study on visual impairment (Example 1.6 and Example 3.13), the primary goal is to identify the influence of demographic variables, such as age, race and education, on visual impairment. Data are available on both eyes, and the association between both responses has to be taken into account for correct analyses. However, one is not primarily interested in this association or its relation with demographic variables. Similar situations occur in other applications, e.g., twin studies and dentistry.

Marginal models were first proposed by Liang and Zeger (1986), Zeger and Liang (1986) in the closely related context of longitudinal data with many short time series (see also Section 6.2.2). Their modelling and estimation approach has subsequently been adopted, modified and extended in various ways, and it is still an area of active research. Our presentation is mainly based on the original work cited above and on Liang, Zeger and Qaqish (1992). We consider the situation where all responses are binary. It is important to note that the dimension of y may vary in marginal models: In contrast to conditional models, they are "reproducible" and parameter interpretation does not depend on the dimension of y.

In marginal models, the effect of covariates on responses and the association between responses is modelled separately. This is in contrast to conditional models, which are defined by only one equation, the conditional mean structure. Let $y_i = (y_{i1}, \ldots, y_{im_i})$ be the vector of binary responses and $x_i = (x_{i1}, \ldots, x_{im_i})$ the vector of covariates from cluster i, $i = 1, \ldots, n$. Here the term cluster is used instead of observation to emphasize that in many situations the components of y_i are correlated observations on the same type of variable. Generally, "cluster size" or the size m_i of the vector $y_i = (y_{i1}, \ldots, y_{im_i})$ may depend on the cluster. In the Baltimore Eye Survey example (Example 1.6) when measuring the performance of two eyes one always has $m_i = m = 2$. However, if measurements are made in families of various sizes, m_i may stand for the family size: Each cluster (family) involves a possibly differing number of correlated binary responses. The covariate components may vary across units within a cluster or may be constant within clusters, i.e., $x_{i1} = \ldots = x_{im_1}$. As long as no confusion arises, we will drop the index i. Then $y = (y_1, \ldots, y_j, \ldots, y_m)$, $x = (x_1, \ldots, x_j, \ldots, x_m)$ denote responses and covariates in a single cluster. Specification of marginal models is as follows.

(i) The *marginal means* or *response probabilities* $P(y_j = 1 | x_j) = E(y_j | x_j)$, $j = 1, \ldots, m$, are specified by common binary response models

$$\pi_j(\beta_j) = E(y_j | x_j) = h_j(z_j' \beta_j), \qquad (3.5.7)$$

where h_j is a response function, e.g., a logit function and $z_j = z_j(x_j)$ is an appropriate design vector.

(ii) The *marginal variance* depends on the marginal mean $\pi_j = \pi_j(\beta_j)$ by the variance function

$$\text{var}(y_j | x_j) = v(\pi_j) = \pi_j(1 - \pi_j). \qquad (3.5.8)$$

(iii) The *covariance* between y_j and y_k is a function of the marginal means $\pi_j = \pi_j(\beta_j)$ and perhaps of additional association parameters α,

$$\text{cov}(y_j, y_k) = c(\pi_j, \pi_k; \alpha), \qquad (3.5.9)$$

with a known function c.

Responses from different clusters are assumed to be independent. Note that parameters $\beta = (\beta_1, \ldots, \beta_j, \ldots, \beta_m)$ and α are the same for each cluster. Therefore marginal models are appropriate for analyzing "population–averaged" effects. This is in contrast to random effects models (Chapter 7), where effects vary from cluster to cluster.

An important feature of marginal models is the following: Marginal effects β can be consistently estimated even if the covariance function $c(\pi_j, \pi_k; \alpha)$ is incorrectly specified. This corresponds to quasi–likelihood models in Section 2.3.1, where the variance function can also be incorrectly specified, while the parameters β can still be estimated consistently, however, with some loss of efficiency, as long as the mean function is correctly specified.

Since the primary scientific objective is often the regression relationship, it is natural to spend more time in correct specification of the marginal mean structure (3.5.7) than in the covariance structure. It is therefore assumed that (3.5.7) is correctly specified, while $c(\pi_j, \pi_k; \alpha)$ is a *working covariance function* for the association between y_i and y_k. Together with the variance function (3.5.8), one obtains a *working covariance matrix*

$$\mathrm{cov}(y) = \Sigma(\beta, \alpha),$$

which depends on β and perhaps on α.

Two main approaches for specifying working covariances have been considered in the literature. Liang and Zeger (1986) and Prentice (1988) use correlations as a measure for associations: The variance structure (3.5.8) is supplemented by a *working correlation matrix* $R(\alpha)$, so that the working covariance matrix is of the form

$$\Sigma(\beta, \alpha) = A^{1/2}(\beta) \, R(\alpha) \, A^{1/2}(\beta),$$

where

$$A(\beta) = \mathrm{diag}\big[\mathrm{var}(y_j|x_j)\big] = \mathrm{diag}\big[\pi_j(\beta)\big(1 - \pi_j(\beta)\big)\big].$$

Various choices for $R(\alpha)$ are suggested by Liang and Zeger (1986) within a longitudinal data context, see also Chapter 6. The simplest choice is a *working independence model*, i.e.,

$$R(\alpha) = I,$$

where I is the identity matrix. Another one is the equicorrelation model with

$$\mathrm{corr}(y_j, y_k) = \alpha$$

for all $j \neq k$. If enough data are available one may leave $R(\alpha)$ completely unspecified, i.e., α has the elements $\alpha_{jk} = \mathrm{corr}(y_j, y_k)$, $j < k$.

An alternative measure of association between pairs of binary responses is the odds ratio. It is easier to interpret and has some desirable properties.

The odds ratio parameterization is suggested by Lipsitz, Laird and Harrington (1991). The odds ratio components $y_j, y_k, j, k = 1, \ldots, m, j \neq k$, are defined by

$$\gamma_{ik} = \frac{P(y_j = 1, y_k = 1)\, P(y_j = 0, y_k = 0)}{P(y_j = 1, y_k = 0)\, P(y_j = 0, y_k = 1)}.$$

Further it is assumed that the vector of pairwise odds ratios can be parameterized by $\gamma = \gamma(\alpha)$. The simplest parameterization is obtained if all odds ratios are taken to be equal. Alternatively, α might be a function of cluster–specific covariates. A common parameterization is to model log–odds ratios as a linear function of covariates as in the visual impairment data example at the end of this section. From the relation (e.g., Lipsitz et al., 1991, Liang et al., 1992)

$$P(y_j = y_k = 1) = E(y_j y_k) =$$
$$= \begin{cases} \dfrac{1 - (\pi_j + \pi_k)(1 - \gamma_{jk}) - s(\pi_i, \pi_j, \gamma_{jk})}{2(\gamma_{jk} - 1)}, & \gamma_{jk} \neq 1 \\ \pi_j \pi_k, & \gamma_{jk} = 1 \end{cases} \quad (3.5.10)$$

with $s(\pi_i, \pi_j, \gamma_{jk}) = [\{1 - (\pi_j + \pi_k)(1 - \gamma_{jk})\}^2 - 4(\gamma_{jk} - 1)\gamma_{jk}\pi_j\pi_k]^{1/2}$, it is seen that the covariance matrix of $y = (y_1, \ldots, y_m)$ can be expressed as a function $\Sigma = \Sigma(\beta, \alpha)$ of the parameters (β, α).

One advantage of the odds ratio is that it is not constrained by the means π_j. In contrast, correlations are constrained by means:

$$\mathrm{corr}(y_j, y_k) = \frac{P(y_j = y_k = 1) - \pi_j \pi_k}{\left[\pi_j(1 - \pi_j)\pi_k(1 - \pi_k)\right]^{1/2}},$$

where $P(y_j = y_k = 1)$ is constrained by $\max(0, \pi_j + \pi_k - 1) \leq P(y_j = y_k = 1) \leq \min(\pi_j, \pi_k)$. This may considerably narrow the range for admissible correlations. However, if correlation between pairs is weak, the correlation parameterization will do as well.

The choice of the working covariance should be a compromise between simplicity and loss of efficiency due to incorrect specification. If association is only a nuisance or of secondary interest, simple working models will suffice. The study of McDonald (1993) favors the independence model, whenever association is a nuisance. If association is also of interest a good specification of odds ratios will be desirable. If in doubt, one will try several models and compare results.

Statistical inference

Now let $y_i = (y_{i1}, \ldots, y_{im_i})$, $x_i = (x_{i1}, \ldots, x_{im_i})$ be the observations for cluster i, $i = 1, \ldots, n$, and $\pi_i(\beta) = (\pi_{i1}(\beta_1), \ldots, \pi_{im_i}(\beta_{m_i}))$, $\Sigma_i(\beta, \alpha)$ be the corresponding marginal mean vectors and working covariance matrices for y_i as described above. Full likelihood estimation is not feasible, since the joint distribution is generally not determined by the mean and covariance structure alone. Therefore a generalized estimation approach is proposed. Keeping the association parameters α fixed for the moment, the generalized estimating equation (GEE) for effect β is

$$s_\beta(\beta, \alpha) = \sum_{i=1}^n Z_i' \, D_i(\beta) \, \Sigma_i^{-1}(\beta, \alpha) \left(y_i - \pi_i(\beta)\right) = 0, \qquad (3.5.11)$$

with design matrices

$$Z_i' = (z_{i1}, \dots, z_{im_i})$$

and diagonal matrices $D_i(\beta) = \mathrm{diag}\big(D_{ij}(\beta_j)\big)$, $D_{ij}(\beta_j) = \partial h_j / \partial \eta_{ij}$ evaluated at $\eta_{ij} = z_{ij}' \beta_j$.

Equation (3.5.11) is a multivariate version of the GEE (2.3.2) in Section 2.3.1, with a correctly specified mean $E(y_i | x_i) = \pi_i(\beta)$ and a possibly misspecified covariance matrix $\mathrm{cov}(y_i | x_i) = \Sigma_i(\beta, \alpha)$. For fixed α, consistent estimates $\hat{\beta}$ are obtained from (3.5.11) by the usual (quasi–) Fisher scoring iterations. Under regularity assumptions this estimator is consistent and assymptotically normal even when α is replaced by a consistent estimator $\hat{\alpha} = \hat{\alpha}(\hat{\beta})$ that converges to some fixed value:

$$\hat{\beta} \overset{a}{\sim} N\big(\beta, \hat{F}^{-1} \, \hat{V} \, \hat{F}^{-1}\big),$$

with the "sandwich" matrix $\hat{F}^{-1} \, \hat{V} \, \hat{F}^{-1}$ defined by

$$\hat{F} = \sum_{i=1}^n Z_i' \, \hat{D}_i \, \hat{\Sigma}_i^{-1} \, \hat{D}_i' \, Z_i,$$

$$\hat{V} = \sum_{i=1}^n Z_i' \, \hat{D}_i \, \hat{\Sigma}_i^{-1} \, (y_i - \hat{\pi}_i)(y_i - \hat{\pi}_i)' \, \hat{\Sigma}_i^{-1} \, \hat{D}_i' \, Z_i,$$

where "^" denotes evaluation at $\hat{\beta}, \hat{\alpha}$.

For the case of an independence working model, $R_i(\alpha) = I$, the working covariance is $\Sigma_i(\beta) = \mathrm{diag}\big[\pi_{ij}(\beta_j)\big(1 - \pi_{ij}(\beta_j)\big)\big]$. The generalized estimating equations then have the usual form of scoring equations, as if observations were independent, and no association parameter α has to be estimated along with β. Compared to the nominal covariance matrix $\hat{F}$ of an independence model, the adjusted sandwich covariance matrix $\hat{F}^{-1} \, \hat{V} \, \hat{F}^{-1}$ protects against assessing false standard deviations for estimated effects $\hat{\beta}$. For the case of positively correlated observations within clusters, standard deviations of effects for covariates that are constant within clusters, as in the visual impairment example, are generally too small if they are based on $\hat{F}$ instead of $\hat{F}^{-1} \, \hat{V} \, \hat{F}^{-1}$. On the other side, standard deviations are too large for effects of subject–specific covariates. If observations within clusters are negatively correlated, "too large" and "too small" have to be exchanged.

If unknown association parameters α are present, several ways of estimating them have been suggested. Liang and Zeger (1986) use a method of moments based on Pearson residuals, see also Section 6.2.2. For the equicorrelation model

$$\text{corr}(y_{ij}, y_{ik}) = \alpha$$

an estimate is

$$\hat{\alpha} = \frac{1}{N-p} \sum_{i=1}^{n} \sum_{j=1}^{m_i} \frac{(y_{ij} - \hat{\pi}_{ij})^2}{\hat{\pi}_{ij}(1 - \hat{\pi}_{ij})}, \tag{3.5.12}$$

with

$$N = \frac{1}{2} \sum_{i=1}^{n} m_i(m_i - 1)$$

and $\hat{\pi}_{ij} = \pi_{ij}(\hat{\beta})$ evaluated at a current iterate $\hat{\beta}$. Cycling to convergence between Fisher scoring steps for $\hat{\beta}$ according to (3.5.11) and estimation of $\hat{\alpha}$ by (3.5.12) leads to a consistent estimation of β. (Note that $\hat{\alpha}$ need only be consistent for *some* α, not a *true* α^0, which does not exist for working correlations).

A totally unspecified $R = R(\alpha)$ can be estimated by

$$\hat{R} = \frac{1}{n} \sum_{i=1}^{n} \hat{A}_i^{-1/2} (y_i - \hat{\pi}_i)(y_i - \hat{\pi}_i)' \hat{A}_i^{-1/2},$$

$\hat{A}_i = \text{diag}(\hat{\pi}_{ij}(1 - \hat{\pi}_{ij}))$, if cluster sizes are all equal to m. This method of moments can lead to considerably biased estimates of α, which is less important if one is only interested in estimating effects β consistently.

Better estimates of α can be obtained if the GEE for β is supplemented by a second GEE for α as in Prentice (1988), Liang et al. (1992) and others. For each cluster define the vector $w_i = (y_{i1}y_{i2}, y_{i1}y_{i3}, \ldots, y_{im_i-1}y_{im_i})$ of pair–wise products $y_{ij}y_{ik}$. Let $\theta_i = E(w_i)$ denote the corresponding vector of second moments $\theta_{ijk} = E(y_{ij}y_{ik}) = P(y_{ij} = y_{ik} = 1)$, which depends on β and α through (3.5.10). The GEE for α is then

$$s_\alpha(\beta, \alpha) = \sum_{i=1}^{n} \frac{\partial \theta_i}{\partial \alpha} C_i^{-1} (w_i - \theta_i) = 0,$$

with a working covariance matrix C_i for $\text{cov}(w_i)$. Complete specification of $\text{cov}(w_i)$ will involve third and fourth moments. A convenient choice is $C_i = \text{diag}\{\text{var}(y_{i1}y_{i2}), \ldots, \text{var}(y_{im_i-1}y_{im_i})\}$. With binary responses, θ_i and C_i are fully determined by the mean and (working) covariance structure. No additional assumptions on higher moments are needed. (Note: Prentice (1988) suggests working with residuals $r_{ij} = (y_{ij} - \pi_{ij})/\{\pi_{ij}(1 - \pi_{ij})\}^{-1/2}$ instead of y_{ij}).

TABLE 3.12. Visual impairment data (from Liang et al., 1992)

Visual impairment	White Age				Black Age				Total
	40–50	51–60	61–70	70+	40–50	51–60	61–70	70+	
Left eye									
Yes	15	24	42	139	29	38	50	85	422
No	617	557	789	673	750	574	473	344	4777
Right eye									
Yes	19	25	48	146	31	37	49	93	448
No	613	556	783	666	748	575	474	226	4751

This GEE approach treats β and α as if they were orthogonal. Zhao and Prentice (1990) suggest a simultaneous GEE for β and α, which involves the (joint) covariance matrix $\text{cov}(y_i, w_i)$. In contrast to the earlier GEE approach, this simultaneous GEE requires correct specification of the covariance structure, making the procedure less robust. Thus simultaneous estimation is only recommended if estimation of association is of primary importance.

A further approach, which is useful when marginal odds ratios are used to model association, has been introduced by Carey, Zeger and Diggle (1993).

Example 3.13: Visual impairment
For 5199 individuals bivariate binary responses were observed, indicating whether or not an eye was visually impaired (VI). Covariates include age in years centered at 60 (A), race (R: 0–white, 1–black), and education in years centered at 9 (E). Table 3.12 gives the data of VI of right and left eyes for race × age combinations. The age–specific risks for whites increase clearly with age. Blacks are more affected than whites, and the discrepancy increases with age. Risks are quite similar for both eyes, as one might expect.

The main objective is to analyze the influence of age and race on visual impairment controlling for education, which serves as a surrogate for socioeconomic status. Therefore Liang et al. (1992) fitted a marginal logistic model

$$\log\left(\frac{\pi_1}{1-\pi_1}\right) = \log\left(\frac{\pi_2}{1-\pi_2}\right) = = \beta_0 + \beta_1 A + \beta_2 A^2 + \beta_3 R$$
$$+\beta_4(A \times R) + \beta_5(A^2 \times R) + \beta_6 E$$

that an eye was visually impaired. For the odds ratio, a log–linear regression model

$$\log \gamma_{12} = \alpha_0 + \alpha_1 R$$

TABLE 3.13. Estimation results for visual impairment data

Covariate	Marginal m.	Conditional m.	Ordinary m.
Intercept	-2.83 (-37)	-3.16 (-43)	-2.82
Age in years (centered at 60)	0.049 (7.1)	0.038 (6.6)	0.049
$(Age-60)^2$	0.0018 (5.3)	0.0012 (4.5)	0.0018
Race (0–white; 1–black)	0.33 (3.2)	0.099 (0.9)	0.33
Age × Race interaction	0.00066 (0.07)	-0.0039 (-0.5)	0.0011
$(Age-60)^2$ by race interaction	-0.0010 (-2.1)	-0.00075 (-2.0)	-0.0011
Education in years (centered at 9)	-0.060 (-0.35)	-0.045 (-3.8)	-0.059
Log odds ratio intercept	2.3 (8.7)	2.3 (13)	—
Log odds ratio race	0.54 (1.3)	0.55 (2.2)	—

Table entries are: parameter (parameter/standard error).

was used. For comparison this is contrasted with a corresponding conditional model, given the response of one eye, and an ordinary logistic model assuming all binary observations are independent.

For the conditional model the linear predictor included, in addition to that of the marginal model above, the response for the other eye and its interaction with race.

Table 3.13 reproduces parameter estimates for the marginal model, the conditional model (with α_0, α_1 estimated by a second GEE) and an ordinary logistic model. Estimates of the latter and the marginal model are quite close, while effects in the conditional model tend to be attenuated compared to the marginal effects because of explicit control for the other eye. For example, the marginal effects for race indicate significant influence of race on visual impairment, while conditional effects do not. In this context, marginal modelling is more sensible.   □

The focus in this section has been on modelling and estimating marginal means of correlated binary data by simple first–order generalized estimating equations (GEE1). Association is considered as a nuisance, and some *working* association model is introduced, mainly to improve efficiency. The approach has been extended to other types of response, in particular multicategorical responses. Also joint modelling and estimation of first and second order marginal moments by simultaneous estimating equations (GEE2) and full likelihood approaches have been developed. Liang et al. (1992) make a suggestion on how to extend the GEE method to multicategorical data using odds ratios. Prentice and Zhao (1991) develop estimating equations for multivariate discrete and continuous responses. Marginal GEE1 models for ordinal responses are described by Miller, Davis and Landis (1993), using correlations as measures for association, and by Williamson, Kim and Lipsitz (1995), Fahrmeir and Pritscher (1996), based on global odds ratios parametrizations. Heagerty and Zeger (1996) propose and compare GEE1, GEE2 and the alternating logistic regression approach of Carey et al. (1993). Maximum likelihood procedures, with different types of parametrization for higher order moments, have been developed by Fitzmaurice and Laird (1993), Fitzmaurice, Laird and Rotnitzky (1993) for binary responses and by Molenberghs and Lesaffre (1992, 1994), Glonek (1996), Glonek and McCullagh (1996) and Heumann (1996) for multivariate categorical responses. Most of these approaches are also appropriate for dealing with correlated repeated outcomes as considered in Section 6.2.2.

4
Selecting and checking models

Fitting data by a certain generalized linear model means choosing appropriate forms for the predictor, the link function and the exponential family or variance function. In the previous chapters Pearsons's χ^2, the deviance and, in the multinomial case, the power divergence family were introduced as general goodness–of–fit statistics. This chapter considers more specific tools to select and check models. Section 4.1 deals with variable selection, i.e., which variables should be included in the linear predictor. Diagnostic methods based on the hat matrix and on residuals are described in Section 4.2, and Section 4.3 is on general misspecification tests, such as Hausman–type tests and tests for non–nested models. We do not treat tests for specific directions, such as testing the correct form of the link function by embedding it in a broader parametric class of link functions. A survey of tests of this type is contained in chapter 11.4 of McCullagh and Nelder (1989). In addition to the methods of this chapter, nonparametric approaches, as in Chapter 5, may also be used to check the adequacy of certain parametric forms.

4.1 Variable selection

Regression analysis is often used in situations where there is a catalogue of many potentially important covariates. Variable selection methods aim at determining submodels with a moderate number of parameters that still fit the data adequately. For normal linear models, various selection methods ranging from traditional stepwise approaches to "all–subsets" methods (e.g., Furnival and Wilson, 1974), are available. The relative merits and drawbacks of stepwise procedures, lower computational costs versus sub-optimality, have been mainly discussed within the linear regression context (e.g., Hocking, 1976, Seeber, 1977, Miller, 1984, 1989). For non–normal regression models, Lawless and Singhal (1978, 1987) developed efficient screening and all–subsets procedures based on various selection criteria. Yet stepwise variable selection is a useful data analytic tool within these more complex models: Maximum likelihood estimates have to be computed iteratively, so that all subset selection by likelihood–based criteria like AIC can become computationally critical in situations with large covariate sets. Secondly, the problem of nonexistence of estimates, which is usually neg-ligible for classical linear models, becomes much more serious for some non–normal models involving a large number of parameters, in particular for models with multicategorical variables. In this case, a complete model search, or even a stepwise backward selection, may break down from the beginning since the full model cannot be fitted, whereas stepwise forward–backward selection (e.g., Fahrmeir and Frost, 1992) is still applicable. Step-wise selection can also be useful for choosing a good initial model in refined selection procedures (Edwards and Havranek, 1987) or in combination with multiple test procedures.

Though we have generalized linear models in mind, the methods apply to any parametric regression models with log–likelihoods of the general form

$$l(\beta; \theta) = \sum_{i=1}^{n} l_i(\eta_i; \theta),$$

where $l_i(\eta_i; \theta)$ is the log–likelihood contribution of (y_i, x_i). The predictor

$$\eta_i = Z_i \beta$$

is linear in the parameters β, with design matrices $Z_i = Z_i(x_i)$ as functions of the covariate vector, in particular $Z_i = (1, x_i)$ for univariate models. The additional parameter vector θ may , e.g., contain an unknown dispersion parameter ϕ, but it may also be void. Variable selection is then understood in the sense of selecting associated sets of subvectors of β. If θ is nonvoid, it is always included in the model. It may be estimated by a method of moments, as ,e.g., the dispersion parameter ϕ in generalized linear models, or by simultaneous ML estimation together with β.

4.1.1   Selection criteria

A submodel corresponds to a subvector, say β_1, of β. Without loss of gen-erality let β be partitioned as (β_1, β_2). The adequacy of a certain submodel can be formally tested by

$$H_0 : \beta_2 = 0, \ \beta_1, \theta \ \text{unrestricted}$$

against

$$H_1 : \beta_1, \beta_2, \theta \ \text{unrestricted},$$

where H_1 stands for the "full" model, or, as in stepwise procedures, for some supermodel. Let the score function $s(\beta)$ for β, the expected (or observed) information matrix $F(\beta)$ and its inverse $A(\beta) = F^{-1}(\beta)$ be partitioned in conformity with the partitioning of β:

$$s = \left[\begin{array}{c} s_1 \\ s_2 \end{array} \right], \quad F = \left[\begin{array}{cc} F_{11} & F_{12} \\ F'_{12} & F_{22} \end{array} \right], \quad A = \left[\begin{array}{cc} A_{11} & A_{12} \\ A'_{12} & A_{22} \end{array} \right].$$

In the following $\hat{\beta} = (\hat{\beta}_1, \hat{\beta}_2)$ denotes the unrestricted MLE under H_1 whereas $\tilde{\beta} = (\tilde{\beta}_1, 0)$ is the MLE under the restriction of H_0. Correspondingly, $\hat{\theta}$ and $\tilde{\theta}$ are consistent estimators of θ under H_0 and H_1. The three common test statistics (compare Section 2.2.2) are the *likelihood ratio statistic*

$$\lambda = -2\{l(\tilde{\beta}_1, 0, \tilde{\theta}) - l(\hat{\beta}_1, \hat{\beta}_2, \hat{\theta})\},$$

the *Wald statistic*

$$w = \hat{\beta}'_2 \hat{A}_{22}^{-1} \hat{\beta}_2, \tag{4.1.1}$$

and the *score statistic*

$$u = \tilde{s}'_2 \tilde{A}_{22} \tilde{s}_2, \tag{4.1.2}$$

where " ^ " resp. " ~ " means evaluation at $(\hat{\beta}, \hat{\theta})$ resp. $(\tilde{\beta}_1, 0, \tilde{\theta})$. As a fourth test statistic one may consider a *modified likelihood ratio statistic*

$$\lambda_m = -2\{l(\bar{\beta}_1, 0, \bar{\theta}) - l(\hat{\beta}_1, \hat{\beta}_2, \hat{\theta})\},$$

where $\bar{\beta}_1$ is a first–order approximation to $\tilde{\beta}$, given by

$$\bar{\beta}_1 = \hat{\beta}_1 - \hat{A}'_{12} \hat{A}_{22}^{-1} \hat{\beta}_2. \tag{4.1.3}$$

Under H_0, all test statistics are asymptotically $\chi^2(r), r = \dim(\beta_2)$, provided that appropriate regularity conditions hold. Generally, the likelihood ratio statistic λ is preferred for a comparably small number of covariates and moderate sample size. For larger sample sizes, the test statistics tend to agree closely, and it is reasonable to use the statistics w, u and λ_m because they are easier and faster to compute. This can be of considerable importance in selection procedures for larger models.

The Wald statistic w is of advantage if the unrestricted MLE has already been computed, as in all–subsets procedures or stepwise backward procedures. The statistic λ_m can provide a better approximation to λ, but requires more computation. The score statistic u is quite useful if a restricted model has been fitted and shall be tested against a more complex model, e.g., as in a stepwise forward selection procedure.

To compare models with different numbers of parameters in an all–subset search, Akaikes Information Criterion

$$AIC = -2l(\tilde{\beta}_1, 0, \tilde{\theta}) + 2(r + s), \qquad (4.1.4)$$

$r = \dim(\tilde{\beta}_1), s = \dim(\tilde{\theta})$, penalizes models with many parameters. Alternatively, one may consider Schwarz' criterion SC (Schwarz, 1978). Since the overall maximum log–likelihood $\hat{l}$ is a constant, one can replace $-2l(\tilde{\beta}_1, 0, \tilde{\theta})$ in (4.1.4) by w, u or λ_m and use the resulting criteria for model comparison.

4.1.2   Selection procedures

All–subsets selection

Lawless and Singhal (1987) adopt and extend Furnival and Wilson's algorithm to generalized linear models, making efficient use of symmetric "sweeps" on (inverse) information matrices. Sweeps are the necessary algebraic manipulations to compute the relevant parts of (inverse) information matrices F and A for submodels defined by $\beta_2 = 0$. For example, sweeping on A_{22} in the augmented tableau

$$\begin{pmatrix} A_{11} & A_{12} & \beta_1 \\ A_{12}' & A_{22} & \beta_2 \\ \beta_1' & \beta_2' & 0 \end{pmatrix}$$

produces A_{22}^{-1}, and it transforms 0 to the Wald statistic w in (4.1.1). As a by–product one obtains the one–step approximation $\bar{\beta}_1$ to $\tilde{\beta}_1$ in (4.1.3). Submodels are tested against the full model, and they are labeled as good if the chosen test statistic (λ, w or λ_m) has a small value. They propose two options for screening models: (i) The "best" m models for fixed r (dimension of β_1) are determined by the models having the m smallest values of the specified test statistics. (ii) The "best" m models overall (i.e., of any dimension) are those with the m smallest modified AIC–values.

Stepwise backward and forward selection

For large models and large data sets, stepwise procedures are a useful additional tool for the reasons mentioned at the beginning of the section. Stepwise selection based on Wald and score tests is described in Fahrmeir and Frost (1992).

Backward steps use Wald tests, starting from a maximal model M, e.g., the "full" model containing all covariates. It is tested against a set U of admissible submodels (e.g., all submodels of M obtained by removing parameters associated with certain variables or all hierarchical submodels). Performing sweeps as in the tableau above, Wald statistics and p–values α_L of the models $L \in U$ are computed. Then the submodel L_0 with

$$\alpha_{L_0} = \max_{L \in U} \alpha_L \quad \text{and} \quad \alpha_{L_0} > \alpha_{out},$$

α_{out} a prechosen exclusion level, is selected. Then the MLE $\hat{\beta}$ for L_0 and its covariance matrix are computed, using the first approximation (4.1.3) as a starting value for the iterations. Since $\bar{\beta}$ is already a good approximation to $\hat{\beta}$, only one or two iterations are required, or they may even be omitted for some of the selection steps. Then L_0 is redefined as M, and the backward procedure is applied iteratively to admissible submodels. It terminates if there is no p–value $\alpha_L > \alpha_{out}$.

Forward steps use score tests starting from a minimal model L, e.g., the model without covariates. It is tested against a list V of admissible supermodels, obtained, e.g., by adding covariates not contained in L. Score statistics and p–values α_M for all $M \in V$ are computed by efficient sweeps; see Fahrmeir and Frost (1992) for details. The supermodel $M_0 \in V$ with

$$\alpha_{M_0} = \min_{M \in V} \alpha_M \quad \text{and} \quad \alpha_{M_0} < \alpha_{in},$$

α_{in} a prechosen inclusion level, is selected. Then M_0 is redefined as L, and forward steps are iterated until there is no p–value smaller than α_{in}.

Combining forward steps with backward steps in the usual way as , e.g., in linear regression or discriminance analysis, yields forward/backward model selection algorithms. Pure forward or forward/backward stepwise selection is quite useful in situations with a large number of variables (parameters) and with comparably sparse data. Typical examples are categorical response models or log–linear models for high–dimensional contingency tables. If many categorical covariates together with higher–order interaction terms are contained in the "full" model M and if data are sparse in some categories of the response variable or in some of the cells, a finite ML estimate will probably not exist, i.e., ML iterations will not converge and no "full" model can be fitted. Thus, all subset selection or even stepwise pure backward selection, which both need the fitted full model, break down.

Selection procedures based on score and Wald tests can also easily be adapted to quasi–likelihood models, replacing w and u by their modified counterparts w_m and u_m defined in Chapter 2, Section 2.3.1, whereas the likelihood ratio statistic may not even be properly defined.

TABLE 4.1. Logit model fit to credit–scoring data

Variable	ML estimate	p–value
Grand mean	− 0.188121	0.614225
X1[1]	+ 0.634647	0.000321
X1[2]	− 1.317027	0.000000
X3	+ 0.035027	0.000008
X4	+ 0.000032	0.330905
X5[1]	− 0.988369	0.000093
X6[1]	− 0.474398	0.003113
X7[1]	+ 0.223511	0.311471
X8[1]	− 0.385423	0.078926

Example 4.1: Credit–scoring (Examples 2.2, 2.5 continued)
This data set, which is contained in Fahrmeir and Hamerle (1984, see p.334 ff. and p.751 ff.), has been analyzed by Kredler (1984) and is reanalyzed for the purpose of comparison. The sample consists of 1000 consumer's credits, the binary response variable "creditability" and 20 covariates, which are assumed to influence creditability. Based on a binary logit model with y = creditworthy as the reference category and with main effects of the covariates only, pure backward selection (exclusion level 0.05) and pure forward selection (inclusion level 0.05) eliminate the same nine variables as in Fahrmeir and Hamerle (1984, p.284).

A full subset selection among 2^{20} = 1048576 possible models becomes computationally infeasible. To compare all subset selection with stepwise selection, analysis was confined to a logit model with the seven covariates X1, X2, X4, X5, X6, X7 and X8, which are given in Example 2.2. The estimation results for the full model containing all main effects are given in Table 4.1. With exclusion/inclusion levels of 0.05, backward and forward selection eliminate the variables X4 and X7. All subset selections with the Wald–, log–likelihood–, Akaike's AIC– and Schwarz' SC–criterion lead to the same best model (see Frost, 1991, for more details).   □

4.2 Diagnostics

Diagnostic tools for assessing the fit of a classical linear regression model are in common use. They are designed to detect discrepancies between the data and the fitted values as well as discrepancies between a few data and the rest. Most of these techniques are based on graphical presentations of residuals, hat matrix and case deletion measures (see, e.g., Belsley, Kuh and Welsch, 1980, and Cook and Weisberg, 1982). As an example consider a plot of residuals against fitted values. If the pattern of residuals increases or decreases with the fitted values some model departure, e.g., a wrong variance function and/or omitted covariates, is to be suspected. On the other hand, if only few residuals are far from the rest, the corresponding data points represent outliers and should be checked further. More precisely, it has to be examined whether fit or parameter estimates greatly change by omission of an outlying observation, and if so, whether the outlier is caused by an extreme response and/or extreme covariate values. Beside displays of residuals, plots of hat matrix and case deletion measures provide information about this question. Note, however, that a model departure indicated by some diagnostic plot may be caused by a faulty choice of the variance function. But it may also arise because one or more covariates are missing, or because of some outliers. That means the source of deviation cannot be determined exactly. Therefore, one has to be careful in the interpretation of such plots. Nevertheless diagnostic methods are an important tool to reveal the influence of single observations or observation sets on the fit.

For non–normal regression models diagnostics are based on extended versions of residuals, hat matrix and case deletion measures. In this section we report on such developments within the context of generalized linear models. The methods can be easily extended to quasi–likelihood and nonlinear nonexponential family models.

We do not report on diagnostics for detecting misscaled covariates or covariates that should be included into the linear predictor. In classical linear regression such methods are known as *partial residual* and *added variable plots* (see ,e.g., Cook and Weisberg, 1982). Landwehr, Pregibon and Shoemaker (1984) extended the partial residual plot to logistic regression. An improved partial residual plot, the so–called *constructed variable plot*, for the generalized linear model has been derived by Wang (1987). It is based on an added variable plot for generalized linear models that was introduced by Wang (1985).

4.2.1   Diagnostic tools for the classical linear model

Consider the classical linear model

$$y = X\beta + \varepsilon$$

where $\varepsilon \sim N(0, \sigma^2 I)$ is normally distributed, X is an $(n \times p)$–matrix of covariates and β is a $(p \times 1)$–vector of coefficients.

Based on the least–squares estimate $\hat{\beta} = (X'X)^{-1}X'y$ the *residual vector* is given by

$$r = y - \hat{y}$$

where $\hat{y} = X\hat{\beta}$ is the vector of fitted values. The residual vector shows which points are ill–fitting.

Another quantity that determines how much influence or leverage the data have on the fitted values is the *hat matrix*

$$H = X(X'X)^{-1}X'.$$

It is called hat matrix because one has $\hat{y} = Hy$. Thus the matrix H maps y into $\hat{y}$. Consider the p–dimensional subspace $\{X\beta, \beta \in \mathbb{R}^p\}$. Then $\hat{y}$ is the perpendicular projection of y into this subspace. The matrix H acts as a projection matrix, which means H is symmetric and idempotent, i.e., $H^2 = H$ holds.

From $\hat{y} = Hy$ it is seen that the element h_{ij} of the hat matrix $H = (h_{ij})$ shows the amount of leverage or influence exerted on $\hat{y}_i$ by y_j. Since H depends only on X this influence is due to the 'design' not to the dependent variable. The most interesting influence is that of y_i on the fitted value $\hat{y}_i$, which is reflected by the diagonal element h_{ii}. For the projection matrix H one has

$$\text{rank}(H) = \sum_{i=1}^{n} h_{ii} = p$$

and $0 \leq h_{ii} \leq 1$. Therefore p/n is the average size of a diagonal element. As a rule of thumb an x–point for which $h_{ii} > 2p/n$ holds is considered a high–leverage point (e.g., Hoaglin and Welsch, 1978).

In addition to the basic residual vector r several standardized versions that have a constant variance are in common use. If $r = (r_1, \ldots, r_n)$ denotes the basic residual vector the standardization

$$r_i^* = r_i/\sqrt{1 - h_{ii}}$$

has variance σ^2, which can be estimated consistently by $\hat{\sigma}^2 = r'r/(n-p)$. The standardized residual r_i^* is scaled to variance one by dividing r_i^* by $\hat{\sigma}$.

The role of residual vector and hat matrix may also be seen by looking at the effect of omitting single observations. If the ith observation is omitted the change in LS estimates is given by

$$\Delta_i \hat{\beta} = \hat{\beta} - \hat{\beta}_{(i)} = (X'X)^{-1}x_i r_i/(1 - h_{ii})$$

where $\hat{\beta}_{(i)}$ is the LS estimate when the ith observation is omitted. The change $\Delta_i \hat{\beta}$ increases with increasing residual r_i and increasing diagonal element h_{ii}.

4.2.2 Generalized hat matrix

Let us now consider a (univariate or multivariate) generalized linear model which for the ith q–dimensional observation has the form

$$\mu_i = h(Z_i \beta).$$

For univariate GLMs one has $q = 1$; for multicategorical responses as considered in Chapter 3 one has $q = k - 1$ where k is the number of categories. As is common for simple goodness–of–fit statistics (Sections 2.2.2, 3.4.2) data should be grouped as far as possible. Therefore, the number g of grouped observations will be used in the following. In this section we will consider a generalized form of the hat matrix based on ML estimation. The hat matrix yields a measure for the leverage of data and is a building block of regression diagnostics in the sense of Pregibon (1981). It is useful to remember that $A^{1/2}(A^{T/2})$ denotes a left (the corresponding right) square root of matrix A such that $A^{1/2}A^{T/2} = A$ holds. The inverse matrices are denoted by $A^{-1/2} = (A^{1/2})^{-1}$ and $A^{-T/2} = (A^{T/2})^{-1}$.

The iterative procedure for maximum likelihood estimates is given in Sections 2.2.1 and 3.4.1. At convergence the estimate has the form

$$\hat{\beta} = (Z'W(\hat{\beta})Z)^{-1}Z'W(\hat{\beta})\tilde{y}(\hat{\beta})$$

where $\tilde{y}(\hat{\beta}) = Z\hat{\beta} + D^{-1}(\hat{\beta})(y - \mu(\hat{\beta}))$. The estimate $\hat{\beta}$ is a weighted least squares solution of the linear problem $\tilde{y}(\hat{\beta}) = Z\beta + \varepsilon$ or an unweighted least squares solution of the linear problem

$$\tilde{y}_0(\hat{\beta}) = Z_0\beta + \tilde{\varepsilon}$$

where $\tilde{y}_0(\hat{\beta}) = W^{T/2}(\hat{\beta})\tilde{y}(\hat{\beta})$, $Z_0 = W^{T/2}(\hat{\beta})Z$. The hat matrix corresponding to this model has the form

$$
\begin{aligned}
H &= Z_0(Z_0'Z_0)^{-1}Z_0' = W^{T/2}(\hat{\beta})Z(Z'W(\hat{\beta})Z)^{-1}Z'W^{1/2}(\hat{\beta}) \\
&= W^{T/2}(\hat{\beta})ZF^{-1}(\hat{\beta})Z'W^{1/2}(\hat{\beta}).
\end{aligned}
$$

From $Z_0\hat{\beta} = H\tilde{y}_0(\hat{\beta})$ it is seen that H maps the observation $\tilde{y}_0(\hat{\beta}) = W^{T/2}(\hat{\beta})\tilde{y}(\hat{\beta})$ into the "fitted" value $W^{T/2}(\hat{\beta})Z\hat{\beta}$.

The $(gq \times gq)$–matrix H is idempotent and symmetric and therefore may be viewed as a projection matrix for which $\text{tr}(H) = \text{rank}(H)$ holds. For univariate response ($q = 1$) the diagonal elements h_{ii} of the matrix $H = (h_{ij})$ are determined by $0 \le h_{ii} \le 1$ and high values of h_{ii} (close to 1) correspond to extreme points in the design space. However, in contrast to the classical linear model the hat matrix does not only depend on the design matrix but also on the fit. Therefore extreme points in the design space do not necessarily have a high value of h_{ii}. For multivariate responses it is useful to consider the blocks H_{ij} of the matrix $H = (H_{ij}), i, j = 1, ..., g$, where H_{ij} is a $(q \times q)$ matrix. The $(q \times q)$ submatrix H_{ii} corresponds to the ith observation, and $\det(H_{ii})$ or $\text{tr}(H_{ii})$ may be used as an indicator for the leverage of y_i.

Example 4.2: Vaso–constriction
The data taken from Finney (1947) were obtained in a carefully controlled study in human physiology where a reflex "vaso–constriction" may occur

TABLE 4.2. Vaso–constriction data

Index	Volume	Rate	Y	Index	Volume	Rate	Y
1	3.70	0.825	1	20	1.80	1.800	1
2	3.50	1.090	1	21	0.40	2.000	0
3	1.25	2.500	1	22	0.95	1.360	0
4	0.75	1.500	1	23	1.35	1.350	0
5	0.80	3.200	1	24	1.50	1.360	0
6	0.70	3.500	1	25	1.60	1.780	1
7	0.60	0.750	0	26	0.60	1.500	0
8	1.10	1.700	0	27	1.80	1.500	1
9	0.90	0.750	0	28	0.95	1.900	0
10	0.90	0.450	0	29	1.90	0.950	1
11	0.80	0.570	0	30	1.60	0.400	0
12	0.55	2.750	0	31	2.70	0.750	1
13	0.60	3.000	0	32	2.35	0.030	0
14	1.40	2.330	1	33	1.10	1.830	0
15	0.75	3.750	1	34	1.10	2.200	1
16	2.30	1.640	1	35	1.20	2.000	1
17	3.20	1.600	1	36	0.80	3.330	1
18	0.85	1.415	1	37	0.95	1.900	0
19	1.70	1.060	0	38	0.75	1.900	0
				39	1.30	1.625	1

in the skin of the digits after taking a single deep breath. The response y is the occurrence ($y = 1$) or nonoccurrence ($y = 0$) of vaso–constriction in the skin of the digits of one subject after it inhaled a certain volume of air at a certain rate. Available are the responses of three subjects. The first contributed 9 responses, the second contributed 8 responses, and the third contributed 22 responses. The $i = 1, \ldots, 39$ observations are listed in Table 4.2.

Although the data represent repeated measurements, an analysis that assumes independent observations may be applied, as claimed by Pregibon (1981). Therefore we use the binary logit model

$$\log(P(y = 1)/P(y = 0)) = \beta_0 + \beta_1 \, \log(\text{volume}) + \beta_2 \, \log(\text{rate})$$

to analyze the effect of volume of air and inspiration rate on the occurrence of vaso–constriction. The deviance of the fit is 29.23 on 36 degrees of freedom. Pearson's χ^2 takes value 34.23. Both statistics are less than the expectation 36 of an asymptotic $\chi^2(36)$–distribution, so that the fitted logit model seems to be adequate. The MLE's for β, their estimated standard errors and the p–values for testing $H_0 : \beta_r = 0, r = 0, 1, 2$, are given in Table 4.3. The MLE's for β_1 and β_2 show that with increasing volume as well

TABLE 4.3. Logit model fit to vaso–constriction data

	MLE	Standard error	p–value
β_0	-2.875	1.319	0.029
β_1	5.179	1.862	0.005
β_2	4.562	1.835	0.013

as with increasing rate the probability for occurrence of vaso–constriction becomes higher. In addition, all three parameters are highly significant as can be seen from the p–values.

The leverage of the observation $y_i, i = 1, \ldots, 39$, can be examined by an index plot of the diagonal elements h_{ii} of the hat matrix H. Figure 4.1 gives such a plot. Observation 31 has the highest value of h_{ii}. Since its value is higher than $2p/n = 0.1533$ one would suspect that observation 31 corresponds to a high–leverage point. However, as can be seen from Table

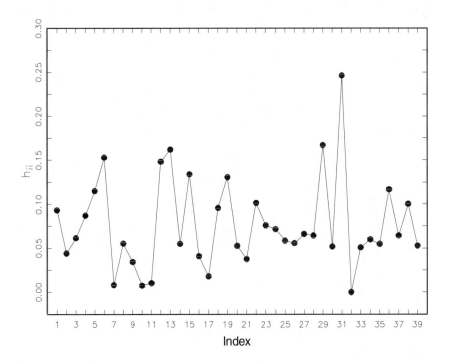

FIGURE 4.1. Index plot of h_{ii} for vaso–constriction data.

4.2 observation 31 is not extreme concerning the values of "volume" and "rate," and its effect on the fit is minor in comparison to other observations identified by residual analytic tools (see Example 4.4).   □

Example 4.3: Job expectation (Examples 3.3, 3.5 continued)
Recall the grouped data for job expectation of students of psychology given in Table 3.2. In Example 3.5 the relationship between the trichotomous response "job expectation" (Y) and "age" was analyzed by the cumulative logistic model

$$P(Y \leq r)|\text{age}) = F(\theta_r + \gamma \log(\text{age})), \quad r = 1, 2.$$

Results of the fit are given in Table 3.4. Now we are interested in the leverage of the ith observation group, $i = 1, \ldots, 13$. Note that these observation groups differ in the only available covariate "age," and in the local sample size n_i as can be seen from Table 3.2. Since the response variable is three–categorical we use $\det(H_{ii})$ resp. $\operatorname{tr}(H_{ii})$ as a measure for leverage. Figure 4.2 gives both measures plotted against the index i. In contrast to the values of $\operatorname{tr}(H_{ii})$, which only take into account the diagonal elements of H_{ii}, the values of $\det(H_{ii})$ are based on all elements of H_{ii} and are much smaller. Both measures, however, identify the observation groups 2 and 3 as those having the highest leverage values. These values, however, are primarily caused by the relative large local sample sizes n_i and not by an extremeness in the design space.   □

4.2.3   Residuals and goodness–of–fit statistics

The quantities considered in residual analysis help to identify poorly fitting observations that are not well explained by the model. Moreover, they should reveal the impact of observations on the goodness–of–fit. Systematic departures from the model may be checked by residual plots. The methods considered in the following are based on ML estimates for generalized linear models with q–dimensional observations y_i.

Generalized residuals should have properties similar to standardized residuals for the normal linear model. In particular the following forms of generalized residuals are widely used.

The Pearson residual for observation y_i is based on the *Pearson goodness–of–fit statistic*

$$\chi^2 = \sum_{i=1}^{g} \chi_P^2(y_i, \hat{\mu}_i)$$

with the ith component

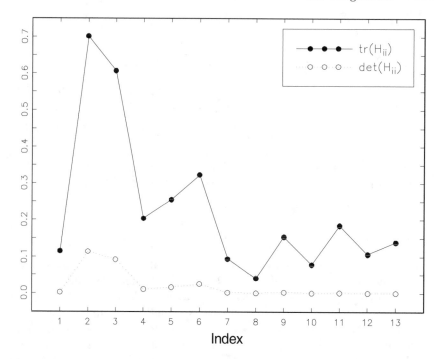

FIGURE 4.2. Index plot of $\text{tr}(H_{ii})$ and $\det(H_{ii})$ for grouped job expectation data.

$$\chi_P^2(y_i, \hat{\mu}_i) = (y_i - \hat{\mu}_i)'\Sigma_i^{-1}(\hat{\beta})(y_i - \hat{\mu}_i)$$

where $\Sigma_i(\hat{\beta})$ denotes the estimated covariance matrix of y_i and $\hat{\mu}_i = \mu_i(\hat{\beta})$.

For the standardized form of the residual one needs a standardization of $y_i - \mu_i(\hat{\beta})$. The *Pearson residual* is given by

$$r_i^P = \Sigma_i^{-1/2}(\hat{\beta})(y_i - \hat{\mu}_i)$$

which is a vector of length q. The Pearson residual may be considered a square root of $\chi_P^2(y_i, \hat{\mu}_i)$ since $\chi_P^2(y_i, \hat{\mu}_i) = (r_i^P)'r_i^P$ holds. For increasing local sample size $n_i \to \infty$ as well as $n \to \infty$ and the assumption of nearly independent y_i and $\hat{\mu}_i$ the covariance of $y_i - \hat{\mu}_i$ has the form

$$\text{cov}(y_i - \hat{\mu}_i) = \Sigma_i^{1/2}(\hat{\beta})(I - H_{ii})\Sigma_i^{T/2}(\hat{\beta}),$$

where H_{ii} denotes the ith $(q \times q)$–matrix in the diagonal of the hat matrix H and I denotes the identity matrix. This is easily seen from the asymptotic covariance

$$\text{cov}(\hat{\mu}_i) = D_i(\hat{\beta})'Z_i\text{cov}(\hat{\beta})Z_i'D_i(\hat{\beta})$$

and

$$\Sigma_i^{1/2}(\hat{\beta})H_{ii}\Sigma_i^{T/2}(\hat{\beta}) = D_i(\hat{\beta})'Z_iF^{-1}(\hat{\beta})Z_i'D_i(\hat{\beta}),$$

where $W_i^{1/2}(\hat{\beta}) = D_i(\hat{\beta})\Sigma_i^{-T/2}(\hat{\beta})$ has been used. Therefore a *studentized version* of the *Pearson residual* r_i^P is given by

$$r_{i,s}^P = (I - H_{ii})^{-1/2}r_i^P = (I - H_{ii})^{-1/2}\Sigma_i^{-1/2}(\hat{\beta})(y_i - \hat{\mu}_i)$$

which for large n_i should be approximately multinormal.

For univariate models the Pearson residual takes the simple form

$$r_i^P = \frac{y_i - \hat{\mu}_i}{\sqrt{\widehat{\mathrm{var}(y_i)}}}.$$

Since these residuals are skewed and thus for small n_i cannot be considered approximately normally distributed, transformed values $t(y_i)$ and $E(t(y_i))$ may be used instead of y_i and $\mu_i(\beta) = E(y_i)$. McCullagh and Nelder (1989) (Section 2.4.2) give transformed *Anscombe residuals* for important cases. Alternative forms of transformations aim at the vanishing of the first–order asymptotic skewness when the local sample size increases are given by Pierce and Schafer (1986). They give for the Poisson distribution the Anscombe residual

$$r_i^A = \frac{3}{2}\frac{y_i^{2/3} - (\hat{\mu}_i^{2/3} - \hat{\mu}_i^{-1/3}/9)}{\hat{\mu}_i^{1/6}}$$

and for the binomial distribution $y_i \sim B(n_i, \pi_i)$ the Anscombe residual

$$r_i^A = \sqrt{n_i}\frac{t(y_i) - [t(\hat{\pi}_i) + (\hat{\pi}_i(1 - \hat{\pi}_i))^{-1/3}(2\hat{\pi}_i - 1)/6n_i]}{\hat{\pi}_i(1 - \hat{\pi}_i)^{1/6}}$$

where $t(u) = \int_0^u s^{-1/3}(1 - s)^{-1/3}\,ds$. For alternative variance–stabilizing residuals see Pierce and Schafer (1986).

Another type of residual is based on the *deviance*

$$D = \sum_{i=1}^g \chi_D^2(y_i, \hat{\mu}_i)$$

with components

$$\chi_D^2(y_i, \hat{\mu}_i) = 2(l_i(y_i) - l_i(\hat{\mu}_i))$$

where

$$l_i(\mu_i) = [y_i'\theta_i - b(\theta_i)]\omega_i/\phi + c(y_i, \phi_i)$$

is the likelihood of μ_i based on observation y_i and $\hat{\mu}_i = \mu_i(\hat{\beta})$. If θ_i is degenerate $l_i(\mu_i)$ is defined by $l_i(\mu_i) = 0$. For univariate response models the deviance residual is given by

$$r_i^D = \text{sign}(y_i - \hat{\mu}_i)\sqrt{\chi_D^2(y_i, \hat{\mu}_i)}.$$

Explicit forms for the various distributions are easily derived from Table 2.1 in Chapter 2. For example, one gets for the Poisson distribution the deviance residual

$$r_i^D = \text{sign}(y_i - \hat{\mu}_i)\{2[y_i \log(y_i/\hat{\mu}_i) - (y_i - \hat{\mu}_i)]\}^{1/2}.$$

Adjusted deviances, which show better approximation to the normal distribution, are given by

$$r_i^{AD} = r_i^D + \rho(\theta_i(\hat{\mu}_i))/6$$

where $\rho(\theta) = E_\theta\{[(y - \mu)/s(y)]^3\}$ with $s(y) = \sqrt{\text{var}(y)}$. Pierce and Schafer (1986) give the explicit forms for several distributions:

$$y \sim B(n, \pi) : \rho(\theta) = (1 - 2\pi)/\{n\pi(1 - \pi)\}^{1/2}$$
$$y \sim P(\lambda) : \rho(\theta) = 1/\sqrt{\lambda}$$
$$y \sim \Gamma(\mu, \nu) : \rho(\theta) = 2/\sqrt{\nu}.$$

For these cases they compare the true tail probabilities and normal approximations based on several types of residuals. Anscombe residuals and adjusted deviance residuals, which are very nearly the same, yield rather good approximations with a slight preference for the adjusted deviances. For the case of integer–valued y Pierce and Schafer (1986) suggest making a continuity correction by replacing y in the formulas by $y \pm 1/2$ toward the center of distribution.

For a nonparametric approach to assessing the influence of observations on the goodness–of–fit, see Simonoff and Tsai (1991).

Example 4.4: Vaso–constriction (Example 4.2 continued)
Figure 4.3 gives an index plot of the Pearson residuals r_i^P, the standardized forms $r_{i,s}^P$ and the deviance residuals r_i^D. All residuals are based on the fit of the logistic model given in Example 4.2. Although the response is strictly binary the three residuals behave quite similarly, observations 4 and 18 show the largest values for $r_i^P, r_{i,s}^P$ as well as for r_i^D. Obviously, the standardization of r_i^P by $(1 - h_{ii})^{1/2}$, which yields $r_{i,s}^P$, can be neglected since the leverage values h_{ii} are rather balanced as can be seen from Figure 4.1.

Observations 4 and 18 are also identified as outliers in a normal probability plot of the standardized Pearson residuals $r_{i,s}^P$ that can be seen in Figure

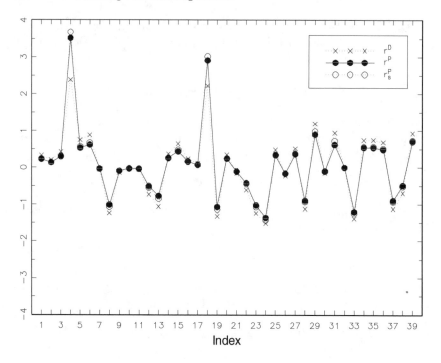

FIGURE 4.3. Index plot of $r_i^P, r_{i,s}^P$ and r_i^D for vaso–constriction data.

4.4. In Figure 4.4 the ordered values of $r_{i,s}^P, i = 1, \ldots, 39$, are plotted against the order statistic of an $N(0,1)$–sample. In the case of approximately $N(0,1)$–distributed residuals such a plot shows approximately a straight line as long as model departures and/or outliers are absent. Clearly, since the response is strictly binary the residuals $r_{i,s}^P$ are not $N(0,1)$–distributed. However, the plot can be used to detect badly fitted observations that are far away from the rest.   □

Example 4.5: Job expectation (Examples 3.3, 3.5, 4.3 continued)
For illustration of residuals in the case of a multicategorical response we refer to the job expectation data. Figure 4.6 shows an index plot of the squared standardized Pearson residuals $(r_{i,s}^P)'(r_{i,s}^P)$ that are based on the cumulative logit model given in Example 4.3. Note that the index $i, i = 1, \ldots, 13$, stands for the ith observation group as given in Table 3.2. The plot in particular identifies observation group 10 as an outlying point. From Table 3.2 it can be seen that this observation group only contains one observation that is 29 years old and does not expect to find an adequate job after getting the degree in psychology. Such a response–covariate combination, however, does not fit into the estimated cumulative logit model, which suggests that the probability of not finding an adequate job decreases with

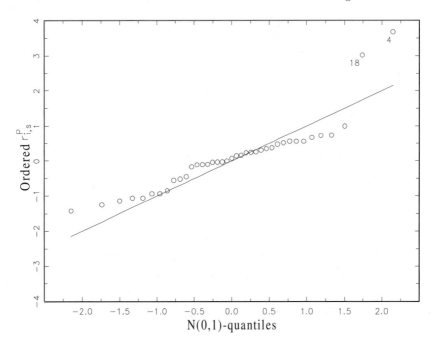

FIGURE 4.4. $N(0,1)$–probability plot of $r_{i,s}^P$ for vaso–constriction data.

increasing age. This is demonstrated by Figure 4.5, which shows the relative frequencies and probability for the fitted model. A similar argument holds for observation group 7 (25 years of age). A look at Figure 4.5 shows that observations 12 (31 years of age) and 13 (34 years of age) are also far from the expected value. However, the local sample size for both observations is only one.

Since the local sample sizes n_i of most observation groups are larger than 1, one may compare the squared standardized Pearsons residuals $(r_{i,s}^P)'(r_{i,s}^P)$ with a χ^2–distribution. More specifically, a $\chi^2(2)$–probability plot is carried out by plotting the ordered values $(r_{i,s}^P)'(r_{i,s}^P)$ against the order statistics of a $\chi^2(2)$–sample. As can be seen in Figure 4.7 the plot approximates a straight line rather well with the exception of observation 10. This leads one to suppose that the "bad" fit given in Table 3.4 is mainly caused by the outlying observation 10 (25 years). □

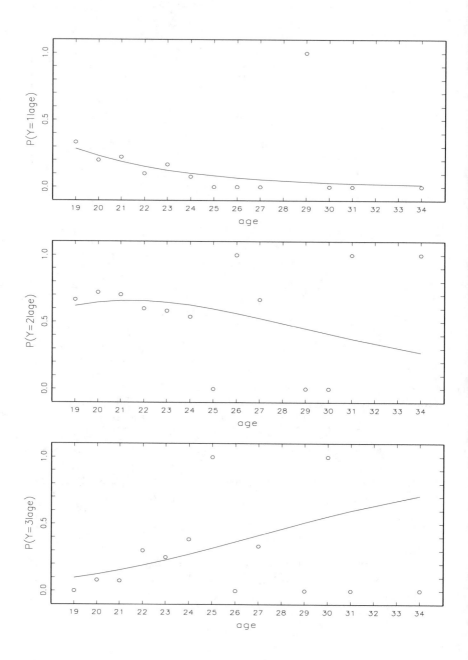

FIGURE 4.5. Relative frequencies and response curve of the fitted cumulative logistic model (responses are "don't expect adequate employment," "not sure," "expect employment immediately after getting the degree").

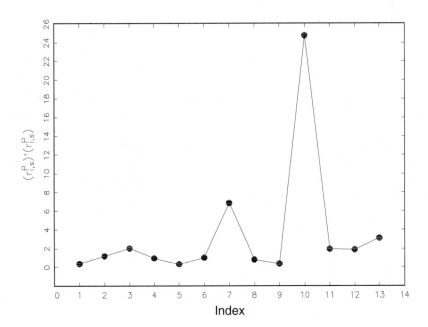

FIGURE 4.6. Index plot of $(r_{i,s}^P)'(r_{i,s}^P)$ for grouped job expectation data.

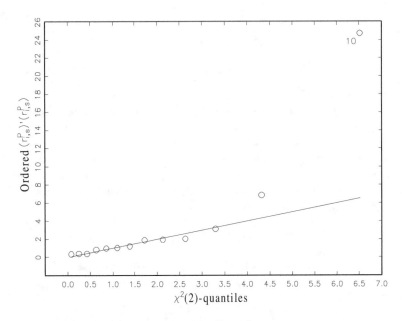

FIGURE 4.7. $\chi^2(2)$–probability plot of $(r_{i,s}^P)'(r_{i,s}^P)$ for grouped job expectation data.

4.2.4   Case deletion

An indicator for the influence of the ith observation (y_i, Z_i) on the vector $\hat{\beta}$ can be calculated by the difference $(\hat{\beta} - \hat{\beta}_{(i)})$, where $\hat{\beta}_{(i)}$ is the MLE obtained from the sample without observation (y_i, Z_i) and $\hat{\beta}$ is the MLE from all observations. If $\hat{\beta}_{(i)}$ is substantially different from $\hat{\beta}$ observation (y_i, Z_i) may be considered influential. Measures of this type have been given by Cook (1977) for linear regression models. Since the estimation of unknown parameters requires an iterative procedure it is computationally expensive to subsequently delete each observation and refit the model. A one–step approximation of $\hat{\beta}_{(i)}$ is obtained by performing just one step of the iterative process when $\hat{\beta}$ is the starting value. The one–step estimate is given by

$$\hat{\beta}_{(i),1} = F_{(i)}^{-1}(\hat{\beta}) Z_{(i)}' W_{(i)}(\hat{\beta}) \tilde{y}(\hat{\beta})$$

where the reduced Fisher matrix is given by

$$F_{(i)}(\hat{\beta}) = F(\hat{\beta}) - Z_i' W_i(\hat{\beta}) Z_i$$

and

$$Z_{(i)}' W_{(i)}(\hat{\beta}) \tilde{y}(\hat{\beta}) = Z' W(\hat{\beta}) \tilde{y}(\hat{\beta}) - Z_i' W_i(\hat{\beta}) \tilde{y}_i(\hat{\beta}).$$

A simpler form of $\hat{\beta}_{(i),1}$ given by Hennevogl and Kranert (1988) is

$$\hat{\beta}_{(i),1} = \hat{\beta} - F^{-1}(\hat{\beta}) Z_i' W_i^{1/2}(\hat{\beta}) (I - H_{ii})^{-1} \Sigma_i^{-1/2}(\hat{\beta}) (y_i - \mu_i(\hat{\beta})) \quad (4.2.1)$$

where H_{ii} is the ith block diagonal of the hat matrix evaluated at $\hat{\beta}$. The difference between the one–step estimate $\hat{\beta}_{(i),1}$ and the original estimate $\hat{\beta}$ may be used as an indicator for the impact of observation (y_i, Z_i) on the estimated parameter.

To determine the influence of observations on the estimate $\hat{\beta}$ one has to consider all the components of $\hat{\beta}$. Therefore it is often useful to have an overall measure as considered by Cook (1977). An asymptotic confidence region for β is given by the log–likelihood distance

$$-2\{l(\beta) - l(\hat{\beta})\} = c$$

which is based on its asymptotic $\chi^2(p)$ distribution. Approximation of $l(\beta)$ by a second–order Taylor expansion yields an approximate confidence region given by

$$(\beta - \hat{\beta})' \text{cov}(\hat{\beta})^{-1} (\beta - \hat{\beta}) \approx c. \quad (4.2.2)$$

If β is replaced by the one–step estimate $\hat{\beta}_{(i),1}$ one gets

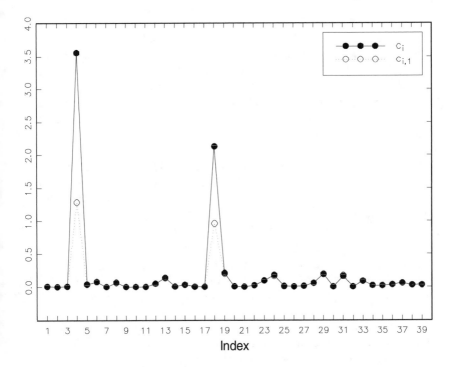

FIGURE 4.8. Index plot of $c_{i,1}$ and c_i for vaso–constriction data.

$$
\begin{aligned}
c_{i,1} &= (\hat{\beta}_{(i),1} - \hat{\beta})'\operatorname{cov}(\hat{\beta})^{-1}(\hat{\beta}_{(i),1} - \hat{\beta}) \\
&= (y_i - \mu_i(\hat{\beta}))'\Sigma_i^{-T/2}(\hat{\beta})(I - H_{ii})^{-1}H_{ii} \\
&\quad (I - H_{ii})^{-1}\Sigma_i^{-1/2}(\hat{\beta})(y_i - \mu_i(\hat{\beta})) \\
&= (r_i^P)'(I - H_{ii})^{-1}H_{ii}(I - H_{ii})^{-1}r_i^P. \qquad (4.2.3)
\end{aligned}
$$

Pregibon (1981) refers to $c_{i,1}$ as a confidence interval displacement diagnostic. The quantity $c_{i,1}$ measures the displacement of $\hat{\beta}$ by omitting observation (y_i, Z_i). As is seen from (4.2.3) $c_{i,1}$ is composed from previously defined diagnostic elements. Again the generalized hat matrix H_{ii} for the ith observation plays an important role. Large values in H_{ii} will produce large values $c_{i,1}$. The same holds for the Pearson residual r_i^P. The original Cook measure as proposed by Cook (1977) for the normal linear model makes use of the leaving–one–out estimate $\hat{\beta}_{(i)}$ instead of the one–step approximation $\hat{\beta}_{(i),1}$. This corresponds to replacing β by $\hat{\beta}_{(i)}$ in (4.2.2), yielding $c_i = (\hat{\beta}_{(i)} - \hat{\beta})\operatorname{cov}(\hat{\beta})^{-1}(\hat{\beta}_{(i)} - \hat{\beta})$. Applications for the generalized linear model show that the use of $\hat{\beta}_{(i),1}$ often tends to underestimate the original Cook measure. The quantities c_i and $c_{i,1}$ assess the effect of omitting observation (y_i, Z_i) on the estimate of the whole parameter vector β. Other quantities that measure the change in various other aspects of the fit are given by Pregibon (1981), Williams (1987), Lee (1988) and Lesaffre and Albert (1989). For example, Williams (1987) measures the change in likelihood ratio statistics, and Lee (1988) derives influence quantities that measure the effect of deleting observation (y_i, Z_i) on the estimates of a subset or single components of the parameter vector β.

Example 4.6: Vaso–constriction (Examples 4.2, 4.4 continued)
Figure 4.8 gives an index plot of the approximate Cook distances $c_{i,1}$ as well as of the exact measures c_i for the logistic model given in Example 4.2. The approximate values $c_{i,1}$ are quite similar to the exact values c_i. Only for observations having an undue influence on the MLE $\hat{\beta}$, i.e., observations 4 and 18, the approximative measure $c_{i,1}$ underestimates the exact value c_i. The large influence of observations 4 and 18, given by $c_{4,1}$ and $c_{18,1}$, is primarily caused by large residuals, as can be seen from Figure 4.3, and not by large leverage values, as can be seen from Figure 4.1.   □

Example 4.7: Job expectation (Examples 3.3, 3.5, 4.3, 4.5 continued)
Approximate Cook distances c_i for the grouped job expectation data of Table 3.2 are given in Figure 4.9. The values c_i were obtained by fitting the cumulative logit model of Example 4.3 to the reduced data that do not contain observation group i. In terms of c_i observations 3, 7, 10 and 13 have the strongest influence. Observation 10 and (somewhat weaker) observations 7 and 13 show high residuals (Figure 4.6) whereas observation 3 shows a large value for $\mathrm{tr}(H_{ii})$. The one–step approximation does not work so well. In particular observations 2 and 10 show a quite different value. Since $c_{i,1}$ is a direct function of residuals and hat–matrix the high value of $\mathrm{tr}(H_{ii})$ of observation 2 may cause this effect.   □

4.3   General tests for misspecification*

Throughout this section we maintain the basic assumption of (conditionally) independent responses. This assumption will generally be violated in the time series and longitudinal data situation. Consequences of its violation can be studied along the lines of White (1984). The tests for misspecification described in the sequel should be of potential usefulness for testing whether a model is misspecified for one of the following reasons:

(i) The linear predictor does not reflect the influence of covariates correctly, e.g., due to omitted covariates, a wrong design matrix in

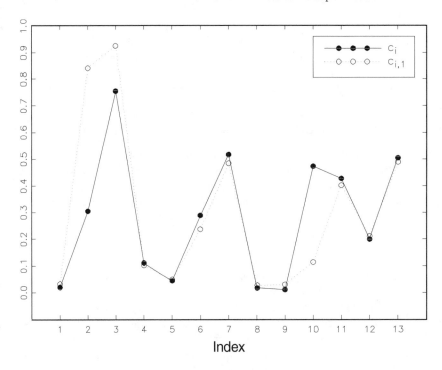

FIGURE 4.9. Index plot of $c_{i,1}$ and c_i for grouped job expectation data.

multicategorical models, or nonlinearities not covered by the chosen predictor.

(ii) Link or response function violation occurs, e.g., when a logit model is used instead of another binary model.

(iii) The exponential family or the variance function is incorrectly specified.

A number of tests of deviations in one of these particular directions has been developed in the literature. Most of them include further parameters in the model, defining, e.g., a generalized family of link functions as in Pregibon (1980), Stukel (1988) and Czado (1992), and test whether this inclusion significantly improves the fit or not. For a survey of tests of this type the reader is referred to chapter 11.4 of McCullagh and Nelder (1989). A nonparametric approach for checking the adequacy of the link function is proposed in Azzalini, Bowman and Härdle (1989) and Horowitz and Härdle (1992).

In Sections 4.3.2 and 4.3.3 we will take a look at some alternative general tests for misspecification that have mainly been discussed in the econometric literature. Most of these tests were developed, at least in the beginning, with a view toward linear and nonlinear regression models for metric responses, but they are applicable, in principle, to any maximum likelihood–type regression problem. We start with a short discussion of the consequences of model assumption violations, adapting White's (1982, 1984) work to generalized linear models. A more detailed presentation can be found in Fahrmeir (1987b, 1990).

4.3.1   Estimation under model misspecification

In the following, terms like E, cov, etc. refer to "true" expectations, co-variances, etc., corresponding to the "true" probabilistic process P generating the data $(y_1, x_1), ..., (y_n, x_n)$. After choosing a certain GLM or quasi–likelihood model, estimation is based on the *quasi–score function*

$$s(\beta) = \sum_{i=1}^{n} M_i(\beta)\Sigma_i^{-1}(\beta)(y_i - \mu_i(\beta)), \qquad (4.3.1)$$

where $\mu_i(\beta), \Sigma_i(\beta)$ is the mean and variance structure implied by the model specification, and $M_i(\beta) = \partial\mu_i(\beta)/\partial\beta = Z_i'D_i(\beta)$. Compared to the quasi–likelihood modelling considered in Section 2.3.1 not only the variance function but also the mean may be misspecified, i.e., there is no "true" β_0 in the admissible set B of parameters such that $E(y_i|x_i) = \mu_i(\beta_0), i = 1, ..., n$, for the *true mean* $m_i = E(y_i|x_i)$. A quasi–MLE $\hat{\beta}$ is computed by the usual algorithms as a root of $s(\beta)$. What can we say about such a QMLE? An investigation of its properties is based on the heuristic idea that a local QMLE should be near to a root β^* of the *expected quasi–score function*

$$s^*(\beta) = Es(\beta) = \sum_{i=1}^{n} M_i(\beta)\Sigma_i^{-1}(\beta)(m_i - \mu_i(\beta)). \qquad (4.3.2)$$

Such a root β^* of (4.3.2) may be termed a quasi–true parameter, since it plays a similar role as the "true" parameter β_0 in correctly specified models. Comparing with (4.3.1) it is seen that $E\,s(\beta)$ is obtained from $s(\beta)$ simply by replacing responses y_i by their true means $m_i = E(y_i|x_i)$. In the case where $s(\beta)$ is obtained as the derivative of a proper log–likelihood model $l(\beta), \beta^*$ can be interpreted as a minimizer of the Kullback–Leibler distance between $l(\beta)$ and the true log–likelihood.

Under certain regularity assumptions (White, 1982; Gourieroux et al., 1984, Fahrmeir, 1990), which are similar but somewhat sharper than in the case of correctly specified models, the QMLE $\hat{\beta}$ is asymptotically normal,

$$\hat{\beta} \stackrel{a}{\sim} N(\beta^*, H^*(V^*)^{-1}H^*)$$

where $H(\beta) = -\partial s(\beta)/\partial\beta'$ is the quasi–information matrix,

$$V(\beta) = \text{cov}\,s(\beta) = \sum_{i=1}^{n} M_i(\beta)\Sigma_i^{-1}(\beta)S_i\Sigma_i^{-1}(\beta)M_i'(\beta),$$

with $S_i = \text{cov}(y_i|x_i)$ as the true covariance matrix of y_i, and "$*$" means evaluation at $\beta = \beta^*$.

Remarks:

(i) If the model is correctly specified, then $\beta^* = \beta_0$ (the "true" parameter), and $S_i = \Sigma_i(\beta_0)$ so that $V(\beta)$ reduces to the common expected information $F(\beta)$. Moreover $H(\beta)$ is asymptotically equivalent to $F(\beta)$, so that c) boils down to the usual asymptotic normality result for GLM's.

(ii) If only the mean is correctly specified, i.e.,

$$m_i = h(Z_i'\beta_0), \qquad i = 1, 2, \dots \qquad \text{for some } \beta_0,$$

then we still have $\beta^* = \beta_0$, since β_0 is a root of (4.3.2). Estimating S_i by $(y_i - \hat{\mu}_i)(y_i - \hat{\mu}_i)'$ and H^* by $\hat{F}$, c) essentially reduces to the asymptotic normality result for quasi–likelihood estimation in Section 2.3.1.

(iii) A special but important case of mean misspecification occurs if the linear predictor is still correctly specified by

$$\eta = \alpha + x'\gamma$$

say (only univariate models are considered), but the response function $h(\cdot)$ is wrong. Extending previous work on estimation under link violation, Li and Duan (1989) show the following: In the population case where (y_i, x_i) are i.i.d. replicates of (y, x) the slope vector γ can still be estimated consistently up to a scale factor, i.e.,

$$\hat{\gamma} \to \gamma^0 = c\gamma,$$

provided that h is a response function that leads to a concave log–likelihood, and that the regressor vector x is sampled randomly from a continuous distribution with the property that the conditional expectation $E(x'a|x'\gamma)$ is linear in $x'\gamma$ for any linear combination $x'a$. The latter condition is fulfilled for elliptically symmetric distributions, including the multivariate normal. This result indicates that estimates $\hat{\beta}$ can still be meaningful even if the response function is grossly misspecified but the linear predictor is correctly specified: We still can estimate the ratios γ_i/γ_k of the slope vector consistently, and these are the key quantities for interpretation of the effects of covariates x_i and x_k. In addition, $\hat{\gamma}$ is asymptotically normal (Li and Duan, 1989, p. 1030).

(iv) For applications, an estimator of the asymptotic covariance matrix is needed. Whereas H^* can be estimated by $\hat{H}$ or $\hat{F}$, where " $\hat{\ }$ " means evaluation at $\hat{\beta}$, no general solution seems to be available for V^* yet, since it contains the unknown covariance matrices S_i. If the mean is not too grossly misspecified in the sense that $\hat{\mu}_i$ fits y_i not too badly, it seems reasonable to use the estimator

$$\hat{V} = \sum_{i=1}^{n} \hat{M}_i \hat{\Sigma}_i^{-1}(y_i - \hat{\mu}_i)(y_i - \hat{\mu}_i)'\hat{\Sigma}_i^{-1}\hat{M}_i'.$$

4.3.2   Hausman–type tests

Hausman tests

The original proposal of Hausman (1978) was made in the context of linear regression, with primary interest in developing a test for misspecification that detects correlation between covariates and errors caused, e.g., by omitted variables. However, the test can also be sensitive to other alternatives, and the basic approach can be extended beyond the frame of linear models. The general idea is to construct two alternative estimators $\hat{\beta}$ and $\tilde{\beta}$ that are consistent for the "true" parameter β_0 under the null hypothesis H_0 of correct specification and that do not converge to the same limit if the model is misspecified. (In the latter case both estimators may be inconsistent.) Furthermore both estimators have to be asymptotically normal. A Hausman test statistic of the Wald type is then defined by (see also Hausman and Taylor, 1981, Holly, 1982)

$$w_h = (\hat{\beta} - \tilde{\beta})' \hat{C}(\hat{\beta}, \tilde{\beta})^- (\hat{\beta} - \tilde{\beta}), \qquad (4.3.3)$$

where $C(\hat{\beta}, \tilde{\beta})$ is the asymptotic covariance matrix of $\hat{\beta} - \tilde{\beta}$ under H_0, and $\hat{C}(\hat{\beta}, \tilde{\beta})$ is a consistent estimate of C. If $\hat{\beta}$ is chosen as the MLE or as any other asymptotically efficient estimator, then

$$\hat{C}(\hat{\beta}, \tilde{\beta}) = \hat{V}(\hat{\beta}) - \tilde{V}(\tilde{\beta})$$

where $\hat{V}$ and $\tilde{V}$ are the asymptotic covariance matrices of $\hat{\beta}$ and $\tilde{\beta}$. Generally, the asymptotic covariance matrix C can be obtained from the joint asymptotic distribution of $(\hat{\beta}, \tilde{\beta})$; see, e.g., Arminger and Schoenberg (1989).

Given certain additional regularity conditions, the Wald–type Hausman statistic w_h has an asymptotic χ^2–distribution with r = rank(C) degrees of freedom under H_0. Large values of w_h should therefore indicate model misspecification. In its general formulation the Hausman test is more a principle than a specific test. Some important questions in particular applications of the basic idea are still left open.

A main problem is the choice of alternative estimators $\hat{\beta}$ and $\tilde{\beta}$. Taking $\hat{\beta}$ as the MLE or any other efficient estimator, one has to find an inefficient alternative estimator $\tilde{\beta}$ such that w_h has enough power to detect various sources of misspecification. For testing correlation among regressors and errors in linear models, estimators $\tilde{\beta}$ based on instrumental variables or weighted least squares have been proposed. For generalized linear models there are more possible candidates, e.g., unweighted least squares estimators or QMLE's $\tilde{\beta}$ maximizing weighted quasi–log–likelihoods

$$l_w(\beta) = \sum_{i=1}^{n} w_i l_i(\beta).$$

In the context of nonlinear regression, White (1981) suggested choosing weights w_i in such a way that the region of the regressor space where the $y_i's$ are fitted poorly is heavily weighted compared to regions with a good fit. Arminger and Schoenberg (1989) adapt this proposal to quasi–likelihood models.

A further problem is: How well does the χ^2–approximation work for finite sample sizes? Two questions arise in this context. First, can the shape of the finite sample distribution adequately be approximated by a χ^2–distribution, in particular in the tails? Monte Carlo studies would provide some evidence, but there seems to be a lack of such investigations in the published literature. Secondly, as first observed by Krämer (1986) for linear regression models, the determination of rank(C) by the rank of the estimate $\hat{C}$ may be misleading: Despite the singularity of C, the estimate $\hat{C}$ may be nonsingular or have higher rank than C. Then wrong inferences will be drawn due to wrong degrees of freedom. Moreover, even the asymptotic χ^2–distribution can be in question: A sufficient and necessary condition that $\hat{C}^-$ is a consistent estimate of C^- is rank $(\hat{C}) \to$ rank(C) in probability as $n \to \infty$, see also Andrews (1987). There are counterexamples where that latter condition is violated, implying that $\hat{C}^-$ is not a consistent estimate of C^- although $\hat{C}$ is a consistent estimate of C. Since consistency of $\hat{C}^-$ is crucial in conventional proofs, even the asymptotic χ^2–distribution can be in question, a fact that has often been overlooked in earlier work.

In a simulation study Frost (1991) investigated several versions of Hausman tests, including unweighted LSE's and weighted QMLE's $\hat{\beta}$, for categorical regression models. The results were disappointing: Various versions of the test often failed to hold its asymptotic significance level and had insufficient power.

Information matrix test

White (1982) proposed comparing estimates $\hat{F}$ and $\hat{V}$ of the (expected) quasi–information $F(\beta) = EH(\beta)$ and of the covariance matrix $V(\beta)$ given in the previous section. Under correct specification the usual regularity assumptions of ML estimation imply $F = V, \hat{F} - \hat{V} \xrightarrow{p} 0$, while $F \neq V$ under misspecification. The information matrix test based on this idea can be interpreted as a Hausman–type test: The difference of alternative estimators is replaced by the vectorized difference

$$\hat{d} = \text{vec}(\hat{V} - \hat{F}),$$

and the Hausman–type statistic for the information matrix test is

$$i = \tilde{d}'\hat{C}^{-1}\tilde{d},$$

where $\tilde{d}$ is a subvector of $\hat{d}$ selected such that its asymptotic covariance matric C is nonsingular. Under H_0 it should be asymptotically χ^2 with

$r = \text{rank}(C)$ degrees of freedom. The main problems connected with its implementation are selection of an appropriate subvector $\tilde{d}$ of $\hat{d}$, and finding consistent and numerically stable estimators $\hat{C}$ for C.

In a number of simulation studies we found that the information matrix test behaved unsatisfactorily. In particular, the probability for falsely rejecting a correctly specified model was often too high. Similar observations have also been made by Andrews (1988).

To summarize: Although Hausman–type tests are implemented in some program packages, in our experience one should be careful when using current versions of Hausman–type tests as a general tool for detecting misspecification in GLM's. It seems that further investigation is necessary here.

4.3.3   Tests for non–nested hypotheses

Hausman–type tests are pure significance tests in the sense that there is no particular alternative H_1 to the model H_0. The tests considered in this section require a specific alternative model H_1. In contrast to classical tests null and alternative hypotheses need not be nested. For binomial responses typical non–nested hypotheses are, e.g.,

H_0: logit model with predictor $\eta = z'\beta$

H_1: linear model with the same predictor,

or

H_0: logit model with predictor $\eta_0 = z_0'\beta_0$

H_1: logit model with different predictor $\eta_1 = z_1'\beta_1$,

or a combination of both. Generally two models H_0 and H_1 are said to be non–nested if H_0 is not nested within H_1, and H_1 is not nested within H_0.

Much of the theoretical literature on non–nested hypotheses testing has its roots in the basic work of Cox (1961, 1962), who proposed a modified likelihood ratio statistic. However "Cox–type tests" are often difficult to implement, and a number of authors proposed simplified versions; see McKinnon (1983) for a survey. A major class of models that is more convenient to deal with is based on the idea of embedding H_0 and H_1 in a supermodel.

Tests based on artificial nesting

For linear models Davidson and McKinnon (1980) formulate an artificial supermodel

$$y = (1 - \lambda)z_0'\beta_0 + \lambda z_1'\beta_1 + \varepsilon$$

in order to test $H_0 : \lambda = 0$ against $H_1 : \lambda \neq 0$. For identifiability reasons, β_1 has to be replaced by some estimator $\hat{\beta}_1$, e.g., the OLSE of the alternative model $y = z_1' \beta_1 + \varepsilon_1$. Davidson and McKinnon suggest a common t–test (i.e., a Wald test) for testing $H_0 : \lambda = 0$, which requires joint estimation of $b = (1 - \lambda)\beta_0$ and λ. Their 'J–test' can be carried out easily with standard software for linear models.

For generalized linear models, Gourieroux (1985) derives a score statistic for testing $\lambda = 0$. Assuming that both models belong to the same exponential family, as in the examples for binomial responses earlier, the test is based on artificial models of the form

$$Ey = \mu = (1 - \lambda)h_0(x_0'\beta_0) + \lambda h_1(x_1'\beta_1),$$

where $h_0(x_0'\beta_0)$ and $h_1(x_1'\beta_1)$ stand for the null and alternative models. For the resulting score statistic we refer the reader to Gourieroux (1985).

Generalized Wald and score tests

A different approach, which is closer to the idea of Hausman's test, has been taken by Gourieroux et al. (1983b). Let

$$s_0(\beta_0) = \sum_{i=1}^{n} M_{i0}\Sigma_{i0}^{-1}(y_i - h_0(Z_{i0}'\beta_0))$$

and

$$s_1(\beta_1) = \sum_{i=1}^{n} M_{i1}^{-1}\Sigma_{i1}^{-1}(y_i - h_1(Z_{i1}'\beta_1))$$

denote the (quasi–) score functions under H_0 and H_1. Based on the theory of estimation under misspecification (Section 4.3.1), the idea is to compare the (quasi–) MLE $\hat{\beta}_1$, i.e., a root of $s_1(\beta_1)$, with the corresponding root $\hat{\beta}_{10}$ of

$$\hat{s}_1(\beta_1) = \sum_{i=1}^{n} M_{i1}\Sigma_{i1}^{-1}(\hat{h}_{i0} - h_1(Z_{i1}'\beta_1))$$

where $\hat{h}_{i0} = h_0(Z_{i0}'\hat{\beta}_0)$ and $\hat{\beta}_0$ is a root of $s_0(\beta_0)$. Compared to the expected (under H_0) score function $E_0 s_1(\beta_1)$, see (4.3.2), the expectation $m_{i0} = E_0 y_i$ is replaced by its estimate $\hat{h}_{i0}$. Therefore it is plausible that under H_0 both $\hat{\beta}_1$ and $\hat{\beta}_{10}$ should tend to the quasi–true value β_1^*. As a consequence of the asymptotic results in Section 4.3.1, it can be conjectured that the difference $\hat{\beta}_1 - \hat{\beta}_{10}$ is asymptotically normal. Gourieroux et al. (1983) give a formal proof of this conjecture. A Wald statistic of the form

$$w = (\hat{\beta}_1 - \hat{\beta}_{10})' \hat{C}_w^- (\hat{\beta}_1 - \hat{\beta}_{10}),$$

where $\hat{C}_w$ is an estimate of the possibly singular asymptotic covariance matrix of $\hat{\beta}_1 - \hat{\beta}_{10}$, can be used to test H_0 versus H_1. Large values of $\hat{\beta}_1 -$

$\hat{\beta}_{10}$ and w are in favour of H_1. Alternatively, the following asymptotically equivalent score statistic can be derived:

$$s = \hat{s}'_{10} \hat{C}_s^- \hat{s}_{10}, \qquad (4.3.4)$$

with $\hat{s}_{10} = s_1(\hat{\beta}_{10})$ and the estimated asymptotic covariance matrix

$$\hat{C}_s = \hat{C}_{11} - \hat{C}_{10} \hat{C}_{00}^{-1} \hat{C}_{01},$$

$$\hat{C}_{00} = \sum_{i=1}^{n} \hat{M}_{i0} \hat{\Sigma}_{i0}^{-1} \hat{M}'_{i0},$$

$$\hat{C}_{01} = \sum_{i=1}^{n} \hat{M}_{i0} \hat{\Sigma}_{i1}^{-1} \hat{M}'_{i1}, \hat{C}_{10} = \sum_{i=1}^{n} M_{i1} \hat{\Sigma}_{i0}^{-1} \hat{M}'_{i0} = \hat{C}'_{01},$$

$$\hat{C}_{11} = \sum_{i=1}^{n} \hat{M}_{i1} \hat{\Sigma}_{i1}^{-1} \hat{\Sigma}_{i0} \hat{\Sigma}_{i1}^{-1} \hat{M}'_{i1},$$

where " ^ " means evaluation at $\beta_0 = \hat{\beta}_0$ or $\beta_1 = \hat{\beta}_{10}$ respectively. Compared to the Wald statistic, $\hat{C}_s$ is easier to implement and numerically more stable than $\hat{C}_w$. If the H_0 model is true, s is asymptotically χ^2–distributed with rank (C_s) degrees of freedom.

In a number of simulation experiments Frost (1991) compared generalized score and Wald tests with the Gourieroux test based on artificial nesting. All in all, the results are in favour of the generalized score test.

To decide between two rival models in applications, the test has to be carried out twice by changing the role of H_0 and H_1. Correspondingly there are four possible results:

(i) H_0 not rejected, H_1 rejected: choice of H_0,

(ii) H_0 rejected, H_1 not rejected: choice of H_1,

(iii) H_0 and H_1 rejected: both models rejected,

(iv) H_0 nor H_1 rejected: no conclusion possible.

Example 4.8: Credit–scoring (Examples 2.2, 2.5, 4.1 continued)
As an illustration we use the credit–scoring data already analyzed previously. We will compare non–nested rival logit models characterized by different covariate vectors using the generalized score statistic (4.3.4). With the same abbreviations as in Example 2.2, we first test

$$H_0 \quad : \quad \text{Model including X1, X3, X5,}$$
$$H_1 \quad : \quad \text{Model including X4, X6, X7, X8.}$$

The resulting score statistic is $s = 20.30$, with df $= 4$ degrees of freedom. The p–value $\alpha = 0.00043$ leads to rejection of H_1. Interchanging H_0 and H_1, one obtains $s = 150.49$, df $= 4$, $\alpha = 0.00$, so that the model including X1, X3, X5 is rejected as well. This suggests that important covariates are omitted in both models.

Enlarging the covariate vector of the models to X1, X3, X5, X6 resp. X4, X5, X6, X7, X8 and testing again leads to rejection of both models. Finally, let us test

$$H_0 \quad : \quad \text{Model including X1, X3, X5, X6, X8}$$
$$H_1 \quad : \quad \text{Model including X3, X4, X5, X6, X7, X8.}$$

One obtains $s = 2.66$, df $= 6$, $\alpha = 0.85$, so that H_0 is not rejected. Interchanging H_0 and H_1 leads to rejection of the model including X3, X4, X5, X6, X7, X8 in favour of the model including X1, X3, X5, X6, X8. This conclusion is consistent with the results obtained by variable selection in Example 4.1. $\square$

5
Semi– and nonparametric approaches to regression analysis

In this chapter developments are given that lead beyond the framework of parametric models. Instead of assuming a functional form that specifies how explanatory variables determine dependent variables, the functional form is assumed to be in some way smooth, and the data are allowed to determine the appropriate functional form under weak restrictions.

In the first section the case of a continuous metric response variable is considered. With no parametric model in mind the estimates of expected responses are purely nonparametric. Only two concepts are developed more thoroughly for the familiar case of continuous responses because they are basic for extensions including the categorical case. The first one is the concept of smoothing by splines, which is extended to smoothing in generalized linear models in Section 5.3. The second concept refers to kernel smoothing. Although usual techniques still apply for dichotomous responses, the multicategorical case makes some modifications necessary. In this case the conditional probability function of a multicategorical response has to be estimated instead of the expectation of a univariate response. Estimation procedures based on categorical kernels are outlined in Section 5.2.

More extensive treatments of special approaches may be found in the literature. Härdle (1990a) considers nonparametric smoothing for continuous responses, Eubank (1988) gives a thorough treatment of splines and Hastie and Tibshirani (1990) show the possibilities of generalized additive modelling. Härdle and Turlach (1992) survey recent developments.

5.1 Smoothing techniques for continuous responses

In this section a short survey of smoothing techniques for the continuous case is given. Observations are bivariate data $(y_i, x_i), i = 1, \ldots, n$, where the response y_i and the explanatory variable x_i are measured on interval scale level. It is assumed that the dependence of y_i on x_i is given by

$$y_i = f(x_i) + \varepsilon_i$$

where f is an unspecified smooth function and ε_i is a noise variable with $E(\varepsilon_i) = 0$. A scatterplot smoother $\hat{f}$ is a smooth estimate of f based on observations (y_i, x_i), $i = 1, \ldots, n$. Most of the smoothers considered in the following are linear. That means if one is only interested in the fit at an observed value x_i the estimate has the linear form

$$\hat{f}(x_i) = \sum_{j=1}^{n} s_{ij} y_j .$$

The weights $s_{ij} = s(x_i, x_j)$ depend on the target point x_i where the response is to be estimated and on the value x_j where the response y_j is observed.

5.1.1 Simple neighbourhood smoothers

A simple device for the estimation of $f(x_i)$ is to use the average of response values in the neighbourhood of x_i. These *local average estimates* have the form

$$\hat{f}(x_i) = \text{Ave}_{j \in N(x_i)}(y_j)$$

where Ave is an averaging operator and $N(x_i)$ is a neighbourhood of x_i. The extension of the neighbourhood is determined by the *span* or *window size* w, which denotes the proportion of total points in a neighbourhood.

Let the window size w be from (0,1) and let the integer part of wn, denoted by $[wn]$, be odd. Then a symmetric neighbourhood may be constructed by

$$N(x_i) = \{\max\{i - \frac{[wn] - 1}{2}, 1\}, \ldots, i-1, i, i+1, \ldots \min\{i + \frac{[wn] - 1}{2}, n\}\}.$$

$N(x_i)$ gives the indices of the ordered data $x_1 < \ldots < x_n$. It contains x_i and $([wn] - 1)/2$ points on either side of x_i if x_i is from the middle of the data. The neighbourhood is smaller near the end points, e.g., at x_1, x_n, which leads to quite biased estimates at the boundary. Of course w determines the smoothness of the estimate. If w is small the estimate is very rough; for large w the estimate is a smooth function.

A special case of a local average estimate is the *running mean* where Ave stands for arithmetic mean. Alternatively the mean may be replaced by the *median*, yielding an estimate that is more resistant to outliers. A drawback of the median is that the resulting smoother is nonlinear.

Another simple smoother is the *running–line smoother*. Instead of computing the mean a linear term

$$\hat{f}(x_i) = \hat{\alpha}_i + \hat{\beta}_i x_i$$

is fitted where $\hat{\alpha}_i, \hat{\beta}_i$ are the least–squares estimates for the data points in the neighbourhood $N(x_i)$. In comparison to the running mean, the running–line smoother reduces bias near the end points. However, both smoothers, running mean and running line may produce curves that are quite jagged.

If the target point x is not from the sample $x_1, \ldots, x_n$ one may interpolate linearly between the fit of the two $\hat{y}_i$ values, which are observed at predictor values from the sample adjacent to x. Alternatively, one may use varying nonsymmetric neighbourhoods. For $k \in \mathbb{N}$ the *k–nearest neighbourhood (k–NN)* estimate is a weighted average based on the varying neighbourhood

$$N(x) = \{i | x_i \text{ is one of the } k\text{–nearest observations to } x\}$$

where near is defined by a distance measure $d(x, x_i)$. Then the degree of smoothing is determined by k. The proportion of points in each neighbourhood is given by $w = k/n$. For example, the linear k–NN estimate has the form

$$\hat{f}(x) = \sum_{i=1}^{n} s(x, x_i) y_i$$

with weights

$$s(x, x_i) = \begin{cases} 1/k, & x_i \in N(x) \\ 0, & \text{otherwise.} \end{cases}$$

The weights fulfill the condition $\Sigma_i s(x, x_i) = 1$. The estimate is not restricted to unidimensional x–values. By appropriate choice of a distance measure, e.g., the Euclidian distance $d(x, x_i) = \|x_i - x\|^2$, the method may be applied to vector–valued x.

5.1.2   Spline smoothing

Cubic smoothing splines

Smoothed estimators may be considered compromises between faith with the data and reduced roughness caused by the noise in the data. This view is made explicit in the construction of smoothing splines. The starting point is the following minimization problem: Find the function f that minimizes

$$\sum_{i=1}^{n}(y_i - f(x_i))^2 + \lambda \int_{-\infty}^{\infty} (f''(u))^2 du \qquad (5.1.1)$$

where f has continuous first and second derivates f', f'', and f'' is quadratically integrable. The first term in (5.1.1) is the residual sum of squares, which is used as a distance function between data and estimator. The second term penalizes roughness of the function by taking the integrated squared second derivative $\int f''(u)^2 du$ as a global measure for curvature or roughness. The parameter $\lambda \geq 0$ is a smoothing parameter that controls the trade–off between the smoothness of the curve and the faith with the data: Large values of the smoothing parameter λ give large weight to the penalty term, therefore enforcing smooth functions with small variance but possibly high bias. For rather small λ, the function f will nearly interpolate the data.

The solution $\hat{f}_\lambda$ of the penalized least–squares estimation problem (5.1.1) is a *cubic smoothing spline* with knots at each distinct x_i (Reinsch, 1967). That means $\hat{f}_\lambda$ is a cubic polynomial between successive x–values and first and second derivatives are continuous at the observation points. At the boundary points the second derivative is zero. Since these piecewise cubic polynomials are defined by a finite number of parameters, optimization of (5.1.1) with respect to a set of functions actually reduces to a finite–dimensional optimization problem. It can be shown (Reinsch 1967; de Boor, 1978, ch. 14) that minimization of (5.1.1) is equivalent to minimizing the *penalized least–squares criterion*

$$PLS(f) = (y - f)'(y - f) + \lambda f'Kf \qquad (5.1.2)$$

where $y = (y_1, \ldots, y_n)$ are the data and $f = (f(x_1), \ldots, f(x_n))$ are the fitted y–values for $x_1 < \ldots < x_n$, i.e., the vector of evaluations of the function $f(x)$. The penalty matrix K has a special structure and is given by

$$K = D'C^{-1}D \qquad (5.1.3)$$

where
$D = (d_{ij})$ is a $(n-2, n)$ upper tridiagonal matrix,

$$D = \begin{pmatrix} \frac{1}{h_1} & -(\frac{1}{h_1} + \frac{1}{h_2}) & \frac{1}{h_2} & & & \\ & \frac{1}{h_2} & -(\frac{1}{h_2} + \frac{1}{h_3}) & \frac{1}{h_3} & & \\ & & \ddots & \ddots & \ddots & \\ & & & \frac{1}{h_{n-2}} & -(\frac{1}{h_{n-2}} + \frac{1}{h_{n-1}}) & \frac{1}{h_{n-1}} \end{pmatrix},$$

$$h_i := x_{i+1} - x_i,$$

and $C = (c_{ij})$ is an $(n-2, n-2)$ tridiagonal symmetric matrix,

$$
C = \frac{1}{6}
\begin{pmatrix}
2(h_1 + h_2) & h_2 & & & \\
h_2 & 2(h_2 + h_3) & h_3 & & \\
& & \ddots & & \\
& \ddots & & \ddots & \\
& & & & h_{n-2} \\
& & h_{n-2} & 2(h_{n-2} + h_{n-1})
\end{pmatrix}.
$$

The minimizing function $\hat{f}_\lambda$ can be obtained by equating the vector $-2(y - f) + 2\lambda K f$ of first derivates of $PLS(f)$ to zero. This yields the *linear smoother*

$$
\hat{f}_\lambda = (I + \lambda K)^{-1} y, \tag{5.1.4}
$$

with smoothing matrix $S_\lambda = (I + \lambda K)^{-1}$, I denoting the identity matrix.

For computational reasons, $\hat{f}_\lambda$ and the smoothing matrix S_λ are generally not computed directly by inversion of $I + \lambda K$ (note that S_λ is an $(n \times n)$– matrix). Instead, $\hat{f}_\lambda$ is computed indirectly in two steps, making efficient use of the band matrices D and C; see Reinsch (1967) and de Boor (1978, ch. 14) for details.

In (5.1.2) the distance between data and estimator is measured by a simple quadratic function. More generally a weighted quadratic distance may be used. For given diagonal weight matrix W a weighted penalized least squares criterion is given by

$$
(y - f)' W (y - f) + \lambda f' K f. \tag{5.1.5}
$$

The solution is again a cubic smoothing spline, with the vector $\hat{f}_\lambda$ of fitted values now given by

$$
\hat{f}_\lambda = (W + \lambda K)^{-1} W y. \tag{5.1.6}
$$

Thus, only the identity matrix I is replaced by the diagonal weight matrix W. Again the solution $\hat{f}_\lambda$ of (5.1.5) is computed indirectly. The weighted spline version (5.1.5), (5.1.6) forms a basic tool for simple spline smoothing in generalized linear models (Section 5.3).

Regression splines

An alternative variant of spline smoothing is based on so–called it regression splines. Instead of solving a minimization problem like (5.1.1) one has to choose a sequence of breakpoints or knots $\xi_1 < \ldots < \xi_s$ from (x_1, x_n) where $x_1 < \ldots < x_n$. Moreover, one has to specify the degree of the polynomial that is to be fitted piecewise in the range $[\xi_i, \xi_{i+1})$ where ξ_0 and ξ_{s+1} denote additional boundary knots. In addition, the polynomials are supposed to join smoothly at the knots. For the case of cubic splines that is considered here the polynomials are constrained to have continuous first and second derivatives at the knots.

For a given set of knots a possible representation is given by

$$\hat{f}(x) = \delta_0 + \delta_1 x + \delta_2 x^2 + \delta_3 x^3 + \sum_{i=1}^{s} \theta_i (x - \xi_i)_+^3 \qquad (5.1.7)$$

where $(x - \xi_i)_+ = \max\{0, x - \xi_i\}$. It is easily seen that $\hat{f}(x)$ is a cubic polynomial in $[\xi_i, \xi_{i+1})$ with continuous first and second derivatives. The smooth function (5.1.7) is a weighted sum of the $s + 4$ functions $P_0(x) = 1; P_1(x) = x, ..., P_{s+3}(x) = (x - \xi_s)_+^3$. These functions are referred to as basis functions used in the expansion. The particular choice used in (5.1.7) is the so–called truncated power series basis.

Expression (5.1.7) is linear in the $s+4$ parameters $\delta_0, ..., \delta_3, \theta_1, ..., \theta_s$, and may be fitted by usual linear regression. However, numerically superior alternatives are available by using alternative basis functions called B–splines (de Boor, 1978).

Cubic splines are a common choice. More difficult is the selection of knots. The number and position of knots strongly determines the degree of smoothing. The position of knots may be chosen uniformly over the data (cardinal splines), at appropriate quantiles or by more complex data–driven schemes. For a detailed discussion of these issues see Friedman and Silverman (1989).

5.1.3   Kernel smoothing

Kernel smoothing is based on the linear form

$$\hat{f}_\lambda(x) = \sum_{i=1}^{n} s(x, x_i) y_i \qquad (5.1.8)$$

where the local weights $s(x, x_i), i = 1, ..., n$, at x are determined by a density function adjusted by a scale parameter λ. The kernel or density function K is a continuous symmetric function with

$$\int K(u) du = 1. \qquad (5.1.9)$$

Widely used kernel functions are the Gaussian kernel where K is the standard Gaussian density or the Epanechnikov kernel

$$K(u) = \begin{cases} \frac{3}{4}(1 - u^2), & \text{for } |u| \leq 1 \\ 0, & \text{otherwise} \end{cases} \qquad (5.1.10)$$

(Epanechnikov, 1969) or the minimum variance kernel

$$K(u) = \begin{cases} \frac{3}{8}(3 - 5u^2), & \text{for } |u| \leq 1 \\ 0, & \text{otherwise.} \end{cases} \qquad (5.1.11)$$

The basic idea in kernel smoothing comes from naive histogram smoothing where a rectangular window (uniform kernel) is moved on the x–axis. In order to get smoother estimates one replaces the uniform kernel by smoother kernels. The weights $s(x, x_i)$ in (5.1.8) are chosen to be proportional to

$$K \left(\frac{x - x_i}{\lambda} \right).$$

That means that for reasonable kernel functions K (e.g., a Gaussian density) the weights will decrease with increasing distance $|x - x_i|$. The window–width or bandwidth λ determines how fast the decrease of weights actually is. For small λ only values in the immediate neighbourhood of x will be influential, for large λ values that are more distant from x may also have influence upon the estimate.

Nadaraya (1964) and Watson (1964) proposed the commonly used so–called Nadaraya–Watson weights

$$s(x, x_i) = \frac{\frac{1}{n\lambda} K \left(\frac{x - x_i}{\lambda} \right)}{\sum_{j=1}^{n} \frac{1}{n\lambda} K \left(\frac{x - x_j}{\lambda} \right)}. \tag{5.1.12}$$

Here the denominator

$$\hat{g}(x) = \sum_{j=1}^{n} \frac{1}{n\lambda} K \left(\frac{x - x_j}{\lambda} \right),$$

which may be considered an estimate of the density of x–values, yields the normalizing of weights

$$\sum_{i=1}^{n} s(x, x_i) = 1.$$

An alternative weight function has been considered in a series of papers by Gasser and Müller (1979, 1984), Müller (1984), Gasser, Müller and Mammitzsch (1985). It is based on ordered values $0 = x_1 < \ldots < x_n = 1$ with $s_i \in [x_i, x_{i+1}]$, $s_o = 0$, $s_n = 1$, chosen between adjacent x–values. The *Gasser–Müller weights* given by

$$s(x, x_i) = \frac{1}{\lambda} \int_{s_{i-1}}^{s_i} K(\frac{x - u}{\lambda}) du \tag{5.1.13}$$

again sum up to one. Other weight functions have been given by Priestley and Chao (1972) and Benedetti (1977).

Relation to other smoothers

Smoothing procedures like nearest neighbourhood estimates and spline smoothing are strongly related to kernel smoothing.

Let the x–values be equidistant with $x_i = i/n, i = 1, \ldots, n$. Then in the middle of the data the nearest neighbourhood estimate is equivalent to a kernel estimate with weights (5.1.12) based on the uniform kernel $K(u) = 1$ for $u \in [-0.5, 0.5)$. The smoothing parameter has to be chosen by $\lambda = k/n$ where k is the number of neighbourhood values.

A combination of kernels and neighbourhood estimates is given by using, e.g., the Nadaraya–Watson weights with smoothing parameter $\lambda_x = d(x, x^{(k)})$ where $d(x, x^{(k)})$ is the distance between the target point x and its kth–nearest neighbour. Then the bandwidth is locally chosen with small values if the data are dense and strong smoothing if the data are sparse (see Stone, 1977, Mack, 1981). Alternative approaches to locally varying bandwidths are considered by Müller and Stadtmüller (1987) and Staniswalis (1989).

Instead of varying smoothing parameters the distance may be defined alternatively. The distance $(x - x_i)$ in the kernel may be substituted by $\hat{F}(x) - \hat{F}(x_i)$ where $\hat{F}$ denotes the empirical distribution function (see Yang, 1981; Stute, 1984).

Although it is not obvious (cubic) spline smoothing yields linear estimates $\hat{f}_\lambda(x) = \sum_i s(x, x_i) y_i$ with "spline weights" s. For $x = x_i$ this is already seen from (5.1.4). Silverman (1984) showed that under regularity conditions the weights may be approximated by kernel weights based on a symmetric kernel K_s with locally varying bandwidths.

For plots of the kernel K_s and effective spline weight functions see Silverman (1984) and Engle et al. (1986).

Bias–variance trade–off

With smoothed regression the bias of the estimate is not to be neglected. The principle of trade–off between the bias and the variance of the estimate is reflected by the squared error

$$E(\hat{f}_\lambda(x) - f(x))^2 = \operatorname{var}\hat{f}_\lambda(x) + [E\hat{f}_\lambda(x) - f(x)]^2. \qquad (5.1.14)$$

It is of interest to look at the effect of smoothing on these components. Gasser and Müller (1984) show under regularity conditions including the consistency condition $n \to \infty, h \to 0, nh \to \infty$ that for fixed variables x and weights (5.1.13) variance and bias^2 are given by

$$\operatorname{var}\hat{f}_\lambda(x) \approx \frac{\sigma^2}{n\lambda} \int K^2(u)du$$

$$[E\hat{f}_\lambda(x) - f(x)]^2 \approx \frac{\lambda^4}{4}[f''(x)][\int u^2 K(u)du]^2$$

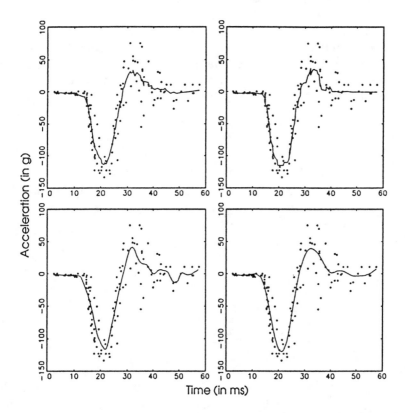

FIGURE 5.1. Smoothed estimates for motorcycle data showing time ($x - axis$) and head acceleration ($y - axis$) after a simulated impact. Smoothers are running lines (top left), running medians (top right), Epanechnikov kernel (bottom left) and cubic splines (bottom right).

where $\sigma^2 = \mathrm{var}(\varepsilon_i)$ and f'' denotes the second derivative of f. That means decreasing λ increases the variance but decreases the bias. The bias becomes large in particular at points with large values of the second derivative.

For the trade–off between variance and bias when alternative weights are used see Lai (1977), Mack (1981) and Härdle (1990a).

Example 5.1: Motorcycle data
For illustration let us consider an example with continuous response. In a data set from Härdle (1990a, table 1) the head acceleration after a simulated impact with motorcycles is investigated in dependence on time (in millisec-

onds) after the impact. Figure 5.1 shows the data as dots and smoothed estimates. The top left shows the running line smoother ($k = 20$), the top right the running median smoother ($k = 34$), below the kernel smoother based on the Epanechnikov kernel ($\lambda = 3.0$) and the estimate resulting from cubic splines ($\lambda = 10$) are given. The smoothing parameter for the splines has been chosen by generalized cross validation (see Section 5.1.4) whereas the other smoothing parameters have been chosen by vision. Kernel and spline methods yield much smoother curves. In particular the running median smoother gives a quite jagged curve.   □

5.1.4   Selection of smoothing parameters*

The mean squared error (5.1.14) is a local criterion considered for fixed point x. Global distance measures are the *average squared error*

$$ASE(\lambda) = \frac{1}{n} \sum_{i=1}^{n} \{\hat{f}_\lambda(x_i) - f(x_i)\}^2 \qquad (5.1.15)$$

and the *mean average squared error*

$$MASE(\lambda) = \frac{1}{n} \sum_{i=1}^{n} E\{\hat{f}_\lambda(x_i) - f(x_i)\}^2. \qquad (5.1.16)$$

The latter may be considered a discrete approximation to the *integrated squared error*

$$ISE(\lambda) = \int (\hat{f}_\lambda(x) - f(x))^2 dx. \qquad (5.1.17)$$

A prediction–oriented measure is the *average predictive squared error*

$$PSE(\lambda) = \frac{1}{n} \sum_{i=1}^{n} E\{\hat{f}_\lambda(x_i) - y_i^*\}^2 \qquad (5.1.18)$$

where y_i^* is a new independent observation at x_i. The connection to $MASE$ is given by $PSE = MASE + \sigma^2$.

Minimization of ASE yields a smoothing parameter that is oriented on the data at hand whereas minimization of $MASE$ aims at the average over all possible data sets. Härdle, Hall and Marron (1988) show for kernel smoothing based on the Priestley–Chao weights that for $n \to \infty$ and equally spaced design points x_i ASE and MASE are very close.

A naive measure that may actually be minimized for the data at hand is the *average squared residual*

$$ASR(\lambda) = \frac{1}{n} \sum_i \{y_i - \hat{f}_\lambda(x_i)\}^2. \tag{5.1.19}$$

However, minimization of ASR does not yield a good approximation of global theoretical measures since the data are used to construct the estimate and simultaneously to assess the properties.

A measure that is more appropriate is the *cross–validation criterion*

$$CV(\lambda) = \frac{1}{n} \sum_{i=1}^{n} \{y_i - \hat{f}_\lambda^{-i}(x_i)\}^2 \tag{5.1.20}$$

where $\hat{f}_\lambda^{-i}(x_i)$ is a version of $\hat{f}_\lambda(x_i)$ computed by leaving out the ith data point. A simple version of the leaving–one–out estimate is reached by correcting the weights computed for the full set of n data points. For linear smoothers $\hat{f}(x_i) = \sum_j s_{ij} y_j$ one may choose

$$\hat{f}_\lambda^{-i}(x_i) = \sum_{j \neq i} \frac{s_{ij}}{1 - s_{ii}} y_j, \tag{5.1.21}$$

where the modified weights $s_{ij}/(1 - s_{ii})$ now sum up to 1. Thus one gets the simple form

$$\hat{f}_\lambda^{-i}(x_i) = \frac{1}{1 - s_{ii}} \hat{f}_\lambda(x_i) - \frac{s_{ii}}{1 - s_{ii}} y_i.$$

Then the essential term $y_i - \hat{f}_\lambda^{-i}(x_i)$ in CV is given by

$$y_i - \hat{f}_\lambda^{-i}(x_i) = \frac{y_i - \hat{f}_\lambda(x_i)}{1 - s_{ii}}$$

and may be computed from the regular fit $\hat{f}_\lambda(x_i)$ based on n observations and the weight s_{ii}. For spline smoothing and kernel smoothing with Nadaraya–Watson weights, (5.1.21) is equivalent to the estimate based on $n - 1$ observations. This does not hold for all smoothers since, for example, a smoother designed for equidistant grid changes when one observation is left out. By using (5.1.21) one gets the criterion

$$CV(\lambda) = \frac{1}{n} \sum_{i=1}^{n} \left\{ \frac{y_i - \hat{f}_\lambda(x_i)}{1 - s_{ii}} \right\}^2.$$

Generalized cross–validation as introduced by Craven and Wahba (1979) replaces s_{ii} by the average $\sum_i s_{ii}/n$. The resulting criterion

$$GCV(\lambda) = \frac{1}{n} \sum_{i=1}^{n} \left\{ \frac{y_i - \hat{f}_\lambda(x_i)}{1 - \frac{1}{n} \sum_j s_{jj}} \right\}^2 = ASR(\lambda)(1 - \frac{1}{n} \sum_i s_{ii})^{-2}$$

is easier to compute since it is the simple averaged squared error corrected by a factor. For weighted cubic spline smoothing as in (5.1.5), the squared "studentized" residuals in $CV(\lambda)$ and $GCV(\lambda)$ have to be multiplied by the corresponding weights w_i of the weight matrix $W = \text{diag}(w_1, ..., w_n)$. For example, $GCV(\lambda)$ is modified to

$$GCV(\lambda) = \frac{1}{n} \sum_{i=1}^{n} w_i \left\{ \frac{y_i - \hat{f}_\lambda(x_i)}{1 - tr(S_\lambda)/n} \right\}^2.$$

Härdle et al. (1988) consider more general bandwidth selectors of the type

$$ASR(\lambda)C(\frac{1}{n\lambda})$$

where $C(\frac{1}{n\lambda})$ is a correction factor. In particular they show that cross validation for special kernel estimators is given by

$$CV(\lambda) = ASR(\lambda)(1 + \frac{2}{n\lambda}K(0) + O_p(n^{-2}\lambda^{-2})).$$

5.2   Kernel smoothing with multicategorical response

For dichotomous response $y \in \{0, 1\}$ the methods of Section 5.1.4 that have been developed for metric responses still apply (see Copas, 1983; Müller and Schmitt, 1988; Staniswalis and Cooper, 1988). However, if y is a categorical response variable with $y \in \{1, \ldots, k\}$ measured on nominal or ordinal scale level weighted averaging of the responses y_i no longer makes sense. Instead of estimating the expectation $E(y|x)$, which is appropriate for continuous or dichotomous 0–1 variables, for a nominal variable with k outcomes the discrete density $p(i|x), i = 1, \ldots, k$, has to be estimated. Therefore in the following, discrete density estimation based on kernels is shortly sketched.

5.2.1   Kernel methods for the estimation of discrete distributions

Let $y \in \{1, \ldots, k\}$ be a multicategorical variable with $\pi_r = P(y = r), r = 1, \ldots, k$. Let $S = \{y_1, \ldots, y_n\}$ denote an i.i.d. sample from this distribution. Then smoothed estimates for π_r as introduced by Aitchison and Aitken (1976) have the form

$$\hat{p}(y|S, \lambda) = \frac{1}{n} \sum_{\tilde{y} \in S} K_\lambda(y|\tilde{y}) \tag{5.2.1}$$

where $K_\lambda(.|\tilde{y})$ is a kernel function and λ is a smoothing parameter.

Equation (5.2.1) has the usual form of density estimates. For the estimation of a continuous variable $K_\lambda(.|\tilde{y})$ may be chosen as a Gaussian kernel $K_\lambda(y|\tilde{y}) = (2\pi)^{-1/2}\lambda^{-1}\exp(-(y-\tilde{y})^2/2\lambda^2)$. For the categorical case considered here K is a discrete kernel. The simplest kernel that is appropriate for nominal response y is the Aitchison and Aitken kernel

$$K_\lambda(y|\tilde{y}) = \left\{ \begin{array}{ll} (1+(k-1)\lambda)/k, & y = \tilde{y} \\ (1-\lambda)/k, & y \neq \tilde{y} \end{array} \right. \tag{5.2.2}$$

where the smoothing parameter λ is from [0,1]. It is easily seen that $\lambda = 1$ yields the simple relative frequency $\hat{p}(y|S,\lambda) = n_y/n$ where n_y is the number of observations having outcome $y \in \{1,\ldots,k\}$. For $\lambda = 0$ (5.2.2) yields the uniform distribution $\hat{p}(y|S,\lambda) = 1/k$ regardless of the observations. That means $\lambda \to 0$ gives an ultrasmooth estimate whereas $\lambda \to 1$ gives the usual ML estimate.

If the response categories $1,\ldots,k$ are ordered, ordinal kernels that take account of the nearness of observation and target value are more appropriate. Discrete ordinal kernels have been proposed by Habbema, Hermans and Remme (1978), Wang and Van Ryzin, (1981), Aitken (1983), Titterington and Bowman (1985). Table 5.1 gives some discrete ordinal kernels. Without loss of generality by rescaling the parameter λ may always be chosen from the unit interval such that $\lambda = 1$ yields the unsmoothed and $\lambda \to 0$ yields the ultrasmooth estimate. Like for continuous kernels the specific form of the ordinal kernel is of minor importance. However, ordinal kernels perform much better than the nominal Aitchison and Aitken kernel in the case of ordinal responses.

If $y = (y_1 \ldots, y_m)$ is a vector of categorical variables with components $y_i \in \{1,\ldots,k_i\}, i = 1,\ldots,m$, kernels may be constructed as product kernels

$$K_\lambda(y|\tilde{y}) = \prod_{i=1}^{m} K_{i,\lambda_i}(y_i|\tilde{y}_i) \tag{5.2.3}$$

where $y = (y_1 \ldots, y_m)$ is the target value, $\tilde{y} = (\tilde{y}_1,\ldots,\tilde{y}_m)$ is the observation and $\lambda = (\lambda_1,\ldots,\lambda_k)$ is a vector–valued smoothing parameter. Component kernels K_i may be chosen according to the scale level of component y_i. Thus for a mixture of nominal and ordinal components a mixture of nominal and ordinal kernels may be used.

It should be noted that discrete kernel smoothing is strongly related to minimum penalized distance estimators with quadratic loss function and Bayesian approaches based on the Dirichlet prior (see Titterington, 1985; Fienberg and Holland, 1973; Leonard, 1977; Titterington and Bowman, 1985; Simonoff, 1983).

Data–driven procedures for the choice of smoothing parameters are based on discrete loss functions. Let $T = \prod_{i=1}^{m}\{1,\ldots,k_i\} = \{z_1,\ldots,z_s\}$ denote the set of possible values of $y = (y_1,\ldots,y_m)$ in the general case of m

TABLE 5.1. Discrete ordinal kernel functions

		Range of admissibility
Habbema kernel	$K_\lambda(x\|y) \propto (1-\lambda)^{\|x-y\|^2}$	$\lambda \in [0,1]$
Uniform kernel	$K_\lambda(x\|y) \propto \begin{cases} (1-\lambda^*)/\|T(y)\| & x \in T(y) \\ \lambda^* & x = y \\ 0 & \text{else} \end{cases}$ where $\quad T(y) = \{z\| \|z-y\| \le k, z \ne y\}$	$\lambda \in [0,1]$ $\lambda^* = \frac{1+k\lambda}{1+k}$
Geometrical kernel	$K_\lambda(x\|y) \propto \begin{cases} \lambda & x = y \\ \frac{1}{2}\lambda(1-\lambda)^{\|x-y\|} & x \ne y \end{cases}$	$\lambda \in (0,1]$
Modified Aitken kernel	$K_\lambda(x\|y) \propto \begin{cases} \lambda^* & x = y \\ -\frac{1-\lambda^*}{2^{\|x-y\|}} & x \ne y \end{cases}$	$\lambda \in [\frac{1}{3}, 1]$ $\lambda^* = \frac{1+2\lambda}{3}$

discrete variables. Let p denote the underlying probability function with $p(z_i) = P(y = z_i)$ and $\hat{p}_\lambda$ denote the estimated probability function with $\hat{p}_\lambda(z_i) = \hat{p}(z_i|S,\lambda)$. Then candidates for the loss function are the quadratic loss

$$\tilde{L}_Q(p, \hat{p}_\lambda) = \sum_{y \in T} (p(y) - \hat{p}_\lambda(y))^2 \qquad (5.2.4)$$

or the Kullback–Leibler distance

$$\tilde{L}_{KL}(p, \hat{p}_\lambda) = \sum_{y \in T} p(y) \log\{\frac{p(y)}{\hat{p}_\lambda(y)}\}. \qquad (5.2.5)$$

With the delta function

$$\delta_{\tilde{y}}(y) = \left\{ \begin{array}{ll} 1, & y = \tilde{y} \\ 0, & \text{otherwise} \end{array} \right.$$

the cross–validation criterion as a widely used criterion for the choice of smoothing parameters is given by

$$CV(\lambda) = \frac{1}{n} \sum_{\tilde{y} \in S} \tilde{L}(\delta_{\tilde{y}}, \hat{p}_{\lambda}(\backslash \tilde{y}))$$

where $\hat{p}_{\lambda}(\backslash \tilde{y})$ is the estimate of p without observation $\tilde{y}$. This criterion may be viewed as the empirical (leaving–one–out) version of loss functions of type (5.2.4) or (5.2.5). For example, the quadratic loss function yields

$$CV(\lambda) = \frac{1}{n} \sum_{\tilde{y} \in S} \sum_{y \in T} (\delta_{\tilde{y}} - \hat{p}_{\lambda}(\backslash \tilde{y}))^2.$$

Choice of smoothing parameters for discrete density estimation has been considered by Aitchison and Aitken (1976), Titterington (1980), Bowman (1980) and Hall (1981). A thorough investigation of general properties has been given by Bowman, Hall and Titterington (1984).

Example 5.2: Memory
Haberman (1978, p.3) considers data from a questionnaire where respondents, who remembered one stressful event in the period from 1 to 18 months prior to the interview, were asked to report the month of occurrence of the event. The number of respondents in months 1 to 18 was 15, 11, 14, 17, 5, 11, 10, 4, 8, 10, 7, 9, 11, 3, 6, 1, 1, 4. Although the probability of a stressful event should be uniformly distributed over months human memory will produce a nonuniform distribution. Haberman (1978) fitted a log–linear model $\log(\pi_t) = \alpha + \beta t$ where π_t is the probability for month $t = 1, \ldots, 18$. In contrast to Haberman's result Read and Cressie (1988) showed that the model does not fit the data sufficiently. We consider nonparametric estimation of the distribution over months.

Figure 5.2 shows the relative frequencies together with a smoothed estimate for the nominal Aitchison and Aitken kernel and the ordinal Habbema kernel after choosing the smoothing parameters by cross–validation. It is seen that the nominal kernel is quite close to the data but is obviously inappropriate since the ordinal nature of the response is not exploited. The ordinal kernel shows some structure of the data; after an initial decrease the probability seems to be quite stable from about 7 to 13 months, and then decreases again.   □

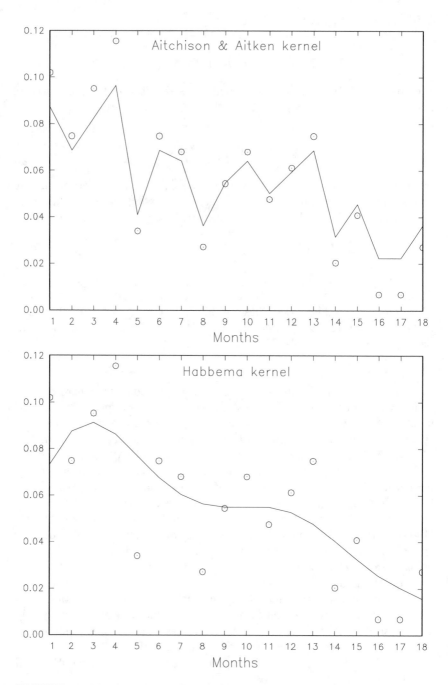

FIGURE 5.2. Memory of stressful event for nominal and ordinal kernel (relative frequencies given as points).

5.2.2 Smoothed categorical regression

Now let $y \in \{1, \ldots, k\}$ denote the categorical response variable and x be a vector of covariates. Let $x_1, \ldots, x_g$ be the observed (or given) values of the predictor variable with $x_i \neq x_j$, $i \neq j$ and $S_i = \{y_1^{(i)}, \ldots, y_{n_i}^{(i)}\}$ denote the local sample of responses observed at the fixed value x_i.

For fixed x_i a simple estimate of the distribution of $y|x_i$ is given by

$$\hat{p}(y|x_i) = \frac{1}{n_i} \sum_{\tilde{y} \in S_i} K_\lambda(y|\tilde{y}) \qquad (5.2.6)$$

where $K_\lambda(\cdot|\tilde{y})$ is a discrete kernel function. The estimate (5.2.6) is the simple density estimate based on the local sample at x_i. Since the number of local observations may be very small it is advantageous to use the information about the response y that is available from observations in the neighbourhood of x_i. Moreover, for continuous x one wants an estimate of $p(y|x)$ for x–values between observed values $x_1, \ldots, x_g$. A smoothed estimate with this property is the *categorical kernel regression estimate* or *direct kernel estimate*

$$\begin{aligned} \hat{p}(y|x, \lambda, \nu) &= \sum_{i=1}^{g} \tilde{s}_\nu(x, x_i)\, \hat{p}(y|x_i) \\ &= \sum_{i=1}^{g} \tilde{s}_\nu(x, x_i) \frac{1}{n_i} \sum_{\tilde{y} \in S_i} K_\lambda(y|\tilde{y}). \end{aligned} \qquad (5.2.7)$$

In (5.2.7) K_λ denotes the discrete kernel and $s_\nu(x, x_i)$ is a weighting function fulfilling

$$\sum_{i=1}^{g} \tilde{s}_\nu(x, x_i) = 1.$$

If x is a unidimensional metric variable $\tilde{s}_\nu(x, x_i)$ may be chosen by $\tilde{s}_\nu(x, x_i) = n_i s_\nu(x, x_i)$, where $s_\nu(x, x_i)$ is the Nadaraya–Watson weight function (5.1.12) or the Gasser–Müller weight function (5.1.13). The kernel regression estimate (5.2.7) is composed from two components, namely, the local density estimate at x_i and the weighting across predictor values. Therefore the smoothness of the estimate is determined by two smoothing parameters. The parameter λ primarily determines the smoothness of the estimated density $p(y|x)$ considered as a function of y where x is fixed, whereas ν primarily determines the smoothness of $p(y|x)$ considered as a function in x where y is fixed. That means an ultrasmooth parameter λ tends to $P(y = 1|x) = \ldots = P(y = k|x)$ whereas an ultrasmooth parameter ν tends to give $P(y = 1|x) = P(y = 1|z)$ for $x \neq z$. The term direct kernel estimate is due to the use of (5.2.7) in discriminant analysis. It serves to distinguish this approach from the "indirect" method of using kernels for the estimation of $p(x|y)$ and transforming them afterward into posterior probabilities $p(y|x)$ (Lauder, 1983; Tutz, 1990a).

A general Nadaraya–Watson weight function allowing for categorical and metric explanatory variables is given by

$$\tilde{s}_\nu(x, x_i) = \frac{n_i K_\nu(x|x_i)}{\sum_{j=1}^{g} n_j K_\nu(x|x_j)}. \tag{5.2.8}$$

When x is categorical K_ν stands for a categorical kernel function like the Aitchison and Aitken kernel (5.2.2). When x is a continuous variable K stands for a continuous kernel like the Epanechnikov kernel. In Section 5.1.4 continuous kernels are written in the usual way as functions $K(.)$ and the smoothing parameter appears in the argument. In order to have a unique form for categorical and continuous kernels one may write, e.g., the Epanechnikov kernel (5.1.10) as

$$K_\nu(x|x_i) = \begin{cases} \frac{3}{4}[1 - \{(x_i - x)/\nu\}^2]/\nu, & |x_i - x|/\nu \le 1 \\ 0 & \text{otherwise.} \end{cases} \tag{5.2.9}$$

For simplicity, both forms, (5.1.9) and (5.2.9), are used for continuous kernels. For vector–valued explanatory variables K_ν stands for the product kernel (5.2.3). Then a multivariate parameter $\nu = (\nu_1, \ldots, \nu_p)$ determines the neighbourhood smoothing. The kernel components have to be chosen according to scale level allowing for a mixture of nominal, ordinal and metric kernels. Alternatively, a mixture of categorical and metric variables may be based on distances between observation and target point that take different scale levels of components into account (see Tutz, 1990b).

When using a continuous kernel like (5.2.9) in (5.2.8) the smoothing parameter is from $(0, \infty)$ with strong smoothing for $\nu \to \infty$. If a discrete kernel is used in (5.2.8) the smoothing parameter is from $[0, 1]$ with strong smoothing at zero (see Table 5.1). Therefore, for continuous kernels the smoothing parameter is sometimes considered in the transformed form $\nu^* = 1 - \exp(-\nu)$ yielding strong smoothing at zero in the same way as for discrete kernels.

Variants of these weight functions in dichotomous regression problems have been used by Copas (1983), Fowlkes (1987), Müller and Schmitt (1988) and Staniswalis and Cooper (1988). The polychotomous case has been considered by Lauder (1983) and Tutz (1990a, 1990b, 1991c).

Example 5.3: Vaso–constriction data (Example 4.2, 4.4 continued)
The objective in this data set is to investigate the effect of rate and volume of air inspired on a transient vaso–constriction in the skin of the digits. For the binary outcome that indicates occurrence or nonoccurrence of vaso–constriction the nominal Aitchison and Aitken kernel is used. As weight

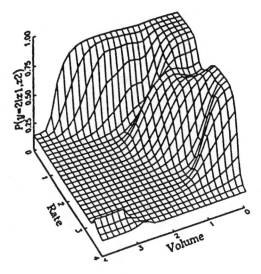

FIGURE 5.3. Kernel–based estimate of the nonoccurrence of vaso–constriction in the skin.

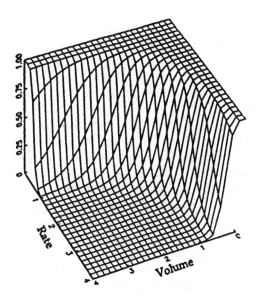

FIGURE 5.4. Response surface for the nonoccurrence of vaso–constriction based on a fitted logit model.

function the Nadaraya–Watson function based on Gaussian kernels is used for the untransformed original variables rate and volume. Cross–validation yields the smoothing parameters $\lambda = 0.720, \nu^* = 0.779$. Figure 5.3 shows the smoothed probability of nonoccurrence of vaso–constriction. The plot suggests that a logistic model may be appropriate. For comparison, in Figure 5.4 the surface is plotted for a logit model based on log–rate and log–volume. The estimated coefficients based on Pregibon's (1982) resistant fit are 7.62 for log–rate and 8.59 for log–volume. Observations 4 and 18, which are not well fitted by the model (see Example 4.2, 4.4, in particular Figure 4.3), show in the smoothed fit. Both observations have volume about 0.8 and rate about 1.5. In Figure 5.3 they appear as a gap near the point (0.8, 1.5) making the surface unsteady. Comparison with Figure 5.4 shows that parametric and nonparametric estimates are roughly comparable. But from the nonparametric fit it is seen where the data produce deviations from the parametric model.   □

Example 5.4: Unemployment data

The data set is a subsample from the socio–economic panel in Germany (Hanefeld, 1987; see also Example 9.1). The response y is multicategorical with $y = 1$ for short–term unemployment (up to 6 months), $y = 2$ for unemployment from 7 to 12 months and $y = 3$ for long–term unemployment (longer than 12 months). The only covariate considered is age (see Table 5.2). For the ordinal response the discrete geometrical kernel is used, for the weighting of log(age) the Gaussian kernel is used. Cross–validation yields $\lambda = 1$ and $\mu = 0.925$. Moreover, a sequential model (see Section 3.3.4) is fitted, namely, the sequential logit model $\log(Y = r | Y \geq r) = F(\theta_r + \gamma \log(age))$ with F denoting the logistic distribution function. The estimates are $\theta_1 = 4.533$, $\theta_2 = 3.819$, $\gamma = -1.286$. The deviance is 128.28 on 83 df. Figure 5.5 shows the nonparametric estimate based on the categorical kernel regression estimate and the response probabilities arising from the sequential model. It is seen that the probability for short–term unemployment ($y = 1$) is decreasing with age and the probability of long–term unemployment ($y = 3$) is increasing with age. The nonparametric fit shows deviations from the parametric model for extreme ages. In particular for high age the probability for short–term unemployment as derived from relative frequencies and kernel regression is lower than the probability estimated by the parametric model. The same holds for category 2 where the parametric model yields a probability estimate that is nearly constant over age whereas the nonparametric fit shows a slightly decreasing probability. These effects sum up to a strong deviation between the nonparametric and the parametric fit for high ages in category 3 (long–term unemployment). It is seen that the nonparametric fit is closer to the data for extreme ages. "Smoothing" by fitting the parametric sequential model is insensitive in these areas. Consequently the deviance signals a bad fit for the simple

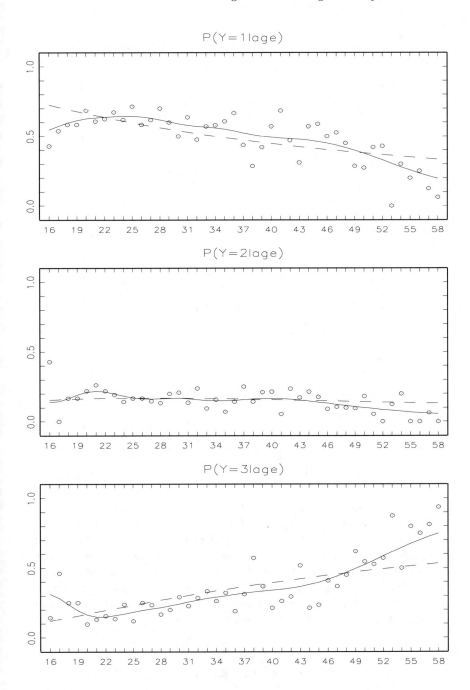

FIGURE 5.5. Categorical kernel regression (——) and sequential logit model (- - -) for unemployment data.

TABLE 5.2. Unemployment in categories $y = 1$ (1–5 months), $y = 2$ (7–12 months), $y = 3$ (> 12 months)

Age in years	1	2	3	Age in years	1	2	3
16	3	3	1	38	4	2	8
17	14	0	12	39	8	4	7
18	28	8	12	40	8	3	3
19	42	12	18	41	13	1	5
20	50	16	7	42	8	4	5
21	51	22	11	43	9	5	15
22	40	14	10	44	8	3	3
23	35	10	7	45	10	3	4
24	26	6	10	46	11	2	9
25	30	7	5	47	10	2	7
26	21	6	9	48	9	2	9
27	21	5	8	49	6	2	13
28	21	4	5	50	3	2	6
29	15	5	5	51	8	1	10
30	12	5	7	52	6	0	8
31	14	3	5	53	0	1	7
32	10	5	6	54	3	2	5
33	12	2	7	55	2	0	8
34	11	3	5	56	1	0	3
35	17	2	9	57	2	1	13
36	14	3	4	58	1	0	15
37	7	4	5				

sequential model where log(age) enters in linear form.

It should be remarked that smoothing over response categories is not necessary here ($\lambda = 1$). Since there are 43 differing ages the effect of smoothing over ages is dominating. Smoothing over the response becomes important if there are many response categories but few differing values of explanatory variables. The extreme case with a fixed value of the explanatory variable (equivalent to no explanatory variable) is considered in Example 5.2 where smoothing over responses is quite helpful.   □

5.2.3   Choice of smoothing parameters*

The smoothness of the regression estimate (5.2.7) is determined by two smoothing parameters, λ and ν. Data–driven procedures for the choice of these parameters may be based on the cross–validation principle. Because the response y is multicategorical, loss functions have to be different from the loss functions used in the case of a metric response (Section 5.1). Estimates are based on the total sample $S = \{(y_j^{(i)}, x_i) | j = 1, \ldots, n_i, \ i = 1, \ldots, g\}$. Now let p denote now the density of (y, x) and $\hat{p}_S$ denote the estimate $\hat{p}_S(y, x) = \hat{p}(y|x, \lambda, \nu)\tilde{p}_S(x)$ where $\tilde{p}_S(x) = n_x/n$ is the relative frequency. A general family of loss functions is given by

$$L(p, \hat{p}_S) = \int p(x)\tilde{L}(p_x, \hat{p}_x)dx \qquad (5.2.10)$$

where $p_x = p(.|x)$ denotes the conditional probability function given x and $\hat{p}_x = \hat{p}(.|x)$ denotes the estimated conditional probability function; $p(x)$ stands for the marginal density of x. The loss function (5.2.10) is a weighted function of the (conditional) loss $\tilde{L}(p_x, \hat{p}_x)$, which is a loss function for the density estimation problem of estimating $p(.|x)$. $\tilde{L}$ has to be specified as one of the loss functions considered in Section 5.2.1, e.g., as quadratic loss (5.2.4) or Kullback–Leibler loss (5.2.5). The family of loss functions (5.2.10) has been called *discriminant loss functions* because of its usefulness in discriminant analysis. When used in discriminant analysis conditional loss functions $\tilde{L}$, which are connected to the classification problem, are more appropriate (see Tutz 1989a, 1991b; Tutz and Groß, 1994). For dichotomous y and quadratic loss function $\tilde{L}$ (5.2.10) is equivalent to the integrated squared error (5.1.7), which is used for metric responses.

As an optimality criterion one may use $EL(p, \hat{p}_S)$ or the prediction–oriented form

$$L^*(p, \hat{p}_S) = E_{y,x}L(\delta_{y,x}, \hat{p}_S)$$

where $\delta_{y,x}$ is the Dirac delta function, i.e., the degenerate distribution with the total mass at (y, x). From $L(\delta_{y,x}, \hat{p}_S) = \tilde{L}(\delta_y, \hat{p}_x)$ one gets

$$L^*(p, \hat{p}_S) = \int p(x) \sum_y p(y|x)\tilde{L}(\delta_y, \hat{p}_x)dx$$

where the sum is over possible values of y. The cross–validation criterion that corresponds to $L^*(p, \hat{p}_S)$ is given by

$$CV(\lambda, \nu) = \frac{1}{n} \sum_{(y,x) \in S} L(\delta_{y,x}, \hat{p}_{S \setminus (y,x)})$$

where $\hat{p}_{S \setminus (y,x)}$ is the estimate with the sample reduced by observation (y, x).

For a specific loss function $\tilde{L}$ the cross–validation function has a simple form. For example, the quadratic loss function yields the quadratic score

$$L(\delta_{y,x}, \hat{p}) = (1 - \hat{p}(y|x))^2 + \sum_{\tilde{y} \neq y} \hat{p}(\tilde{y}|x)^2.$$

TABLE 5.3. Loss functions and empirical loss for observation (y, x)

Conditional loss function $\tilde{L}$	Empirical loss
Bayes risk $\tilde{L}(p_x, \hat{p}_x) = 1 - \sum_y p(y\|x) \cdot$ $\quad Ind_y(\hat{p}(1\|x), \ldots, \hat{p}(k\|x))$	$L(\delta_{y,x}\hat{p}) =$ $\quad 1 - Ind_y(\hat{p}(1\|x), \ldots, \hat{p}(k\|x))$
Quadratic loss $\tilde{L}(p_x, \hat{p}_x) = \sum (p(y\|x) - \hat{p}(y\|x))^2$	Quadratic score $L(\delta_{y,x}\hat{p}) = (1 - \hat{p}(y\|x))^2 +$ $\quad \sum_{\tilde{y} \neq y} \hat{p}(\tilde{y}\|x)^2$
Kullback–Leibler loss $\tilde{L}(p_x, \hat{p}_x) =$ $\sum_y p(y\|x) \log(p(y\|x)/\hat{p}(y\|x))$	Logarithmic score $L(\delta_{y,x}, \hat{p}) = -\log \hat{p}(y\|x)$

Then cross–validation is equivalent to the minimization of the leaving–one–out version of the quadratic score. In particular with prediction in mind an interesting conditional loss function is

$$\tilde{L}(p_x, \hat{p}_x) = 1 - \sum_y p(y|x) \, \mathrm{Ind}_y(\hat{p}(1|x), \ldots, \hat{p}(k|x)) \qquad (5.2.11)$$

where $\mathrm{Ind} = (\mathrm{Ind}_1, \ldots, \mathrm{Ind}_k) : \mathbb{R}^k \to \mathbb{R}$ is the indicator function

$$\mathrm{Ind}_r(y_1, \ldots, y_k) = \begin{cases} 1, & \text{if } y_r > y_i, \ i \neq r \\ 0, & \text{otherwise.} \end{cases}$$

Then one gets

$$L(\delta_{y,x}, \hat{p}) = 1 - \mathrm{Ind}_y(\hat{p}(1|x), \ldots, \hat{p}(k|x))$$

which takes value 0 if for observation (y, x) the estimate $\hat{p}(y|x)$ is the maximal estimated probability and the value 1 otherwise. Therefore $L(\delta_{y,x}, \hat{p})$ is equivalent to the resubstitution error rate where the number of misclassifications in the sample is counted. Consequently minimization of $CV(\lambda, \nu)$ is equivalent to the minimization of the leaving–one–out error rate, which is an often–used measure for the performance of allocation rules in discriminant analysis. Table 5.3 shows some loss functions and their corresponding empirical version. A simple derivation shows that the optimality criterion

$L^*(p, \hat{p}_S)$ is equivalent to the Bayes risk. Minimization of $CV(\lambda, \nu)$ yields consistent estimates $\hat{p}(y|x, S, \lambda, \nu) \to p(y|x)$ and $CV(\lambda, \nu) \to L^*(p, p)$ under regularity conditions for all discrete kernels that are continuous in λ and common conditional loss functions like quadratic or Kullback–Leibler loss (Tutz, 1991c).

5.3   Spline smoothing in generalized linear models

The basic idea for spline smoothing in generalized linear models is to replace linear functions $x\beta$ of covariates by smooth functions $f(x)$. Throughout this section we consider only models for univariate response, but extensions to the multivariate case are possible.

Section 5.3.1 deals with the case where only a single covariate is present. Then the linear predictor $\eta = \alpha + \beta x$ in a generalized linear model is replaced by a spline function $\eta(x) = f(x)$. For more covariates, the predictor may be assumed to be additive, i.e., $\eta(x_1, \ldots, x_p) = f_{(1)}(x_1) + \cdots + f_{(p)}(x_p)$, with smooth functions $f_{(1)}(x_1), \ldots, f_{(p)}(x_p)$. Section 5.3.2 deals with such generalized additive models. The presentation follows mainly Green and Yandell (1985) and Hastie and Tibshirani (1990), which are also recommended for further reading.

5.3.1   Cubic spline smoothing with a single covariate

We consider only the special but important case of cubic spline smoothing. More general forms can be dealt with along similar lines by penalized maximum likelihood estimation as in Green and Yandell (1985) and Green (1987). The basic modification in comparison with Section 5.1.3 is to replace the least–squares fit criterion in (5.1.1), which is equivalent to the log–likelihood for normal data and by log–likelihoods for non–normal data. In the context of generalized linear models the linear predictor $\eta_i = x_i\beta$ is replaced by the more general predictor $\eta_i = f(x_i)$, so that

$$E(y_i|x_i) = h(f(x_i))$$

with known response function h. This leads to the penalized log–likelihood criterion

$$\sum_{i=1}^{n} l_i(y_i; f(x_i)) - \frac{1}{2}\lambda \int (f''(u))^2 du \to \max, \qquad (5.3.1)$$

where l_i is the log–likelihood contribution of observation y_i and f has continuous first and second derivatives f', f'', with f'' quadratically integrable. Compared to the penalized least–squares criterion (5.1.1), the squared Euclidian distance $(y_i - f(x_i))^2$ is replaced by a Kullback–Leibler distance.

For normally distributed observations y_i, the penalized log–likelihood criterion (5.3.1) reduces to the penalized least–squares criterion (5.1.1). As in Section 5.1.2, the penalized log–likelihood criterion explicitly formulates the compromise between faith to the data measured by the first term and roughness of the function expressed by the penalty term, with the smoothing parameter λ controlling this compromise. The solution is again a cubic spline. Parameterizing by the evaluations $f_i = f(x_i)$ of the cubic splines at the observed points $x_1 < \ldots < x_n$, (5.3.1) reduces to

$$PL(f) = \sum_{i=1}^{n} l_i(y_i; f_i) - \frac{1}{2}\lambda f' K f \to \max_{f}, \qquad (5.3.2)$$

with $f = (f_1, \ldots, f_n)$ and K defined as in (5.1.2).

Except for normal likelihoods, this is no longer a quadratic optimization problem with an explicit solution $\hat{f}_\lambda$ as in (5.1.4). The solution can be obtained by Fisher–scoring iterations. In each step, the next iterate, $f^{(k+1)}$ say, is computed as a (weighted) spline smoother (5.1.6) applied to a "working observation" vector $\tilde{y}^{(k)}$, computed from the current iterate $f^{(k)}$. This is in complete analogy to computing maximum likelihood estimates in generalized linear models by iteratively weighted least–squares; compare with Section 2.2.1. Details of this Fisher–scoring algorithm are described in the following.

Fisher scoring for generalized spline smoothing*

The first and expected negative second derivatives of $PL(f)$ are

$$\frac{\partial PL(f)}{\partial f} = s - \lambda K f, \qquad E\left(-\frac{\partial PL(f)}{\partial f \partial f'}\right) = W + \lambda K, \qquad (5.3.3)$$

where s is the vector of first derivatives of the log–likelihood term and W is the diagonal matrix of expected negative second derivatives.

The score function s and the diagonal weight matrix W have the same structure as in common parametric GLM's (compare with Chapter 2):

$$s' = (s_1, \ldots, s_n), \qquad s_i = \frac{D_i}{\sigma_i^2}(y_i - h(f_i)), \qquad (5.3.4)$$

with $D_i = \partial h/\partial f_i$ as the first derivative of the response function and σ_i^2 the variance evaluated at $\mu_i = h(f_i)$,

$$W = \operatorname{diag}(w_i), \qquad w_i = D_i^2/\sigma_i^2. \qquad (5.3.5)$$

The Fisher scoring step to go from the current iterate $f^{(k)}$ to $f^{(k+1)}$, $k = 0, 1, 2, \ldots$, is thus given by

$$(W^{(k)} + \lambda K)(f^{(k+1)} - f^{(k)}) = s^{(k)} - \lambda K f^{(k)}, \qquad (5.3.6)$$

In the same way as for GLM's with linear predictors, the iteration (5.3.6) can be rewritten in iteratively weighted least–squares form:

$$f^{(k+1)} = (W^{(k)} + \lambda K)^{-1} W^{(k)} \tilde{y}^{(k)} = S^{(k)} \tilde{y}^{(k)}, \qquad (5.3.7)$$

with the "working" observation vector

$$\tilde{y}^{(k)} = (W^{(k)})^{-1} s^{(k)} + f^{(k)} = (\tilde{y}_1^{(k)}, ..., \tilde{y}_n^{(k)}),$$

$$\tilde{y}_i^{(k)} = \frac{y_i - h(f_i^{(k)})}{D_i^{(k)}} + f_i^{(k)},$$

and the smoother matrix

$$S^{(k)} = (W^{(k)} + \lambda K)^{-1} W^{(k)}.$$

Comparing with cubic spline smoothing in Section 5.1, it is seen that (5.3.7) is the weighted cubic spline smoother (5.1.7) applied to the working observation vector $\tilde{y}^{(k)}$. Thus in each Fisher–scoring step computationally efficient weighted spline smoothing algorithms can be applied to $\tilde{y}^{(k)}$ to obtain the next iterate $f^{(k+1)}$ in $O(n)$ operations. Iterations are stopped according to a termination criterion, e.g., if

$$\|f^{(k+1)} - f^{(k)}\| / \|f^{(k)}\| < \varepsilon$$

is fulfilled.

Choice of smoothing parameter

Cross–validation or generalized cross–validation may be adapted to the present situation for the data–driven choice of the smoothing parameter λ. O'Sullivan, Yandell and Raynor (1986) replace the usual squared residuals by squared Pearson residuals and define the generalized cross–validation criterion

$$GCV_P(\lambda) = \frac{1}{n} \sum_{i=1}^{n} \left\{ \frac{(y_i - \hat{\mu}_i)/\hat{\sigma}_i}{1 - tr(S_\lambda)/n} \right\}^2,$$

where $\hat{\mu}_i = h(\hat{f}_i)$ and the variances $\hat{\sigma}_i^2$ are evaluated at the last iterate of the scoring algorithm. Another possibility would be to replace Pearson residuals by deviance residuals, compare Hastie and Tibshirani (1990, ch. 6.9). Other approaches for estimating λ would be possible, e.g., an empirical Bayes approach in combination with an EM–type algorithm as in Chapter 8.

Example 5.5: Rainfall data
As an illustration, let us shortly consider smoothing the rainfall data already presented in Chapter 1, Example 1.7. The data, given by the number

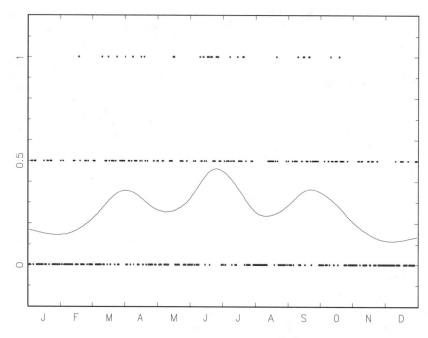

FIGURE 5.6. Smoothed probability of rainfall $\lambda = 4064$.

of occurrences of rainfall in Tokyo for each day during the years 1983–1984, are reproduced in Figures 5.6 and 5.7 as relative frequencies with values 0, 0.5 and 1. To obtain a smooth curve for the (marginal) probability π_t of rainfall at calendar day t, a logistic cubic spline model

$$\pi(t) = \frac{\exp(f(t))}{1 + \exp(f(t))},$$

with $x(t) = t$ as "covariate" was fitted. The smoothing parameter was estimated by generalized cross–validation, resulting in two local minima at $\lambda = 32$ and $\lambda = 4064$; compare with Figure 5.8. The smooth curve in Figure 5.6 ($\lambda = 4064$) shows a clear seasonal pattern, while in Figure 5.7 ($\lambda = 32$) the curve is rougher but closer to the data. The pattern in Figure 5.6, with peaks for wet seasons, nicely reflects the climate in Tokyo. It would be fairly difficult to see this by only looking at the raw data. A further analysis of these data is in Chapter 8 with closely related categorical state–space methods. □

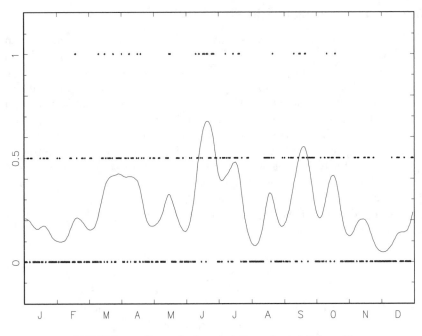

FIGURE 5.7. Smoothed probability of rainfall $\lambda = 32$.

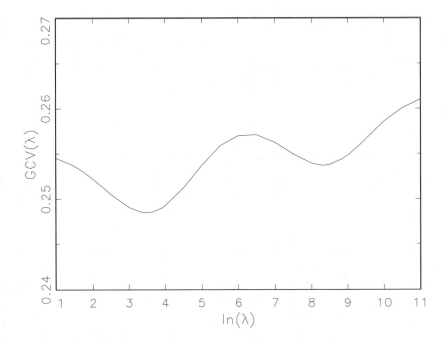

FIGURE 5.8. Generalized cross–validation criterion, with logarithmic scale for λ.

5.3.2   Generalized additive models

To extend nonparametric regression techniques to the case of more than one covariate, one might assume that the predictor is a smooth function $\eta = \eta(x_1, x_2, ...)$ of the covariates and try to estimate it by some surface smoother. Extensions of cubic spline smoothing to two or more dimensions have been considered by O'Sullivan, Yandell and Raynor (1986) and Gu (1990). Such approaches are computationally rather demanding and, more seriously, their performance may suffer from sparse data in dimensions higher than two or three: the curse of dimensionality. As a compromise generalized additive models were introduced in a series of papers by Hastie and Tibshirani (1985, 1986, 1987) and are described in detail in Hastie and Tibshirani (1990). They retain an important feature of GLM's: covariate effects are additive. However, these effects are generally nonlinear, so that the predictor is assumed to have the form

$$\eta = \alpha + \sum_{j=1}^{p} f_{(j)}(x_j), \qquad (5.3.8)$$

where the covariate vector $x = (x_1, ..., x_p)$ can be a function of original covariates, including, for example, interactions.

A special case occurs if one covariate function, say $f_{(1)}(x_1)$, is modelled nonparametrically, while the remaining covariates still enter as a linear combination, say $z'\beta = x_2\beta_2 + \cdots + x_p\beta_p$. Such semi–parametric or partial spline models are considered in Green and Yandell (1985).

Compared to the definition of univariate GLM's in Chapter 2, the distributional assumptions remain the same, while the mean of y_i given $x_i = (x_{i1}, ..., x_{ip})$ is now assumed to be

$$\mu_i = E(y_i|x_i) = h(\eta_i), \quad \eta_i = \alpha + \sum_{j=1}^{p} f_{(j)}(x_{ij}). \qquad (5.3.9)$$

In order to estimate the functions $f_{(1)}, ..., f_{(p)}$ by cubic splines, the criterion (5.3.1) is generalized to

$$\sum_{i=1}^{n} l_i(y_i; \eta_i) - \frac{1}{2} \sum_{j=1}^{p} \lambda_j \int (f''_{(j)}(u))^2 du \rightarrow \max, \qquad (5.3.10)$$

with η_i given by (5.3.9), and with separate penalty terms and smoothing parameters for each function. The maximizing functions are again cubic splines. Parameterizing by the vectors of evaluations

$$f_j = (f_{(j)}(x_{1j}), ..., f_{(j)}(x_{nj}))', \quad j = 1, ..., p,$$

(5.3.10) can be rewritten as the penalized log–likelihood criterion

$$PL(f_1, ..., f_p) = \sum_{i=1}^{n} l_i(y_i; \eta_i) - \frac{1}{2}\sum_{j=1}^{p} \lambda_j f_j' K_j f_j, \qquad (5.3.11)$$

where penalty matrices K_j for each predictor f_j are defined analogously to K in (5.1.3).

Maximization of (5.3.11) can again be carried out by Fisher scoring iterations. Compared to the algorithm in Section 5.3.1, a further inner "backfitting" iteration loop has to be included.

Fisher scoring with backfitting*

Differentiation of $\partial PL/\partial f_j, j = 1, ..., p$, yields the likelihood equations

$$s_1 = \lambda_1 K_1 f_1, ..., s_p = \lambda_p K_p f_p, \qquad (5.3.12)$$

where the derivative $s = (s_1, ..., s_n)$ of the log–likelihood is given by

$$s_i = \frac{D_i}{\sigma_i^2}(y_i - \mu_i),$$

with $D_i = \partial h/\partial \eta$ as the first derivative of the response function and σ_i^2 the variance function evaluated at $\mu_i = h(\eta_i), \eta_i$ given by (5.3.9). The constant term α is omitted here and in the following derivation of the Fisher–scoring iterations. It is used in the backfitting algorithm only in order to guarantee uniqueness of the smoothers by centering the working observations, compare with Hastie and Tibshirani (1990, sections 5.2, 5.3).

With the expected information matrix

$$W = \text{diag}(w_1, ..., w_n), w_i = D_i^2/\sigma_i^2$$

Fisher–scoring iterations are given by

$$\begin{bmatrix} W^{(k)} + \lambda_1 K_1 & W^{(k)} & \cdots & W^{(k)} \\ W^{(k)} & W^{(k)} + \lambda_2 K_2 & \cdots & W^{(k)} \\ \vdots & \vdots & \ddots & \vdots \\ W^{(k)} & W^{(k)} & \cdots & W^{(k)} + \lambda_p K_p \end{bmatrix} \begin{bmatrix} f_1^{(k+1)} - f_1^{(k)} \\ f_2^{(k+1)} - f_2^{(k)} \\ \vdots \\ f_p^{(k+1)} - f_p^{(k)} \end{bmatrix}$$

$$= \begin{bmatrix} s^{(k)} - \lambda_1 K_1 f_1^{(k)} \\ s^{(k)} - \lambda_2 K_2 f_2^{(k)} \\ \vdots \\ s^{(k)} - \lambda_p K_p f_p^{(k)} \end{bmatrix},$$

where $W^{(k)}, s^{(k)}$ are W, s evaluated at $\eta^{(k)} = \eta(f_1^{(k)}, ..., f_p^{(k)})$. Defining the working observation vector

$$\tilde{y}^{(k)} = \eta^{(k)} + (W^{(k)})^{-1}s^{(k)}$$

and the smoother matrices

$$S_j^{(k)} = (W^{(k)} + \lambda_j K_j)^{-1} W^{(k)}$$

the iterations can be written as

$$
\begin{bmatrix}
I & S_1^{(k)} & \cdots & S_1^{(k)} \\
S_2^{(k)} & I & \cdots & S_2^{(k)} \\
\vdots & \vdots & \ddots & \vdots \\
S_p^{(k)} & S_p^{(k)} & \cdots & I
\end{bmatrix}
\begin{bmatrix}
f_1^{(k+1)} \\
f_2^{(k+1)} \\
\vdots \\
f_p^{(k+1)}
\end{bmatrix}
=
\begin{bmatrix}
S_1^{(k)} \tilde{y}^{(k)} \\
S_2^{(k)} \tilde{y}^{(k)} \\
\vdots \\
S_p^{(k)} \tilde{y}^{(k)}
\end{bmatrix}. \tag{5.3.13}
$$

A direct solution of the np–dimensional system (5.3.13) to obtain the next iterate $f^{(k+1)} = (f_1^{(k+1)}, ..., f_p^{(k+1)})$ will be computationally feasible only in special cases. Instead (5.3.13) is rewritten as

$$
\begin{bmatrix}
f_1^{(k+1)} \\
f_2^{(k+1)} \\
\vdots \\
f_p^{(k+1)}
\end{bmatrix}
=
\begin{bmatrix}
S_1(\tilde{y}^{(k)} - \sum_{j \neq 1} f_j^{(k+1)}) \\
S_2(\tilde{y}^{(k)} - \sum_{j \neq 2} f_j^{(k+1)}) \\
\vdots \\
S_p(\tilde{y}^{(k)} - \sum_{j \neq p} f_j^{(k+1)})
\end{bmatrix} \tag{5.3.14}
$$

and solved iteratively by the "backfitting" or Gauss–Seidel algorithm in the inner loop of the following algorithm.

Initialization: $\alpha^{(0)} = g(\sum_{i=1}^n y_i / n), f_1^{(0)}, = ... = f_p^{(0)} = 0$, where $g = h^{-1}$ is the link function.

Scoring steps for $k = 0, 1, 2, ...$: Compute the current working observations

$$\tilde{y}_i^{(k)} = \eta_i^{(k)} + \frac{y_i - h(\eta^{(k)})}{D_i^{(k)}}$$

and the weights $w_i^{(k)} = (D_i^{(k)} / \sigma_i^{(k)})^2$ with $\eta_i^{(k)} = \alpha^{(k)} + \sum_{j=1}^p f_j^{(k)}(x_{ij}), i = 1, ..., n$. Solve (5.3.13) for $f^{(k+1)}$ by the inner *backfitting loop*:

(i) Initialize $\alpha^{(k)} = \frac{1}{n} \sum_{i=1}^n \tilde{y}_i^{(k)}, f_j^0 := f_j^{(k)}, j = 1, ..., p$,

(ii) Compute updates $f_j^0 \to f_j^1, j = 1, ..., p$, in each backfitting iteration:

$$f_j^1 = S_j(\tilde{y} - \alpha - \sum_{h<j} f_h^1 - \sum_{l>j} f_l^0)$$

by application of weighted cubic spline smoothing to the "working residual" $\tilde{y} - \alpha - \sum f_h^1 - \sum f_l^0$. Set $f_j^0 := f_j^1 \; j = 1, ..., p$, after each such iteration. Stop backfitting iterations until $\| f_j^0 - f_j^1 \| \; j = 1, ..., p$ is nearly zero. Set $f_j^{(k+1)} := f_j^1$ for the final iterate.

Stop if some termination criterion, for example,

$$\frac{\sum_{j=1}^{p} \|f_j^{(k+1)} - f_j^{(k)}\|}{\sum_{j=1}^{p} \|f_j^{(k)}\|} \leq \varepsilon,$$

is reached.

Remark: The constant α has been introduced for identifiability reasons. This implies that $\tilde{y}$ has been centered to have mean 0, and it ensures that at every stage of the procedure the $f_j's$ have mean 0.

A data–driven choice of the smoothing parameters $\lambda = (\lambda_1, ..., \lambda_p)$ by generalized cross–validation is possible in principle. For additive models the criterion is

$$GCV_P(\lambda) = \frac{1}{n} \sum_{i=1}^{n} \left\{ \frac{(y_i - \hat{\mu}_i)/\hat{\sigma}_i}{1 - tr(R_\lambda)/n} \right\}^2,$$

where R_λ is the smoother matrix that generates the additive predictor $\hat{\eta} = R_\lambda \tilde{y}$ in the last iteration step. However, optimization of the criterion would require efficient computation of $tr(R_\lambda)$ in each step of a multidimensional search algorithm. Moreover, it may be very difficult to find a global minimum in the multidimensional case. It seems that additional research is necessary for automatic smoothing parameter selection; compare with Hastie and Tibshirani (1990, section 6.9).

Let us finally remark that smoothing in generalized additive models is not restricted to cubic spline smoothing as described earlier. The approach works for other symmetric linear smoothers based on smoother matrices $S_1, ..., S_p$ as well: One only has to define the penalty matrices $K_1, ..., K_p$ in the penalized log–likelihood criterion by

$$K_j = S_j^- - I,$$

where S_j^- is any generalized inverse of S_j. Weighted cubic spline smoothing in the inner loop (ii) of the backfitting algorithm is then substituted by a corresponding symmetric linear smoother, for example, a running line smoother.

Example 5.6: Vaso–constriction data (Example 5.3 continued)
Figures 5.9 and 5.10 show the nonoccurrence of vaso–constriction estimated with an additive logit model. The additive structure resulting from the sum of the component functions is clearly seen.

The effects of the two explanatory variables may be considered separately. The effect of rate on the occurrence of constriction is almost monotone: if rate decreases the probability of nonoccurence increases. The effect of volume is different: there is a decrease in probability if volume decreases between 1.6 and 0.8. For stronger smoothing (Figure 5.10) the decrease is

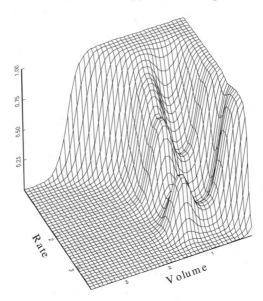

FIGURE 5.9. Nonoccurrence of vaso–constriction of the skin smoothed by an additive model with $\lambda_1 = \lambda_2 = 0.001$.

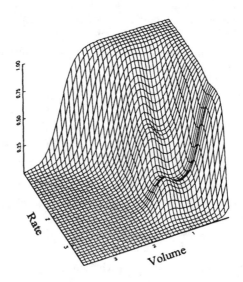

FIGURE 5.10. Nonoccurrence of vaso–constriction of the skin smoothed by an additive model with $\lambda_1 = \lambda_2 = 0.003$.

almost a plateau. A comparison with the kernel–based smoother in Figure 5.3 shows the difference between approaches: data that indicate non–monotone behaviour influence the behaviour of the smoothing components in the additive model whereas they yield more local deviation concerning both components in the case of kernel smoothing. □

5.4   Further developments

Although locally weighted regression has already been considered by Stone (1977) and Cleveland (1979), in recent years extensions of the method have found much attention. The basic idea is to fit for each point of interest x_0 low–order polynomials in x locally at x_0. Observations in the neighborhood are weighted down e.g. by kernel weights possibly in combination with a nearest neighborhood rule. The estimate of $f(x_0)$ is taken from the fitted polynomial at x_0. The Nadaraya–Watson estimator considered in Section 5.1.3 may be considered as a special case of constant fitting, i.e. using polynomials of degree zero. Hastie and Loader (1993) give a nice introduction to this method of local polynomial regression and demonstrate the advantages over kernel smoothing. Local regression techniques reduce automatically bias, particularly in boundary regions, they are less sensitive to the design and one of the biggest advantages is when the predictor is two or three dimensional. Then kernel estimates suffer from boundary effects over much of the domain. Problems with local regression are variances for random designs and possible numerical instability (see e.g. Seifert and Gasser, 1996). Asymptotic properties of local polynomials demonstrating minimax efficiency have been given by Fan (1992, 1993), Fan and Gijbels (1996). In the case of binary or Poisson response local polynomial regression becomes local polynomial likelihood fitting. Instead of computing a local weighted least squares solution one maximizes the corresponding local likelihood, again fitting polynomials at the target point. Fan, Heckman and Wand (1995) give some asymptotic properties for this case, see also Fan and Gijbels (1996).

There is a growing body of literature which uses nonparametric fitting techniques as diagnostic tool for linear or generalized linear models, see e.g. Azzalini, Bowman and Härdle (1989), Staniswalis and Severini (1991), Firth, Glosup and Hinkley (1991), Le Cessie and van Houwelingen (1991), Eubank and Hart (1993), Azzalini and Bowman (1993).

An interesting class of flexible models, called varying–coefficient model, has been introduced by Hastie and Tibshirani (1993). They consider models that are linear in the regressors but their coefficients are allowed to change smoothly with the values of other variables called effect modifiers. Thus the rather general models are partially nonparametric with generalized linear models, generalized additive models and partial linear models as special cases. Hastie and Tibshirani (1993) give estimation procedures based on the penalized likelihood functions whereas Kauermann and Tutz (1995), Tutz and Kauermann (1997) consider local likelihood estimates which allow the derivation of asymptotic properties.

Since the first printing of this book several very readable books have been published: Green and Silverman (1994) give a very thorough account of the roughness penalty approach to smoothing, Simonoff (1996) introduces to smoothing techniques highlighting the smoothing of sparse categorical data and Fan and Gijbels (1996) demonstrate the usefulness of local polynomial modelling.

6

Fixed parameter models for time series and longitudinal data

The methods of the foregoing sections are mainly appropriate for modelling and analyzing a broad class of non–normal cross–sectional data. Extensions to time–dependent data are possible in a variety of ways. Time series are repeated observations (y_t, x_t) on a response variable y of primary interest and on a vector of covariates taken at times $t = 1, ..., T$. Discrete time longitudinal or panel data are repeated observations (y_{it}, x_{it}) taken for units $i = 1, ..., n$ at times $t = 1, ..., T_i$. The restriction to integral times is made to simplify notation but is not necessary for most of the approaches. Longitudinal data may be viewed as a cross section of individual time series, reducing to a single time series for $n = 1$, or as a sequence of cross–sectional observations where units are identifiable over time. If a comparably small number of longer time series is observed, models and methods will be similar to those for single time series. If, however, many short time series have been observed, models, and often the scientific objective, can be different.

In this chapter, we will consider extensions of generalized linear models for time series (Section 6.1) and longitudinal data (Section 6.2) where parameters or covariate effects are fixed. In each section, we will consider conditional ("observation-driven") and marginal models. This corresponds to the distinction between conditional and marginal approaches in Section 3.5. For longitudinal data, Chapters 8 and 10 in Diggle, Liang and Zeger (1994) provide a detailed introduction to both approaches. Models with parameters varying across units according to a mixing distribution or across time according to a stochastic process are treated in Chapters 7 and 8.

6.1   Time series

While time series analysis for approximately Gaussian data has a long tradition, models for non–Gaussian data have received attention only more recently. Typical examples are binary or multicategorical time series, e.g., daily rainfall with categories no/yes or no/low/high (Example 1.7), and time series of counts, e.g., monthly number of cases of poliomyelitis (Example 1.8), daily number of purchases of some good, etc. Categorical and discrete–valued time series were often analyzed as time homogeneous Markov chains, i.e., Markov chains with stationary transition probabilities. However, without further constraints the number of parameters increases exponentially with the order of the Markov chain, and in many applications nonhomogeneous Markov chains are more appropriate since exogenous variables possibly give rise to nonstationary transition probabilities. For binary time series, Cox (1970) proposed an autoregressive logistic model, where covariates and a finite number of past outcomes are part of the linear predictor. Autoregressive generalized linear models of this kind and extensions are considered in Section 6.1.1 on conditional models for y_t given $y_{t-1}, ..., y_1$ and x_t. In certain applications, the marginal effect of covariates x_t on the response y_t is of primary interest, whereas the dependence of observations is regarded as a nuisance. Then it is more reasonable to base inference on the marginal distributions of y_t given x_t only, since in conditional models the influence of past observations may condition away covariate effects (Section 6.1.3).

6.1.1   Conditional models

A broad class of non–normal time series models can be obtained by the following simple dynamization of generalized linear models or quasi–likelihood models for independent observations: Similarly as in autoregressive models for normal responses, conditional distributions or moments of y_t given the past can be defined by including past values $y_{t-1}, ..., y_{t-l}, ...$ together with covariates.

Generalized autoregressive models

Consider first the case of univariate responses y_t and let

$$H_t = \{y_{t-1}, y_{t-2}, ..., y_1, x_t, x_{t-1}, ..., x_1\}$$

be the "history" of past observations, and present and past covariates. Generalized autoregressive models are characterized by the following structure:

(i) The conditional densities $f(y_t|H_t), t = 1, 2, ...$ are of the exponential family type.

(ii) The conditional expectation $\mu_t = E(y_t|H_t)$ is of the form

$$\mu_t = h(z_t'\beta), \tag{6.1.1}$$

where h is a response function as in Chapter 2 and the p–dimensional design vector z_t is a function of H_t, i.e., $z_t = z_t(H_t)$.

Assumptions (i) and (ii) imply that the conditional variance $\sigma_t^2 = \operatorname{var}(y_t|H_t)$ is given by

$$\sigma_t^2 = v(\mu_t)\phi,$$

where $v(\cdot)$ is the variance function corresponding to the specific exponential family and ϕ is the scale parameter. Thus, the definition is formally identical to that of GLM's for independent observations in Chapter 2; only *conditional* rather than *marginal* distributions and moments are modelled.

Often only a finite number of past observations, $y_{t-1}, ..., y_{t-l}$ say, will be included in the design vector. We will call this a *generalized autoregressive model* or *Markov model of order l*.

Let us consider some examples.

(i) *Binary and binomial time series.*

For a binary time series $\{y_t\}, y_t \in \{0,1\}$, the conditional distribution of y_t given H_t is determined by

$$\pi_t = P(y_t = 1|H_t).$$

If no covariates are observed, then

$$\pi_t = h(\beta_0 + \beta_1 y_{t-1} + \ldots + \beta_l y_{t-l}) = h(z_t'\beta), \quad t > l,$$

with $z_t' = (1, y_{t-1}, ..., y_{t-l}), \beta = (\beta_0, ..., \beta_l)'$, defines a purely autoregressive model of order l, with variance $\sigma_t^2 = \pi_t(1 - \pi_t)$ and $\phi = 1$. Using the logistic response function, this is the Markov chain of order l suggested by Cox (1970). Additional interaction terms such as $y_{t-1}y_{t-2}$, etc., may be included. Incorporation of all possible interactions up to order l yields a saturated model, containing the same number of parameters as the general homogeneous Markov chain.

Inclusion of covariates, e.g., by

$$\begin{aligned} \pi_t &= h(\beta_0 + \beta_1 y_{t-1} + \ldots + \beta_l y_{t-l} + x_t'\gamma), \quad t > l \\ z_t' &= (1, y_{t-1}, ..., y_{t-l}, x_t'), \beta' = (\beta_0, ..., \beta_l, \gamma'), \end{aligned}$$

allows the modelling of *nonstationary Markov chains*. Interaction terms of past observations and covariates, e.g., as in

$$\pi_t = h(\beta_0 + \beta_1 y_{t-1} + \beta_2 x_t + \beta_3 y_{t-1} x_t), \tag{6.1.2}$$

can be useful or even necessary. Model (6.1.2) is equivalent to the following parameterization of nonhomogeneous transition probabilities:

$$
\begin{aligned}
\pi_{i1} &= P(y_t = 1 | y_{t-1} = i, x_t) = h(\alpha_{0i} + \alpha_{1i}x_t), \quad i = 0, 1, \dots \\
\pi_{i0} &= P(y_t = 0 | y_{t-1} = i, x_t) = 1 - \pi_{i1}.
\end{aligned}
$$

Setting $y_{t-1} = 0$ or $= 1$ in (6.1.2), it is easily seen that

$$
\alpha_{00} = \beta_0, \alpha_{01} = \beta_0 + \beta_1, \alpha_{10} = \beta_2, \alpha_{11} = \beta_2 + \beta_3.
$$

Thus, *nonhomogeneous models for transition probabilities*, which have been suggested by some authors, see, e.g., Garber (1989), are contained in the general model $\pi_t = h(z'_t\beta)$ by appropriate inclusion of interaction terms.

Finally, the Markovian property may be dropped completely, as for $z_t = (1, \bar{y}_{t-1})'$, with the arithmetic or some weighted mean $\bar{y}_{t-1}$ of past observations.

Extension to the case where $\{y_t\}$ is a time series of relative frequencies assumed to be (scaled) binomial with repetition number n_t and conditional expectation $\mu_t = E(y_t | H_t)$ is straightforward: Now, $\mu_t = h(z'_t\beta)$ with analogous models as earlier, but $\sigma_t^2 = \mu_t(1 - \mu_t)/n_t$.

(ii) *Count data.*

For counts, log–linear Poisson models for $y_t | H_t$ are reasonable specifications. In complete analogy to the preceding binary models, one may assume

$$
\lambda_t = E(y_t | H_t) = \exp(\beta_0 + \beta_1 y_{t-1} + \dots + \beta_l y_{t-l} + x'_t\gamma).
$$

For $\gamma = 0$, such purely autoregressive models have been considered by Wong (1986). Although this model seems sensible, certain limitations are implied. Consider the case $l = 1$ for simplicity. Then λ_t grows exponentially for $\beta_1 > 0$, while $\beta_1 < 0$ corresponds to a stationary process for $\gamma = 0$. Therefore, other specifications may sometimes be more appropriate. For example, $\lambda_t = \exp(\beta_0 + \beta_1 \log y_{t-1})$ is equivalent to

$$
\lambda_t = \lambda(y_{t-1})^{\beta_1}, \quad \lambda = \exp(\beta_0).
$$

For $\beta_1 = 0$ the rate is constant, $\lambda_t = \lambda$. For $\beta_1 > 0, \lambda_t$ is increased by the previous outcome, while for $\beta_1 < 0$ it is decreased. For $y_{t-1} = 0, \lambda_t = 0$, i.e., $y_{t-1} = 0$ is absorbing. Models of this type, but more general, have been considered by Zeger and Qaqish (1988); see also below.

(iii) *Categorical time series.*
Generalized autoregressive models are easily extended to multivariate responses. For multicategorical time series $\{y_t\}$, where y_t is observed in k categories, let the response be coded by $y_t = (y_{t1}, ..., y_{tq})', q = k - 1$,

$$y_{tj} = \begin{cases} 1, & \text{category } j \text{ has been observed,} \\ 0, & \text{otherwise }, \quad j = 1, ..., q. \end{cases}$$

Correspondingly $\pi_t = (\pi_{t1}, ..., \pi_{tq})$ is the vector of conditional probabilities

$$\pi_{tj} = P(y_{tj} = 1|H_t), \quad j = 1, ..., q.$$

The general *autoregressive model for categorical time series* (e.g., Fahrmeir and Kaufmann, 1987; Kaufmann, 1987) now becomes

$$\pi_t = h(Z_t\beta), \quad t > l$$

where h is a q–dimensional response function as in Chapter 3, but the design matrix $Z_t = Z_t(H_t)$ is now a function of past observations and past and present covariates. In this way all the response models for unordered and ordered categories are easily adapted to the time series situation, admitting a flexible and parsimonious treatment of higher–order dependence and, to some extent, of nonstationarity. With the various possibilities for h and Z_t, a wide variety of models can be covered. Choosing, e.g., a multinomial logit model with $Z_t = \text{diag}\,(z_t')$ and

$$z_t' = (1, y_{t-1,1}, \ldots, y_{t-1,q}, \ldots, y_{t-l,1}, \ldots, y_{t-l,q}, x_t'),$$

a nonhomogeneous Markov chain of order l, with k unordered states, is obtained. Interactions may be included as in the binary case.

Quasi–likelihood models and extensions

Generalized autoregressive models are genuine likelihood models in the sense that the (conditional) distribution of $y_1, ..., y_t$, given the covariates $x_1, ..., x_t$, is completely determined by specification of the exponential family type and the mean structure (6.1.1), thereby implying a certain variance structure. As in the case of independent observations one may wish to separate the mean and variance structure. Also, it may not always be sensible to treat covariate effects and effects of past observations symmetrically as in the simple mean structure model (6.1.1). Zeger and Qaqish (1988) consider the following class of quasi–likelihood Markov models: The conditional mean structure is assumed to be

$$\mu_t = h(x_t'\gamma + \sum_{i=1}^{l} \beta_i f_i(H_t)), \tag{6.1.3}$$

where h is a response function, and the functions f_i are functions of past responses y_{t-i} and, possibly, past linear combinations $x'_{t-i}\gamma$. The conditional variance is assumed to be

$$\text{var}\,(y_t|H_t) = v(\mu_t)\phi, \tag{6.1.4}$$

where v is a variance function and ϕ is an unknown dispersion parameter.

Some examples, not contained in the more restricted framework of generalized autoregressive models, are:

(iv) *Autoregressive conditionally heteroscedastic (ARCH) models for Gaussian responses.*
Linear autoregressive models with covariates are defined by

$$\mu_t = \beta y_{t-1} + x'_t\gamma, \quad v(\mu_t) = 1,$$

setting $l = 1$ for simplicity. Alternatively, for

$$\mu_t = \beta(y_{t-1} - x'_{t-1}\gamma), \quad v(\mu_t) = \mu_t^2/\beta = \beta(y_{t-1} - x'_{t-1}\gamma)^2$$

we get an ARCH–model (e.g., Engle, 1982). Such models can account for time series with periods of increased variability.

(v) *Counts*
As an alternative to Example (ii), consider a log–linear Poisson model with

$$\lambda_t = \exp\{x'_t\gamma + \beta[\log(y^*_{t-1}) - x'_{t-1}\gamma]\}$$

and var $(y_t|y_{t-1}, x_t) = \lambda_t\phi$, where $y^*_{t-1} = \max(y_{t-1}, c), 0 < c < 1$ (Zeger and Qaqish, 1988). This is equivalent to

$$\lambda_t = \exp(x'_t\gamma)\left[\frac{y^*_{t-1}}{\exp(x'_{t-1}\gamma)}\right]^\beta. \tag{6.1.5}$$

Compared to the common log–linear model, the rate at time t is modified by the ratio of the past response and of $\exp(x'_{t-1}\gamma)$. Positive (negative) values of β correspond to positive (negative) autocorrelation. The parameter c determines the probability that $y_t > 0$ in the case of $y_{t-1} = 0$, so that $y_{t-1} = 0$ is no longer an absorbing state.

(vi) *Conditional gamma models.*
In analogy to (iv) and (v), Zeger and Qaqish (1988) propose a model with canonical link

$$\mu_t^{-1} = x'_t\gamma + \sum_{i=1}^{l} \beta_i\left(\frac{1}{y_{t-i}} - x'_{t-i}\gamma\right)$$

and $\sigma_t^2 = \mu_t^2 \phi$. Thus the linear predictor $x_t' \gamma$ at time t is modified by a weighted sum of past "residuals" $1/y_{t-i} - x_{t-i}' \gamma$. They illustrate a simplified version of this model, where $x_t' \gamma = \gamma$, with an analysis of interspike times collected from neurons in the motor cortex of a monkey.

The additional flexibility of models of this type does not come free. Generally they cannot be reformulated in terms of the mean structure (6.1.1), which allows application of the usual fitting procedures of generalized linear models; see Section 6.1.2. However, keeping either β or γ fixed, examples (v) and (vi) fit into the usual framework. Therefore a second level of iteration will be necessary for simultaneously estimating β and γ; see Section 6.1.2.

In the models considered so far, it was assumed that the conditional densities $f(y_t | H_t)$, or at least conditional first and second moments, are correctly specified. In the case of quasi–likelihood models for independent observations (Section 2.3.1), we have seen that consistent parameter estimation, however, with some loss of efficiency, is still possible if the mean is correctly specified whereas the true variance function is replaced by some "working" variance. One may ask if similar results hold for the time series situation. This is indeed the case. For generalized autoregressive models, conditioning in $f(y_t | H_t)$, $E(y_t | H_t)$, etc., is on the complete "history" H_t. If one works with some Markov model of order l, it has to be *assumed* that

$$f(y_t | H_t) = f(y_t | y_{t-1}, ..., y_{t-l}, x_t).$$

If this Markov property holds, then estimation can be based on genuine likelihoods. If this is not the case, or if the correct $f(y_t | H_t)$ cannot be determined, we may *deliberately* condition only on the l most recent observations $y_{t-1}, ..., y_{t-l}$. If $f(y_t | y_{t-1}, ..., y_{t-l}, x_t)$ is correctly specified, e.g., by one of the autoregressive models of order l discussed earlier, then consistent quasi–likelihood estimation is still possible. Compared to the case of independent observations, however, the correction for the asymptotic covariance matrix of the estimators becomes more difficult, and asymptotic theory requires more stringent assumptions; see Section 6.1.2.

One may even go a step further and assume only a correctly specified conditional mean $E(y_t | y_{t-l}, ..., y_{t-l}, x_t)$ together with some appropriate variance and autocorrelation structure. The extreme case, where conditioning is only on x_t and not on past responses, leads to marginal models in Section 6.1.3.

Some final remarks concern extensions to   nonlinear and nonexponential family time series models in analogy to corresponding models for independent observations mentioned in Section 2.3.3. Nonlinear and robust models for metrical outcomes have been considered and discussed by various authors. Several approaches to discrete outcomes, not covered by the conditional models here, may be cast into the more general nonlinear–nonexponential framework. In their development of D(iscrete) ARMA processes, Jacobs and Lewis (1983) have been guided by the autocorrelation structure of ARMA processes for continuous variables. Guan and Yuan (1991) and Ronning and Jung (1992) discuss estimation of integer–valued AR models with Poisson marginals. In other approaches, binary time series $\{y_t\}$ are assumed to be generated by truncation of a latent series $\{\tilde{y}_t\}$ of continuous outcomes (e.g., Gourieroux et al., 1983a; Grether and Maddala, 1982). Pruscha (1993) proposes a categorical time series model that combines the autoregressive model (iii) with a recursion for the probability vector π_t. In the most general model (Heckman, 1981, for longitudinal data), the latent variable $\tilde{y}_t$ is a linear combination of covariates, past values of $\{y_t\}$ and $\{\tilde{y}_t\}$, and an error term. However, fitting such models requires considerable computational effort and no asymptotic theory is available for this general case.

6.1.2   Statistical inference for conditional models

Estimation and testing for generalized autoregressive models can be based on genuine likelihood models. In the case of deterministic covariates, the joint density of the observations $y_1, ..., y_T$ factorizes into a product of conditional densities,

$$f(y_1, ..., y_T|\beta) = \prod_{t=1}^{T} f(y_t|y_{t-1}, ..., y_1; \beta),$$

and the conditional densities are determined by assuming some specific generalized autoregressive model. If covariates are stochastic, the joint density factorizes into

$$f(y_1, ..., y_T, x_1, ..., x_T|\beta) = \prod_{t=1}^{T} f(y_t|H_t; \beta) \prod_{t=1}^{T} f(x_t|C_t)$$

where $H_t = (y_1, ..., y_{t-1}, x_1, ..., x_t)$ and $C_t = (y_1, ..., y_{t-1}, x_1, ..., x_{t-1})$. Assuming that the second product does not depend on β, estimation can be based on the (partial) likelihood defined by the first product. In any case, the log–likelihood of $y_1, ..., y_T$ is given by

$$l(\beta) = \sum_{t=1}^{T} l_t(\beta), \quad l_t(\beta) = \log f(y_t|H_t; \beta),$$

where the conditional densities are given by the definition of generalized autoregressive models. For Markov models of order l, this is, strictly speaking, a conditional log–likelihood, given the starting values $y_0, ..., y_{-l+1}$. The first derivative of $l(\beta)$, the *score function*, is

$$s(\beta) = \sum_{t=1}^{T} Z_t' D_t(\beta) \Sigma_t^{-1}(\beta)(y_t - \mu_t(\beta)), \qquad (6.1.6)$$

where $\mu_t(\beta) = h(Z_t(\beta))$ is the *conditional* expectation, $\Sigma_t(\beta) = \mathrm{cov}\,(y_t|H_t)$ the *conditional* covariance matrix, and $D_t(\beta) = \partial h/\partial \eta$ evaluated at $\eta_t = Z_t\beta$. For univariate responses, Z_t reduces to z_t', and $\Sigma_t(\beta)$ to the conditional variance σ_t^2. Comparing with corresponding expressions for independent observations in Chapters 2 and 3, it is seen that they are formally identical if past responses are treated like additional covariates. Of course, observed information matrices $F_{obs}(\beta) = -\partial^2 l(\beta)/\partial\beta\partial\beta'$ are also of the same form. However, it is generally not possible to write down the *unconditional* expected information matrix $F(\beta) = EF_{obs}(\beta)$ in explicit form. Instead, for dependent observations, the *conditional information matrix*

$$G(\beta) = \sum_{t=1}^{T} \mathrm{cov}\,(s_t(\beta)|H_t),$$

where $s_t(\beta) = \partial \log f(y_t|H_t;\beta)/\partial\beta = Z_t' D_t(\beta) \Sigma_t^{-1}(\beta)(y_t - \mu_t(\beta))$ is the individual score function contribution, plays an important role in computations and asymptotic considerations. In our context it is given by

$$G(\beta) = \sum_{t=1}^{T} Z_t' D_t(\beta_t) \Sigma_t^{-1} D_t(\beta) Z_t \qquad (6.1.7)$$

and has the same form as the expected information for independent observations. Integrating out the observations $\{y_t\}$, we get, in principle, the unconditional expected information $F(\beta) = \mathrm{cov}\, s(\beta)$.

To compute an MLE $\hat{\beta}$ as a root of $s(\beta)$ corresponding to a (local) maximum of $l(\beta)$, we can use the same iterative algorithms as for independent observations, treating past responses just like additional covariates. For the Fisher–scoring algorithm or its equivalent iterative weighted–least–squares procedure this means that conditional information matrices (6.1.7) are used instead of unconditional ones.

Under appropriate "regularity assumptions" the MLE $\hat{\beta}$ is consistent and asymptotically normal,

$$\hat{\beta} \stackrel{a}{\sim} N(\beta, G^{-1}(\hat{\beta})),$$

with the inverse of the conditional information matrix the asymptotic covariance matrix. Since dependent observations contain less information than independent observations, such regularity assumptions are generally somewhat more restrictive. Often stationarity or ergodicity conditions are assumed, allowing the application of limit theorems for stationary sequences of random variables. Certain forms of nonstationarity can be admitted, however, additional mathematical effort has to be invested. Kaufmann (1987) provides rigorous asymptotic estimation theory for categorical time series; see also Fahrmeir and Kaufmann (1987). (As a key tool, martingale limit theory (e.g., Hall and Heyde, 1980) can be applied to the sequence $s_t(\beta) = \partial l_t(\beta)/\partial\beta$ of individual conditional score function contributions, since they form a martingale sequence). To illustrate what kind of nonstationarity can be admitted, consider an autoregressive logit or probit model of the form (6.1.1) and design vector $z_t' = (1, y_{t-1}, \ldots, y_{t-l}, x_t')$ without interaction terms between past responses and covariates. Then the following two conditions are sufficient for consistency and asymptotic normality of the MLE (corollary 1 in Fahrmeir and Kaufmann, 1987):

(i) The covariate sequence $\{x_t\}$ is bounded.

(ii) The empirical covariance matrix

$$S_t = \sum_{s=1}^{t}(x_s - \bar{x})(x_s - \bar{x})'$$

of the regressors diverges, i.e., $\lambda_{\min}S_t \to \infty$ or equivalently $S_t^{-1} \to 0$.

This condition corresponds exactly to (2.2.10) of Chapter 2 for independent observations, after removal of the constant 1. No convergence assumptions on S_t such as $S_t/t \to S$ are required. Although no rigorous proof has been given, we conjecture that growing regressors are admissible with similar growth rates as for the case of independent observations.

Testing linear hypotheses by   likelihood ratio, Wald and score statistics is possible in complete analogy to Section 2.2.2 if unconditional information matrices are replaced by conditional ones. The common asymptotic properties of test statistics remain valid under essentially the same general conditions required for consistency and asymptotic normality of the MLE (e.g., Fahrmeir 1987b, 1988). All other tools for statistical inference that rely on the MLE and test statistics such as goodness–of–fit statistics and variable selection methods may be used in the same way as in Chapters 2, 3 and 4.

For the more general quasi–likelihood models defined by (6.1.3) and (6.1.4), the predictor

$$\eta_t = x_t'\gamma + \sum_{i=1}^{l} \beta_i f_i(H_t)$$

is no longer linear in γ and β if $f_i(H_t)$ depends on γ. The MLE for (β, γ) may in principle be computed by the method of Fisher scoring, however, score functions and conditional information matrices are not of the simple form (6.1.6), (6.1.7) and have to be redefined by evaluating $\partial\mu_t/\partial\gamma$ and

$\partial\mu_t/\partial\beta$. If $f_i(H_t)$ is linear in γ as in Examples (v) and (vi), a two–step iteration procedure alternating between estimates for γ and β may be easier to implement; see Zeger and Qaqish (1988, p. 1024). General asymptotic theory can be applied and the (quasi–)MLE will possess the usual asymptotic properties under appropriate regularity conditions.

Finally, let us briefly discuss the situation considered at the end of Section 6.1.1, where modelling and estimation is based on a Markov model of order l and on the resulting quasi–log–likelihood

$$l(\beta) = \sum_{t=1}^{T} l_t(\beta) = \sum_{t=1}^{T} \log f(y_t|y_{t-1}, ..., y_{t-l}, x_t; \beta).$$

Since $f(y_t|H_t) \neq f(y_t|y_{t-1}, ..., y_1, x_t; \beta)$, this is not the true log–likelihood of the data. Yet β can be estimated consistently and asymptotically normal, but the asymptotic covariance matrix has to be corrected. Due to the dependence of the observations, this correction is not as simple as in the case of independent observations. For stationary processes and under suitable mixing assumptions Azzalini (1983), Levine (1983) and White (1984) give relevant results. However, it seems that one cannot move away too far from stationarity.

Example 6.1: Polio incidence in USA

Table 6.1 lists a time series of the monthly number y_t of cases of poliomyelitis reported by the U.S. Centers for Disease Control for the years 1970 to 1983, $t = 0, \ldots, 167$, taken from Zeger (1988a).

Let us analyse whether polio incidence has been decreasing since 1970. A plot of the time series given in Figure 6.1 reveals some seasonality, but does not provide clear evidence for a long–term decrease in the rate of U.S. polio infection. Thus, we will regress y_t on a linear trend, as well as sine, cosine pairs at the annual and semiannual frequencies. Since the counts y_t and $y_{t+\tau}, \tau > 0$, are not independent we also take into account the effect of past polio counts by conditioning on these. As a basis we use a conditional log–linear Poisson model of the form

$$\lambda_t = E(y_t|y_{t-1}, \ldots, y_{t-l}) = \exp(\alpha + \beta\, t \times 10^{-3} + z_t'\delta + \sum_{j=1}^{l} \gamma_j y_{t-j}),$$

where the term z_t' including cos $(2\pi t/12)$, sin $(2\pi t/12)$, cos $(2\pi t/6)$, sin $(2\pi t/6)$ represents the seasonal part. For the autoregressive model of order $l = 5$ the fitted incidence rate $\hat\mu_t$ can be seen in Figure 6.2. MLE's on which the fitted incidence rate $\hat\mu_t$ is based are given in Table 6.2 together with p–values that already take into account the estimated nuisance–parameter $\hat\phi = 1.761$. A long–term decrease in polio incidence is indicated by the fitted curve in Figure 6.2 as well as by the negative sign of the MLE for the

TABLE 6.1. Monthly number of poliomyelitis cases in USA for 1970 to 1983

	Jan.	Feb.	Mar.	Apr.	May	June	July	Aug.	Sept.	Oct.	Nov.	Dec.
1970	0	1	0	0	1	3	9	2	3	5	3	5
1971	2	2	0	1	0	1	3	3	2	1	1	5
1972	0	3	1	0	1	4	0	0	1	6	14	1
1973	1	0	0	1	1	1	1	0	1	0	1	0
1974	1	0	1	0	1	0	1	0	1	0	0	2
1975	0	1	0	1	0	0	1	2	0	0	1	2
1976	0	3	1	1	0	2	0	4	0	1	1	1
1977	1	1	0	1	1	0	2	1	3	1	2	4
1978	0	0	0	1	0	1	0	2	2	4	2	3
1979	3	0	0	3	7	8	2	4	1	1	2	4
1980	0	1	1	1	3	0	0	0	0	1	0	1
1981	1	0	0	0	0	0	1	2	0	2	0	0
1982	0	1	0	1	0	1	0	2	0	0	1	2
1983	0	1	0	0	0	1	2	1	0	1	3	6

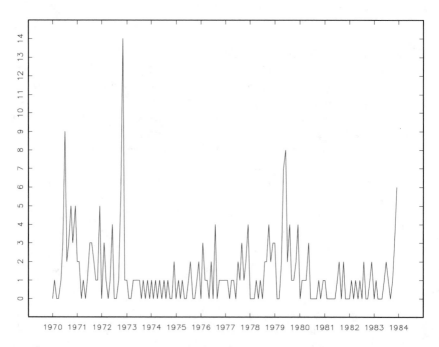

FIGURE 6.1. Monthly number of polio cases in USA from 1970 to 1983.

trend component. This is also confirmed by the Wald test of the hypothesis $H_0\colon \beta = 0$, which cannot clearly be rejected due to the p–value 0.095.

TABLE 6.2. Log–linear AR(5)–model fit to polio data

	MLE	p–value
1	0.160	0.523
$t \times 10^{-3}$	-3.332	0.095
$\cos(2\pi t/12)$	-0.217	0.116
$\sin(2\pi t/12)$	-0.462	0.002
$\cos(2\pi t/6)$	0.128	0.354
$\sin(2\pi t/6)$	-0.372	0.008
y_{t-1}	0.085	0.022
y_{t-2}	0.040	0.445
y_{t-3}	-0.040	0.501
y_{t-4}	0.029	0.532
y_{t-5}	0.081	0.059

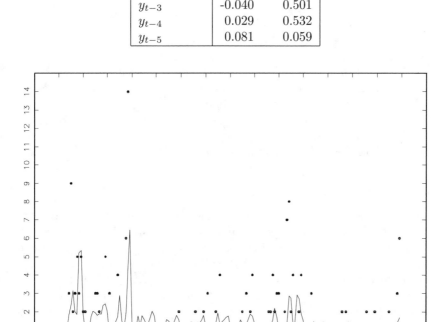

FIGURE 6.2. Predicted polio incidence $\hat{\mu}_t$ based on a log–linear AR(l=5)–model fit.

Concerning time dependence, the autoregressive order $l = 5$ should not be reduced further since the p–value of y_{t-5} indicates a strong time dependence. The effect of successive dropping of autoregressive terms on the intercept α and trend component β is documented in Table 6.3. Obviously

TABLE 6.3. Log–linear AR(1)–model fits to polio data

AR–order	1	$t \times 10^{-3}$
l=1	0.352 (0.193)	-3.934 (1.914)
l=2	0.297 (0.218)	-3.715 (1.955)
l=3	0.335 (0.225)	-3.882 (1.983)
l=4	0.278 (0.242)	-3.718 (2.008)
l=5	0.160 (0.251)	-3.332 (1.997)

MLEs and standard errors (in brackets) of intercept
and trend component

different autoregressive models yield *different* trend component estimates. Such a phenomena is typical for conditional models, which usually are not robust against different forms of time–dependence specification. Moreover, interpretation of the trend components changes from one autoregressive model to the other, since the conditioning on past observations changes. Effects of seasonal components behave quite similarly. Therefore marginal modelling as in the next section may be more sensible if one is interested in the marginal effect of trend and season. □

6.1.3   Marginal models

In a number of applications, the main scientific objective is not to fit a conditional or predictive model for y_t but to express y_t or its expectation as a simple function of x_t. More formally, this means one is rather interested in the marginal distribution $f(y_t|x_t)$ or the marginal expectation $E(y_t|x_t)$ instead of some conditional distribution $f(y_t|y_{t-1}, ..., x_t)$ or conditional expectation. This is in complete analogy to Section 3.5.2, where conditional and marginal models for correlated responses have been contrasted within a cross–sectional context.

For illustration, let us first consider two simple Gaussian models. The conditional model

$$y_t = x_t\beta + \gamma y_{t-1} + u_t, \quad u_t \sim N(0, \sigma^2)$$

has conditional expectation $x_t\beta + \gamma y_{t-1}$, while its marginal expectation is (for $|\gamma| < 1$)

$$Ey_t = (\sum_{s=0}^{\infty} \gamma^s x_{t-s})\beta,$$

which is not easy to interpret. As an alternative consider the model

$$\tilde{y}_t = x_t\beta + \varepsilon_t, \quad \varepsilon_t = \gamma\varepsilon_{t-1} + u_t. \tag{6.1.8}$$

Now the marginal expectation is

$$E\tilde{y}_t = x_t\beta,$$

which is quite easily interpreted.

Generally, the basic idea is to model marginal and not conditional expectations and to supplement the marginal mean structure by a variance–covariance structure. This leads to the following specification of marginal models for univariate generalized time series models (Zeger, Diggle and Yasui, 1990), which is analogous to the definition in Section 3.5. Assume

(i) a *marginal mean structure*

$$E(y_t|x_t) = \mu_t = h(z_t'\beta), \quad z_t = z_t(x_t)$$

   as for independent observations, together with

(ii) a *variance function*

$$\mathrm{var}\,(y_t|x_t) = v(\mu_t)\phi,$$

(iii) and an *(auto–)covariance function*

$$\mathrm{cov}\,(y_t, y_{t+r}) = c(\mu_t, \mu_{t+r}; \alpha),$$

   where α is a further parameter.

The most crucial point is the specification of the covariance function. However, consistent estimation by a generalized estimating approach is still possible under regularity assumptions if $c(\mu_t, \mu_{t+r}; \alpha)$ is only a "working" covariance.

As an example we first consider a regression model for time series of counts (Zeger, 1988a). It is a log–linear model specified in analogy to the linear Gaussian model (6.1.8). Conditional on an unobserved latent process ε_t the counts are assumed to be independent with conditional mean and variance

$$E(y_t|\varepsilon_t) = \mathrm{var}\,(y_t|\varepsilon_t) = \exp(x_t'\beta)\varepsilon_t. \tag{6.1.9}$$

Furthermore suppose that $\{\varepsilon_t\}$ is a stationary process with $E(\varepsilon_t) = 1$, for simplicity, and $cov(\varepsilon_t, \varepsilon_{t+\tau}) = \sigma^2\rho_\varepsilon(\tau)$. Then the marginal moments of y_t are

$$\mu_t = Ey_t = \exp(x_t'\beta), \quad \text{var}\,(y_t) = \mu_t + \sigma^2\mu_t^2, \qquad (6.1.10)$$

$$\text{cov}\,(y_t, y_{t+\tau}) = \mu_t\mu_{t+\tau}\sigma^2\rho_\varepsilon(\tau). \qquad (6.1.11)$$

If the autocorrelation function ρ_ε is fully specified by a further set of parameters α, as, e.g., $\rho_\varepsilon(\tau) = \rho^\tau$ for a first–order autoregressive process ε_t, then the covariance function is in the form (iii) with $\alpha = (\sigma^2, \rho)$.

For binary time series, a marginal logit model

$$\pi_t = E(y_t|x_t) = x_t'\beta, \quad \sigma_t^2 = \text{var}\,(y_t|x_t) = \pi_t(1 - \pi_t),$$

and a specification of the odds ratio

$$OR(y_t, y_{t+\tau}) = \frac{P(y_t = y_{t+\tau} = 1)P(y_t = y_{t+\tau} = 0)}{P(y_t = 1, y_{t+\tau} = 0)P(y_t = 0, y_{t+\tau} = 1)}$$

are suggested. For more parsimonious parameterization, "stationary" and "truncated" (log) odds ratios

$$\log OR(y_t, y_{t+\tau}) = \left\{ \begin{array}{ll} \gamma(\tau), & \tau \le h \\ 0, & \tau > h \end{array} \right.$$

are useful as a "working" autocovariance structure.

Estimation of marginal models

If covariance parameters α or consistent estimates are known, regression parameters β can be estimated by adapting the generalized estimating equation approach in Sections 2.3.1 and 3.5.2 to the time series case. Setting

$$y = (y_1, ..., y_T)', \quad Z = (z_1, ..., z_T)', \quad \mu = (\mu_1, ..., \mu_T)',$$

$$D(\beta) = \text{diag}\,(D_t(\beta)), \quad M(\beta) = Z'D(\beta), \quad \Sigma(\beta) = \text{cov}\,(y),$$

the quasi–score function (generalized estimating function) can be written in matrix form as

$$s(\beta) = Z'D(\beta)\Sigma^{-1}(\beta)(y - \mu).$$

With independent observations $\Sigma(\beta)$ is diagonal. With time series data, $\Sigma(\beta)$ is nondiagonal with elements

$$(\Sigma(\beta))_{st} = \text{cov}\,(y_s, y_t) = c(\mu_s, \mu_t; \alpha).$$

A quasi–MLE $\hat{\beta}$ is obtained as a root of the generalized estimating equation $s(\beta) = 0$ and can, in principle, be computed by the method of Fisher scoring. However, this requires inversion of the $T \times T$–matrix $\Sigma(\beta)$ in each iteration step. To simplify computations, it will be useful to approximate the actual covariance matrix by a simpler "working" covariance matrix Σ_w, e.g., a band diagonal matrix.

Under regularity assumptions that admit the application of limit theorems for dependent data, the quasi–MLE $\hat{\beta}$ is consistent and asymptotically normal with asymptotic covariance matrix

$$\hat{A} = \left(Z'D(\hat{\beta})\Sigma^{-1}(\hat{\beta})D(\hat{\beta})Z \right)^{-1},$$

if $\Sigma(\beta)$ is *correctly specified*. If a working covariance matrix $\Sigma_w(\beta)$ is used to simplify computations, the covariance matrix has to be modified to the sandwich matrix

$$\hat{A} = (\hat{M}\hat{\Sigma}_w^{-1}\hat{M}')^{-1}\hat{M}\hat{\Sigma}_w^{-1}\hat{\Sigma}\hat{\Sigma}_w^{-1}\hat{M}'(\hat{M}\hat{\Sigma}_w^{-1}\hat{M}')^{-1}$$

where " $\hat{\ }$ " means evaluation at $\hat{\beta}$ and $\hat{M} = Z'D(\hat{\beta})$. Of course, direct application of this result requires that the autocovariance structure $\Sigma(\beta)$ is known. The asymptotic normality result remains true if the "nuisance" parameter α is replaced by a consistent estimate $\hat{\alpha}$. Such an estimate can be obtained, e.g., by a method of moments. For example, in the log–linear model for count data $\mathrm{var}\,(y_t) = \mu_t + \sigma^2\mu_t^2$. Hence σ^2 can be estimated by

$$\hat{\sigma}^2 = \sum_t \{(y_t - \hat{\mu}_t)^2 - \hat{\mu}_t\}/\sum_t \hat{\mu}_t^2,$$

and similarly the autocorrelation function can be estimated by

$$\hat{\rho}_\varepsilon(\tau) = \hat{\sigma}^{-2} \sum_{t=\tau+1}^{T} \{(y_t - \hat{\mu}_t)(y_{t-\tau} - \hat{\mu}_{t-\tau})\}/ \sum_{t=\tau+1}^{T} \hat{\mu}_t\hat{\mu}_{t-\tau}.$$

If ρ_ε is itself parameterized by $\rho_\varepsilon(\tau;\alpha)$, then $\hat{\rho}_\varepsilon(\tau,\alpha)$ is solved for α. For example, if ε_t is assumed to be a stationary autoregressive process, then $\hat{\alpha}$ can be obtained by solving the Yule–Walker equations.

Example 6.2: polio incidence in USA (Example 6.1 continued)
Zeger (1988a) applied a marginal log–linear Poisson model (6.1.9) – (6.1.11) to the polio incidence data of Table 6.1, including trend and seasonal components in the linear predictor as in Example 6.1 but omit autoregressive terms. Estimates $\hat{\sigma}^2, \hat{\rho}_\varepsilon$ were obtained as above and a tridiagonal "working" covariance matrix Σ_w was used in the Fisher–scoring steps. The results of the trend and seasonal components are given in Table 6.4.

Compared to conditional modelling in Example 6.1, signs of the parameter estimates remain mostly the same, but absolute values of marginal effects differ significantly from corresponding effects in the conditional autoregressive model of order 5 (Table 6.2). In particular, the trend term in Table 6.4 indicates a long–term decrease in the rate of polio infection more clearly. In contrast, this effect is attenuated with conditional models.   □

TABLE 6.4. Marginal model fit for polio data

	Estimate	Std. error
$t \times 10^{-3}$	-4.35	2.68
$\cos(2\pi t/12)$	-0.11	0.16
$\sin(2\pi t/12)$	-0.48	0.17
$\cos(2\pi t/6$	-0.20	0.14
$\sin(2\pi t/6)$	-0.41	0.14
$\hat{\sigma}^2$	0.77	—
$\hat{\rho}(1)$	0.25	—

6.2 Longitudinal data

In this section we consider longitudinal data in discrete time, where time series data or repeated observations $(y_{it}, x_{it}), t = 1, ..., T_i$, are available for each individual or unit $i = 1, ..., n$ of a population. Linear models for Gaussian outcomes y_{it} have been treated extensively for a long time; see, e.g., Hsiao (1986) and Dielman (1989). Corresponding models for discrete and non–normal outcomes have received attention only more recently, but the gap is rapidly closing. The text of Diggle, Liang and Zeger (1994) is an important contribution with focus on biostatistical applications.

Longitudinal data may be viewed as data with correlated responses within clusters (Section 3.5): The cluster i contains repeated responses y_{it}, $t = 1, ..., T_i$ on the same subject i, $i = 1, ..., n$, and it is very likely that these repeated responses are correlated. Therefore methods for analysing correlated responses will reappear in this section in modified form. For example, due to the longitudinal data situation, special covariance structures will be of interest. As in Section 3.5 and as for the pure time series situation in Section 6.1 it is important to distinguish between conditional and marginal models. For example, consider an epidemiologic study as in Example 1.10, where y_{it} is the health or illness state, measured in categories, of individual i at time t and x_{it} is a vector of risk factors or individual characteristics, possibly changing over time. If one is interested in analysing conditional probabilities for a certain state or transitions to a state given the individual's history and the effect of covariates on these probabilities, then conditional models are needed. If the main scientific objective is the effect of covariates on the state of health, marginal models are appropriate. The distinction is also important from the methodological point of view: Most models for non–normal outcomes are nonlinear so that conditional and marginal models will generally not be of the same form. For example, marginal probabilities calculated from a conditional logistic model will not be of the logistic form.

6.2.1   Conditional models

Notation is simplified if observations at time t are collected in "panel waves"

$$(y_t, x_t) = (y_{1t}, ..., y_{nt}; x_{1t}, ..., x_{nt}), \quad t = 1, ..., T,$$

where $T = \max(T_i)$ is the maximal length of individual time series and y_{it}, x_{it} are omitted if $t > T_i$. The "history" of covariates and past responses at time t is then denoted by

$$H_t = \{x_t, ..., x_1, y_{t-1}, ..., y_1\}. \tag{6.2.1}$$

Initial values $y_0, y_{-1}, ...,$ etc., which are needed in autoregressive models, are assumed to be part of the covariates. Conditional models for time series (Section 6.1.1) are easily extended to panel data if we assume that individual responses y_{it} within y_t given H_t are conditionally independent:

$$(C) \qquad\qquad f(y_t|H_t) = \prod_{i \in R_t} f(y_{it}|H_t), \quad t = 1, ..., T,$$

where R_t is the set of units still observed at time t. This condition is clearly fulfilled if the individual n time series $\{y_{it}, t = 1, ..., T\}$ are totally independent. However (C) is weaker since interaction among units may be introduced via the common history. Condition (C) may be weakened further by conditioning y_{it} additionally on $y_{jt}, j \neq i$, similarly for conditional symmetric models for multivariate cross–sectional observations in Section 3.5.1; see Zeger and Liang (1989). However, estimation becomes more difficult.

Generalized autoregressive models, quasi–likelihood models

The simplest models of generalized autoregressive form are obtained if it is assumed that parameters are constant across time and units and that conditional densities of y_{it} given H_t belong to a uni– or multivariate simple exponential family. Conditional means are supposed to be given by

$$\mu_{it} = E(y_{it}|H_t) = h(Z_{it}\beta), \tag{6.2.2}$$

where the design matrix is a function of H_t, i.e., $Z_{it} = Z_{it}(H_t)$. As for cross–sectional and time series data, various specifications of the response function and the design matrix are possible, in particular for multicategorical responses. As an important subclass, Markov models are obtained if only a finite number of past responses is included in H_t.

Together with a specific exponential family the mean structure (6.2.2) completely specifies the variance function. Separation of the mean and variance structure is possible by adopting the extended models defined by (6.1.3) and (6.1.4) to the panel data situation.

The assumption that parameters β are homogeneous with respect to time and the population can be relaxed, in principle, in a number of ways. Whether this is sensible will also depend on the data available. For example, if we allow for individual–specific parameters β_i, (6.2.2) is replaced by

$$\mu_{it} = h(Z_{it}\beta_i), i = 1, ..., n.$$

If individual time series are long enough, the $\beta_i's$ can be consistently estimated separately. However, if individual time series are short this will lead to serious bias or even nonexistence of estimates. To avoid such problems, other conceivable working hypotheses are homogeneity of parameters in subpopulations or homogeneity with respect to a part of the parameters as, e.g., in models where only the intercept term varies. An attractive alternative is random effects models (Chapter 7). Analogous remarks apply to parameters β_t varying over time. If the population is large enough it may be possible to separately fit a model for each panel wave y_t as in Stram, Wei and Ware (1988). For smaller cross sections one will run into trouble due to bias or nonexistence of estimates, and some kind of smoothing as in Chapter 8 will be useful.

Though the class of conditional models considered so far seems to be quite large, it does not contain all approaches proposed previously in the literature. Heckman (1981) derives dynamic models for panel data with binary responses y_{it} by relating them to latent continuous responses $\tilde{y}_{it}$ via a threshold mechanism: $y_{it} = 1 \Leftrightarrow \tilde{y}_{it} \geq 0, y_{it} = 0$ else. He assumes a linear autoregressive model $\tilde{y}_{it} = \eta_{it} + \varepsilon_{it}$, where η_{it} is a linear combination of covariates x_{it}, past responses and past latent variables. The inclusion of $\tilde{y}_{i,t-1}, \tilde{y}_{i,t-2}, ...$ formalizes the idea that former latent propensities to choose a state influence the probability for the current choice. The conditional probability for y_{it} given the past cannot be expressed in the simple mean structure model above.

Statistical inference

Under the conditional independence assumption (C), the log–likelihood given all data factorizes into individual contributions. Therefore log–likelihoods, score functions and conditional and observed information matrices can be written as a sum of the individual contributions. For generalized autoregressive models with parameters β constant across time and units we obtain the log–likelihood

$$l(\beta) = \sum_{i,t} l_{it}(\beta), \quad l_{it}(\beta) = \log f(y_{it}|H_t, \beta), \tag{6.2.3}$$

the score function

$$s(\beta) = \sum_{i,t} s_{it}(\beta) = \sum_{i,t} Z'_{it} D_{it}(\beta) \Sigma_{it}^{-1}(y_{it} - \mu_{it}(\beta)),$$

and the conditional information matrix

$$G(\beta) = \sum_{i,t} G_{it}(\beta) = \sum_{i,t} Z'_{it} D_{it}(\beta) \Sigma_{it}^{-1} D'_{it}(\beta) Z_{it} \qquad (6.2.4)$$

with the usual notation for individual conditional means μ_{it}, variances Σ_{it}, etc. The modifications necessary for parameters varying over time or units are obvious. MLE's are obtained by the usual iterative methods from $s(\hat{\beta}) = 0$. For quasi–likelihood models one starts directly from the appropriately modified quasi–score function.

For asymptotics, there are three main situations that can be considered:

$$n \to \infty, \qquad t \text{ finite}, \ \beta \text{ constant across units},$$
$$n \text{ finite}, \qquad t \to \infty, \ \beta \text{ constant across time},$$
$$n \to \infty, \qquad t \to \infty.$$

These possibilities correspond to longitudinal data with many short time series, a moderate number of longer time series, and many longer time series. The second situation can be treated similarly as for single time series. For $n \to \infty$ and $t \to \infty$ it may be possible to admit some parameters varying over units and others varying over time, e.g., as for an additive intercept term $\beta_{0it} = \beta_{0i} + \beta_{it}$. Under appropriate regularity assumptions the (quasi–) MLE will be consistent and asymptotically normal, however, to our knowledge rigorous proofs are not available. Yet it seems reasonable to assume that

$$\hat{\beta} \overset{a}{\sim} N(\beta, G^{-1}(\hat{\beta})),$$

and to base test statistics, goodness–of–fit tests, etc., on this working assumption. Anyway, finite sample behaviour requires further investigations.

Transition models

For discrete outcomes y_{it} a main objective often is the modelling of transition probabilities between pairs of categories or "states" at successive occasions $t - 1$ and t. If a first–order Markov assumption holds, transitions may be analyzed with the models and methods above by inclusion of $y_{i,t-1}$ as an additional covariate. It can be preferable, however, to model transition probabilities

$$p_{jk}(x_{it}) = P(y_{it} = k | y_{i,t-1} = j, x_{it}; \beta), \quad t = 2, ..., T,$$

separately, e.g., by a logistic model for all pairs (j, k) and to express the likelihood as the product of transition probabilities of all observed transitions. This direct approach, which has been used, e.g., by Garber (1989) and Hopper and Young (1989), is more convenient and flexible if one has to distinguish between different types of categories, such as transient, recurrent or absorbing states.

If the first–order Markov property holds and transition probabilities are correctly specified usual likelihood theory applies. One may ask, however, whether transition probabilities can be consistently estimated when the Markov property is violated. This is no problem if the population is large enough to estimate separate response models for each of the $t-1$ occasions; see, e.g., Ware, Lipsitz and Speizer (1988). Otherwise the question is analogous to the estimation of marginal models from time series (Section 6.1.3) or longitudinal data (Section 6.2.2). If transition probabilities are correctly specified, then results of White (1984, theorem 3.1) show that consistent parameter estimation is possible for $T \to \infty$ under regularity assumptions. Again little is known about finite sample behaviour, in particular concerning covariance matrix estimation.

Subject–specific approaches and conditional likelihood

A binary logistic model that includes a subject–specific parameter is given by

$$P(y_{it} = 1 | x_{it}) = \frac{\exp(\alpha_i + x_{it}\beta)}{1 + \exp(\alpha_i + x'_{it}\beta)}.$$

The model accounts for heterogeneity by introducing the parameter α_i. Models of this type have been studied in the psychometrics literature in the context of item response theory. The simplest version is the Rasch model where α_i stands for the subject's ability and $x'_{it}\beta$ simplifies to a parameter γ_t that stands for the item difficulty (see Rasch, 1961; Andersen, 1980).

The basic idea proposed by Andersen (1973) is to consider the conditional likelihood given $y_{i\cdot} = \sum_{t=1}^{T} y_{it}$ where it is assumed that the number of observations does not depend on i. From the form

$$P(y_{i1}, \ldots, y_{iT} | y_{i\cdot}) = \frac{\exp(\sum_{t=1}^{T} y_{it} x'_{it}\beta)}{\sum_{\sum_j \tilde{y}_{ij} = y_{i\cdot}} \exp(\sum_{t=1}^{T} \tilde{y}_{it} x'_{it}\beta)}$$

it is seen that the conditional probability given $y_{i\cdot}$ does not depend on the heterogeneity parameter α_i since $y_{i\cdot}$ is a sufficient statistic for α_i. Therefore conditioning on $y_{1\cdot}, \ldots, y_{n\cdot}$ allows conditional inference upon the parameter β. Originally developed for item response models this conditioned likelihood approach has been considered more recently for categorical repeated measurement and panel data (Conaway, 1989; Hamerle and Ronning, 1992). An alternative approach to subject–specific models based on the connection of

TABLE 6.5. Variables and questions of the IFO business test

Variable	Questions
Production plan P	Our production with respect to product X is planned to be in the course of the next three months, corrected for seasonal variations–increased, remained the same, decreased.
Expected business D	Our business conditions for product X are expected to be in the next six months, corrected for seasonal variation–improved, about the same, deteriorated.
Orders in hand O	Our orders in hand (domestic and foreign) for X are in relation to the preceding month–higher, the same, lower.

(Since the variable "expected business condition" is considered as a substitute for "expected demand," we have chosen the mnemonic D).

the Rasch model and quasi–symmetry has been investigated in a series of papers by Agresti (1993a,b) and Agresti and Lang (1993).

Example 6.3: IFO business test
The IFO Institute in Munich collects micro–data of firms each month for its "business tests." The monthly questionnaire contains questions on the tendency of successive change of realizations, plans and expectations of variables like production, orders in hand, demand, etc. Answers are categorical, most of them trichotomous with categories like "increase" $(+)$, "decrease" $(-)$ and "no change" $(=)$. Currently several thousand firms from various industry branches participate in this survey on a voluntary basis. Table 6.5 contains some typical variables and questions.

One of the main objectives of previous research has been to study the dependency of certain variables, such as production plans and prices, on other economic variables, and to check whether the results are in agreement with economic theory; see, e.g., König, Nerlove and Oudiz (1981) and Nerlove (1983). Based on this work, we will analyze the dependency of the variable P_t at month t on the explanatory variables D_t and O_t and on the production plans P_{t-1} in the previous month. In this section, we consider 317 firms of the machine–building industry from January 1980 to December 1990. Previous work of Morawitz and Tutz (1990) showed that simple cumulative models (Section 3.3.1) are not always appropriate for analysing these data, while extended versions with explanatory variables

TABLE 6.6. Estimates of main effects

Category	Parameter	Estimator	p–value
$+$	threshold	-6.51	0.000
$=$	threshold	-1.45	0.000
$+$	P^+_{t-1}	4.34	0.000
$=$	P^+_{t-1}	2.35	0.000
$+$	$P^=_{t-1}$	1.11	0.003
$=$	$P^=_{t-1}$	1.82	0.000
$+$	D^+_t	3.78	0.000
$=$	D^+_t	1.61	0.000
$+$	$D^=_t$	2.13	0.000
$=$	$D^=_t$	1.68	0.000
$+$	O^+_t	3.54	0.000
$=$	O^+_t	1.50	0.000
$+$	O^+_t	1.59	0.000
$=$	$O^=_t$	0.98	0.000
Pearsons's χ^2:		24.17, 16 df,	p–value 0.09
Deviance:		24.05, 16 df,	p–value 0.09

as threshold variables (Section 3.3.2) provide reasonable fits. For work on business tendency surveys, see also Ronning (1980, 1987).   □

For the following each of the trichotomous (k = 3) variables P, D and O is described by two (q = 2) dummy variables, with "decrease" (-) as the reference category. The relevant dummy variables for "increase" (+) and "no change" (=) are shortened by $P^+, P^=, D^+, D^=$ and $O^+, O^=$. Then an extended cumulative model of the form (3.3.10), Chapter 3, where all covariates are threshold variables, is specified by

$$P\left(P_t = \text{“}+\text{”}\right) = F(\beta_{10} + w'\beta_1)$$
$$P\left(P_t = \text{“}+\text{” or “}=\text{”}\right) = F(\beta_{20} + w'\beta_2),$$

where $P\left(P_t = \text{“}+\text{”}\right)$ and $P\left(P_t = \text{“}+\text{” or } P_t = \text{“}=\text{”}\right)$ stand for the probability of increasing and nondecreasing production plans, and F is the logistic distribution function. The parameters (β_{10}, β_1) and (β_{20}, β_2) are category–specific, and the threshold design vector w contains all six main effects and interaction effects if necessary:

$$w = (P^+_{t-1}, P^=_{t-1}, D^+_t, D^=_t, O^+_t, O^=_t, P^+_{t-1} * D^+_t, \ldots).$$

For the data at hand it turned out that the model without three–factor interactions but including all two–factor interactions fits the data reasonably well, while all smaller models are rejected by the Pearson or deviance statistic. Table 6.6 contains estimation results for the main effects of the fitted model. Effects of two–factor interactions are generally quite smaller and are omitted for a simplified presentation. The significant influence of previous production plans on current production plans provides a clear indication for the continuity of production planning. Also, the strong influence of an improved expected business condition indicates high dependency of economic growth on positive expectation, as described in several economic theories, and the effect of "orders in hand" allows similar interpretation.

6.2.2   Marginal models

In many longitudinal studies, data consist of a small or moderate number of repeated observations for many subjects, and the main objective is to analyse the effect of covariates on a response variable, without conditioning on previous responses. Thus one has essentially the same situation as in Section 3.5, but within a longitudinal context. In this setting it is more natural to view the data as a cross section of individual time series

$$y_i = (y_{i1}, ..., y_{iT_i}), \quad x_i = (x_{i1}, ..., x_{iT_i}), \quad i = 1, ..., n,$$

and to assume that, given the covariates, individual time series are mutually independent. Marginal models for non–normal longitudinal data have been introduced by Liang and Zeger (1986) and Zeger and Liang (1986). Although the method can be extended to multicategorical responses, the following presentation is also restricted to the univariate case for simplicity. Similarly as in Sections 3.5.2 and 6.1.3 one assumes:

(i) The *marginal mean* of y_{it} is correctly specified by

$$\mu_{it} = E(y_{it}|x_{it}) = h(z'_{it}\beta),$$

where h is a response function and $z_{it} = z_{it}(x_{it})$ a design vector as described in Chapter 2.

(ii) As in quasi–likelihood models for independent observations, the mean structure assumption is supplemented by describing the *variance* of y_{it} as a function of μ_{it}, i.e.,

$$\sigma^2_{it} = \text{var}(y_{it}) = v(\mu_{it})\phi,$$

where v is a known variance function. Liang and Zeger (1986) assume that the marginal distribution of y_{it} follows a generalized linear model. Then $v(\mu_{it})$ is completely determined by the exponential family assumption. In Zeger and Liang (1986) the exponential family assumption is dropped and (i) and (ii) describe only the first two moments of the marginal distribution.

(iii) To account for within–units dependence in y_i, *working covariances*, possibly depending on an unknown parameter vector α, are introduced:

$$\text{cov}(y_{is}, y_{it}) = c(\mu_{is}, \mu_{it}; \alpha),$$

with a known function c. Together with the variance function, this defines a *working covariance matrix*

$$\text{cov}(y_i) = \Sigma_i(\beta, \alpha)$$

for subject i, which depends on β and perhaps on α.

Since parameters β are constant across subjects, marginal models are appropriate for analyzing "population–averaged" effects. Subject–specific approaches based on conditional likelihoods are mentioned at the end of Section 6.2.1, and random effects models are described in detail in Chapter 7.

As for clustered cross–sectional data, two proposals for specifying working covariances exist. Liang and Zeger (1986) use correlations and assume a $T_i \times T_i$ *working correlation matrix* $R_i(\alpha)$ for y_i.

Setting

$$A_i = \text{diag}(\sigma_{i1}^2, ..., \sigma_{iT_i}^2),$$

this leads to a "working" covariance matrix

$$\Sigma_i(\alpha) = A_i^{1/2} R_i(\alpha) A_i^{1/2} \tag{6.2.5}$$

for y_i. Choice of the working correlation matrix R_i can be guided by some compromise between simplicity and efficiency, it may be supported by assumptions on an underlying latent process, and it will also depend on the amount of data available.

The simplest specification is the working assumption of uncorrelated repeated observations, i.e.,

$$R_i(\alpha) = I. \tag{6.2.6}$$

This leads to the same simple estimating equations as for independent observations.

Another extreme case is applicable if $T_i = T$ and $R_i(\alpha) = R(\alpha)$ for all units. Then one may let $R(\alpha)$ fully unspecified and estimate it by the empirical correlation matrix.

The choice

$$\text{corr}(y_{is}, y_{it}) = (R_i)_{st} = \alpha, \quad s \neq t \tag{6.2.7}$$

corresponds to the exchangeable correlation structure in linear random effects models with a random intercept term; compare with Chapter 7. With

$$(R_i)_{st} = \alpha(|t - s|) \tag{6.2.8}$$

correlations are stationary. The special form $\alpha(|t - s|) = \alpha^{|t-s|}$ mimics the autocorrelation function of a Gaussian AR(1)–process. Further examples can be found in Liang and Zeger (1986), or obtained from an underlying random–effects model as, e.g., in Hamerle and Nagl (1988) and Thall and Vail (1990). For binary responses, odds ratios

$$\gamma_{ist} = \frac{P(y_{is} = y_{it} = 1)P(y_{is} = y_{it} = 0)}{P(y_{is} = 1, y_{it} = 0)P(y_{is} = 0, y_{it} = 1)},$$

with the vector of pairwise odds ratios parameterized by $\gamma = \gamma(\alpha)$, may be a better alternative as measure for association. This is in complete analogy to Section 3.5.2, and we will not go into details in this section.

A closely related class of models for ordered categorical outcomes has been proposed by Stram, Wei and Ware (1988) and Stram and Wei (1988). Admitting time–dependent parameters, they propose to fit a marginal cumulative response model

$$\pi_{it} = h(Z_{it}\beta_t)$$

with time–dependent parameter β_t separately for each occasion. For $n \to \infty$, the combined estimate $\hat{\beta} = (\hat{\beta}_1, ..., \hat{\beta}_T)$ becomes asymptotically normal, and its asymptotic covariance matrix can be estimated empirically. Zeger (1988b) shows that $\hat{\beta}$ can be viewed as the solution of a GEE with working correlation matrix $R(\alpha) = I$, and that the covariance matrix is identical to the one obtained from the GEE approach. A similar approach has been considered by Moulton and Zeger (1989). They combine the estimated coefficients at each time point by using bootstrap methods or weighted–least–squares methods. For applications of these approaches to clinical trials see Davis (1991).

In principle, the estimate $\hat{\beta}$ could be used to examine trends in the sequence β_t, to combine estimates, etc. However, enough data for each occasion are required to obtain reasonable estimates. If this is not the case, models with time–varying parameters according to a stochastic process or a "state space" model (Chapter 8) will be preferable.

Statistical inference

Estimation of β is carried out in complete analogy to Section 3.5.2. According to the model assumptions, the observation vectors $y_i = (y_{i1}, \ldots, y_{iT_i})$, $i = 1, \ldots, n$ are independent and have means $E(y_i|x_i) = \mu_i(\beta) = (\mu_{i1}, \ldots \ldots, \mu_{iT_i})$ and working covariance matrices $\Sigma_i(\beta, \alpha)$. Defining design matrices $Z_i' = (z_{i1}, \ldots, z_{iT_i})$ and diagonal matrices $D_i(\beta) = \mathrm{diag}(D_{it}(\beta))$, $D_{it}(\beta) = \partial h/\partial \eta$ evaluated at $\eta = z_{it}'\beta$, the generalized estimating equation for β is

$$s_\beta(\beta, \alpha) = \sum_{i=1}^{n} Z_i' D_i(\beta)\Sigma_i^{-1}(\beta, \alpha)(y_i - \mu_i(\beta)) = 0. \qquad (6.2.9)$$

In the special case of the working independence assumption the working covariance matrix Σ_i is diagonal and (6.2.9) reduces to the usual form of the score function for independent observations. Generally, Σ_i is a $T_i \times T_i$ matrix that depends on the covariate effects β, the correlation parameter α and the dispersion parameter. To compute $\hat{\beta}$, one therefore has to iterate between a modified Fisher scoring for β and estimation of α and ϕ, e.g., by some method of moments or by a second GEE. In this section we will focus on the method of moments; estimation by a GEE method is in complete analogy to Section 3.5.2. Given current estimates $\hat{\alpha}$ and $\hat{\phi}$, the GEE $s(\beta) = 0$ for $\hat{\beta}$ is solved by the iterations

$$\hat{\beta}^{(k+1)} = \hat{\beta}^{(k)} + (\hat{F}^{(k)})^{-1}\hat{s}^{(k)},$$

with the "working" Fisher matrix

$$\hat{F}^{(k)} = \sum_{i=1}^{n} Z_i D_i(\hat{\beta}^{(k)}) \Sigma_i^{-1}(\hat{\beta}^{(k)}, \hat{\alpha}, \hat{\phi}) D_i(\hat{\beta}^{(k)}) Z_i',$$

where " $\,\hat{}\,$ " means evaluation at $\beta = \hat{\beta}^{(k)}$, and the corresponding quasi–score function $\hat{s}^{(k)} = s(\hat{\beta}^{(k)}, \hat{\alpha}, \hat{\phi})$. Given a current estimate of β, the parameters α and ϕ can be estimated from current Pearson residuals

$$\hat{r}_{it} = \frac{y_{it} - \hat{\mu}_{it}}{(v(\hat{\mu}_{it}))^{1/2}}.$$

The dispersion parameter is estimated consistently by

$$\hat{\phi} = \frac{1}{N - p} \sum_{i=1}^{n} \sum_{t_1}^{T_i} \hat{r}_{it}^2, \quad N = \sum_{i=1}^{n} T_i.$$

Estimation of α depends on the choice of $R_i(\alpha)$. As a general strategy it is proposed to estimate α by a simple function of the Pearson residuals. For the exchangeable correlation model (6.2.7) α can be estimated by

$$\hat{\alpha} = \frac{1}{\hat{\phi}\left\{\sum_{i=1}^{n} \frac{1}{2} T_i(T_i - 1) - p\right\}} \sum_{i=1}^{n} \sum_{t>s} \hat{r}_{it}\hat{r}_{is}.$$

The parameters of the stationary model (6.2.8) can be estimated analogously by averages of the corresponding residuals $\hat{r}_{it}, \hat{r}_{is}$.

A totally unspecified $R = R(\alpha)$ can be estimated by

$$\frac{1}{n\hat{\phi}} \sum_{i=1}^{n} A_i^{-1/2}(y_i - \hat{\mu}_i)(y_i - \hat{\mu}_i)' A_i^{-1/2}$$

if T is small compared to n.

Consistency and asymptotic normality results for $\hat{\beta}$ can be obtained on the lines of asymptotic theory of quasi–MLE's for independent observations if $T_i, i = 1, ..., n$ is fixed and $n \to \infty$. Standard asymptotic theory imposes "mild" regularity conditions leading to $n^{1/2}$–asymptotics as in Theorem 2 of Liang and Zeger (1986): If $\hat{\alpha}$ is $n^{1/2}$–consistent given β, ϕ, and $\hat{\phi}$ is $n^{1/2}$–consistent given β, then $\hat{\beta}$ is $n^{1/2}$–consistent and asymptotically multivariate Gaussian; briefly

$$\hat{\beta} \overset{a}{\sim} N(\beta, F^{-1}VF^{-1}),$$

with

$$F = \sum_{i=1}^{n} Z_i' D_i \Sigma_i^{-1} D_i Z_i, \quad V = \sum_{i=1}^{n} Z_i' D_i \Sigma_i^{-1} \text{cov}\,(y_i) \Sigma_i^{-1} D_i Z_i \,.$$

The asymptotic covariance matrix F^{-1} can be consistently estimated by replacing β, α and ϕ by their estimates and cov (y_i) by $(y_i - \hat{\mu}_i)(y_i - \hat{\mu}_i)'$, i.e.,

$$\hat{A} = \text{cov}\,(\hat{\beta}) \overset{a}{\sim} \hat{F}^{-1} \left\{ \sum_{i=1}^{n} Z_i' \hat{D}_i \hat{\Sigma}_i^{-1} (y_i - \hat{\mu}_i)(y_i - \hat{\mu}_i)' \hat{\Sigma}_i^{-1} \hat{D}_i Z_i \right\} \hat{F}^{-1} \,.$$

(6.2.10)

Based on the preceding asymptotic result, confidence intervals for β and Wald and score tests may be constructed; compare with Section 2.3.1. If the covariance structure Σ_i is correctly specified or consistently estimated as in the totally unspecified case, then estimation is asymptotically efficient. If $R_i(\alpha)$ is just a working assumption, then some loss of efficiency will occur. Liang and Zeger (1986) and Hamerle and Nagl (1988) provide some evidence that this loss can be quite small for simple choices of $R_i(\alpha)$. The study of McDonald (1993) recommends the *independence* working model, whenever correlation is merely regarded as a nuisance.

Approaches that add a second GEE for α to the GEE (6.2.9) for β or that treat (β, α) simultaneously proceed as in Section 3.5.2 and the literature cited there, see also Zeger, Liang and Self (1985) and Thall and Vail (1990) in the longitudinal context.

Example 6.4: Ohio children

Zeger, Liang and Albert (1988) analyzed a subset of data from the Harvard Study of Air Pollution and Health, reported by Laird et al. (1984) and reproduced in Table 6.7. The data consist in reports for 537 children from Ohio, examined annually from age 7 to 10. Responses $y_{it}, i = 1, ..., 537, t = 1, ..., 4$ are binary, with $y_{it} = 1$ for the presence and $y_{it} = 0$ for the absence of respiratory infection. To analyse the influence of mother's smoking status

TABLE 6.7. Presence and absence of respiratory infection

Mother Did Not Smoke					Mother Smoked				
Age of child				frequency	Age of child				frequency
7	8	9	10		7	8	9	10	
0	0	0	0	237	0	0	0	0	118
0	0	0	1	10	0	0	0	1	6
0	0	1	0	15	0	0	1	0	8
0	0	1	1	4	0	0	1	1	2
0	1	0	0	16	0	1	0	0	11
0	1	0	1	2	0	1	0	1	1
0	1	1	0	7	0	1	1	0	6
0	1	1	1	3	0	1	1	1	4
1	0	0	0	24	1	0	0	0	7
1	0	0	1	3	1	0	0	1	3
1	0	1	0	3	1	0	1	0	3
1	0	1	1	2	1	0	1	1	1
1	1	0	0	6	1	1	0	0	4
1	1	0	1	2	1	1	0	1	2
1	1	1	0	5	1	1	1	0	4
1	1	1	1	11	1	1	1	1	7

and of age on children's respiratory disease, we assume that the marginal probability of infection follows a logit model

$$
\log \frac{P(\text{infection})}{P(\text{no infection})} = \beta_0 + \beta_S x_S + \beta_{A1} x_{A1} + \beta_{A2} x_{A2} + \beta_{A3} x_{A3} +
$$
$$
+ \ \beta_{SA1} x_S x_{A1} + \beta_{SA2} x_S x_{A2} + \beta_{SA3} x_S x_{A3} \, ,
$$

where smoking status is coded by $x_S = 1$ for smoking, $x_S = -1$ for non–smoking, x_{A1}, x_{A2}, x_{A3} represent age in effect–coding, and $x_S x_{A1}$, $x_S x_{A2}$, $x_S x_{A3}$ are interaction terms. Table 6.8 shows estimates based on three correlation assumptions: $R = I$ (independence assumption), $R_{st} = \alpha$, $s \neq t$ (exchangeable correlation), and unspecified correlation R. For all three working correlations, estimates are identical for the first relevant digits, so only one column is given for point estimates and robust standard deviations, based on (6.2.10). For comparison naive standard deviations, obtained from the usual ML method for independent observations, are also displayed.

TABLE 6.8. Marginal logit model fits for Ohio children data

Parameter	Effect	Standard Deviation Robust	Naive
$\hat{\beta}_0$	-1.696	0.090	0.062
$\hat{\beta}_S$	0.136	0.090	0.062
$\hat{\beta}_{A1}$	0.059	0.088	0.107
$\hat{\beta}_{A2}$	0.156	0.081	0.104
$\hat{\beta}_{A3}$	0.066	0.082	0.106
$\hat{\beta}_{SA1}$	-0.115	0.088	0.107
$\hat{\beta}_{SA2}$	0.069	0.081	0.104
$\hat{\beta}_{SA3}$	0.025	0.082	0.106

TABLE 6.9. Main effects model fits for Ohio children data

Parameter	Effect Independent	Exchangeable/Unspecified	Standard Deviation Robust	Naive
$\hat{\beta}_0$	-1.695	-1.696	0.090	0.062
$\hat{\beta}_S$	0.136	0.130	0.089	0.062
$\hat{\beta}_{A1}$	0.087	0.087	0.086	0.103
$\hat{\beta}_{A2}$	0.141	0.141	0.079	0.102
$\hat{\beta}_{A3}$	0.060	0.060	0.080	0.103

Looking at the parameter estimate $\hat{\beta}_S$ alone, the results seem to indicate a positive effect of mother's smoking on the probability of infection. The naive standard deviation also slightly supports this finding. However, the correct standard deviation, which is larger since smoking status is a time–independent covariate, shows that the effect of smoking is overinterpreted if analysis is falsely based on common ML estimation for independent observations. The effects $\hat{\beta}_{A1}$ and $\hat{\beta}_{A3}$ of ages 7 and 9 are slightly positive but nonsignificant. The positive effect $\hat{\beta}_{A2}$ of age 8 is slightly more significant. Comparing standard deviations, it is seen that robust estimates are smaller since age is a time–dependent covariate. However, $\hat{\beta}_{A4} = -\hat{\beta}_{A1} - \hat{\beta}_{A2} - \hat{\beta}_{A3} = -2.81$ of age 10 is negative and highly significant (stand. dev. = 0.94). This means that the probability of infection is significantly lower at age 10. It seems that children's constitution is more stable at this age. Interaction effects are nonsignificant on the basis of the estimated standard deviations. Therefore their contribution on the probability of infection should only be interpreted with great caution. This is also supported by the estimates in Table 6.9, obtained from fitting marginal models without interaction effects.   □

As in Section 3.5.2, we focussed on repeated binary outcomes and the so–called GEE1 approach. For extensions and modifications to other types of responses, in particular repeated multicategorical outcomes, and GEE2 or likelihood–based methods we refer readers to the end of Section 3.5.2. An excellent general introduction to the analysis of longitudinal data is given by Diggle, Liang and Zeger (1994). Additional aspects that arise in longitudinal studies are time–varying covariates and incomplete data due to drop–outs. Problems with time–varying covariates are discussed in Fitzmaurice et al. (1993), Pepe and Anderson (1994) and Robins, Rotnitzky and Zhao (1995). Effects of dropouts or missing data are studied by Fitzmaurice, Molenberghs and Lipsitz (1995), Lesaffre, Molenberghs and Dewulf (1996) and Robins et al. (1995). More recently, nonparametric approaches for flexible modelling and exploring mean and association structures have been suggested, see for example Heagerty and Zeger (1997) and Gieger (1997).

7
Random effects models

This chapter is concerned with random effects models for analyzing non–normal data that are assumed to be clustered or correlated. The clustering may be due to repeated measurements over time, as in longitudinal studies, or to subsampling the primary sampling units, as in cross–sectional studies. In each of these cases the data consist of repeated observations (y_{it}, x_{it}), $t = 1, \ldots, T_i$, for each individual or unit $i = 1, \ldots, n$, where y denotes a response variable of primary interest and x a vector of covariates. Typical examples include panel data, where the cluster–specific data

$$(y_i, x_i) = (y_{i1}, \ldots, y_{iT_i}, x_{i1}, \ldots, x_{iT_i})$$

correspond to a time series of length T_i, or large–scale health studies, where (y_i, x_i) represents the data of a primary sampling unit, say a hospital or a geographical region.

Let us consider the Ohio children data (Example 6.4), which are further investigated in this chapter. In this data set the objective is to investigate the effect of mother's smoking status on the presence/absence of respiratory infection. Since children's response is considered at ages 7, 8, 9 and 10 we have repeated measurements. In Example 6.4 a marginal fixed effect approach was used to analyze these data. However, susceptibility to respiratory infection differs in highly individual ways that cannot be traced back to only one covariate and time effects. If the focus is on *individual* risk probabilities, unobservable heterogeneity should be taken into account. Therefore one may consider a logit model with covariates "mother's smoking status," "age" and a random intercept that is allowed to vary randomly across children.

Most of the models considered in previous sections specify effects to be constant across clusters. However, when one is interested in individual risks, a better assumption is that the parameters, e.g., intercept and/or covariate effects, may vary across clusters. Models of the first type are called *population–averaged* whereas the latter approach is *cluster–* or *subject–specific*. Parameters in population–averaged models may be interpreted with respect to the marginal or population–averaged distribution. In subject–specific models the parameters refer to the influence of covariates for individuals. The distinction between these approaches is irrelevant for Gaussian outcomes, but it becomes important for categorical data since the mixture of subject–specific logistic models in general is not a logistic model for the population. For the comparison of these alternative model types, see also Zeger, Liang and Albert (1988), Neuhaus, Kalbfleisch and Hauck (1991) and Agresti (1993b).

The cluster– or subject–specific approach considered in this chapter is based on random effects: Cluster–specific effects are assumed to be independent and identically distributed according to a mixing distribution. In principle, cluster–specific effects can also be treated as fixed. In the fixed effects approach cluster–specific dummy variables have to be introduced such that each of the cluster–specific effects is treated as an unknown fixed parameter. However, the substantial number of parameters to be estimated often gives rise to serious problems, in particular when the cluster sizes T_i are small or moderate. Random effects approaches are much more parsimonious in the parameters and estimation can be carried out even when cluster sizes are small.

For normal data, linear random effects models are commonly used in theory and practice. Estimation of unknown parameters and of cluster–specific random effects is easier since the mixing distribution can be chosen to be conjugate to the data density so that the posterior distribution is available in closed form. Therefore random effects models for Gaussian data are only sketched in Section 7.1. In Section 7.2 random effects generalized linear models are introduced. The statistical analysis of random effects models for non–normal data becomes more difficult due to the lack of analytically and computationally tractable mixing distributions. Therefore, estimation methods based on full marginal and posterior distributions or on posterior means require repeated approximations of multidimensional integrals. In this chapter we report on work in this area that still is in progress. In Section 7.3 an estimation procedure based on posterior modes is given that has been used by Stiratelli, Laird and Ware (1984) and Harville and Mee (1984). In Section 7.4 integration–based two–step procedures are considered. First the fixed effects are estimated by maximizing the marginal likelihood based on Gauss–Hermite quadrature or Monte Carlo techniques. Then an empirical Bayes approach is used to estimate the random effects. Techniques of this type have been used by Hinde (1982), Anderson and Aitkin (1985), Anderson and Hinde (1988) and Jansen (1990). In Section

7.6 a marginal estimation approach due to Zeger, Liang and Albert (1988) is sketched.

7.1   Linear random effects models for normal data

For clustered Gaussian data linear random effects models provide an efficient tool for analyzing cluster–specific intercept and/or covariate effects. Although there is some development of nonlinear models (e.g., Lindstrom and Bates, 1990) we will consider only linear models. This section gives a short review on modelling and estimation techniques. More detailed expositions may be found in Hsiao (1986), Rao and Kleffe (1988), Dielman (1989), Jones (1993), Lindsey (1993).

7.1.1   Two–stage random effects models

Linear random effects models as considered by Laird and Ware (1982) extend the classical linear model for Gaussian response variables. We first define the general model and describe familiar special cases further later.

At the *first stage* the normal responses y_{it} are assumed to depend linearly on unknown population–specific effects β and on unknown cluster–specific effects b_i,

$$y_{it} = z'_{it}\beta + w'_{it}b_i + \varepsilon_{it}, \tag{7.1.1}$$

where z_{it} and w_{it} represent design vectors, w_{it} often being a subvector of z_{it}, and the disturbances ε_{it} being uncorrelated normal random variables with $E(\varepsilon_{it}) = 0$ and $\text{var}(\varepsilon_{it}) = \sigma^2$. The design vector z_{it} and thus w_{it} may depend on deterministic or stochastic covariates and on past responses, as in longitudinal studies, such that

$$z_{it} = z_{it}(x_{it}, y^*_{i,t-1}), \qquad \text{with} \quad y^*_{i,t-1} = (y_{i,t-1}, \dots, y_{i1}).$$

Furthermore, it is assumed that z_{it} contains the intercept term "1".

At the *second stage* the effects b_i are assumed to vary independently from one cluster to another according to a mixing distribution with mean $E(b_i) = 0$. Since the disturbances are Gaussian a normal mixing density with unknown covariance matrix $\text{cov}(b_i) = Q$ is commonly chosen,

$$b_i \sim N(0, Q), \qquad Q > 0, \tag{7.1.2}$$

and the sequences $\{\epsilon_{it}\}$ and $\{b_i\}$ are assumed to be mutually uncorrelated. Although it is possible to use a more general form of covariance structure for $\{\varepsilon_{it}\}$, e.g., first–order autoregression (see Jones, 1993), we will only consider the uncorrelated case. In matrix notation model (7.1.1) takes the form

$$y_i = Z_i\beta + W_ib_i + \varepsilon_i, \tag{7.1.3}$$

where $y_i = (y_{i1}, \ldots, y_{iT_i})$ is the response vector, $Z_i' = (z_{i1}, \ldots, z_{iT_i})$, $W_i' = (w_{i1}, \ldots, w_{iT_i})$ are design matrices and $\varepsilon_i = (\varepsilon_{i1}, \ldots, \varepsilon_{iT_i})$ is the vector of within cluster errors, $\varepsilon_i \sim N(0, \sigma_\varepsilon^2 I)$.

The assumption of Gaussian errors allows us to rewrite the model as a multivariate heteroscedastic linear regression model

$$y_i = Z_i \beta + \varepsilon_i^*, \tag{7.1.4}$$

where the multivariate disturbances $\varepsilon_i^* = (\varepsilon_{i1}^*, \ldots, \varepsilon_{iT_i}^*)$, with components $\varepsilon_{it}^* = w_{it}' b_i + \varepsilon_i$, are independent and normally distributed,

$$\varepsilon_i^* \sim N(0, V_i), \qquad \text{with} \quad V_i = I\sigma_\varepsilon^2 + W_i Q W_i'. \tag{7.1.5}$$

Equation (7.1.4), together with the covariance structure (7.1.5), represents the *marginal version* of the linear random effects model, where the conditioning on the cluster–specific effects b_i is dropped. Marginally, responses are correlated within clusters, as can be seen from (7.1.5).

In the following we discuss some special versions of linear random effects models, which are applied to analyse varying intercepts, (partially) varying slopes or covariate effects, or cluster–specific effects being hierarchically nested.

Random intercepts

In many empirical studies cluster–specific characteristics that possibly determine the response variable in addition to the observed covariates x_{it} have not been collected due to technical or economical circumstances. To take account of such cluster–specific effects a linear model with cluster–specific intercepts τ_i is appropriate,

$$y_{it} = \tau_i + x_{it}' \gamma + \varepsilon_{it}, \tag{7.1.6}$$

where the slope coefficients γ are constant and the intercepts τ_i are assumed to be i.i.d. with unknown parameters $E(\tau_i) = \tau$ and $\text{var}(\tau_i) = \sigma^2$. The unobservable deviations between the population mean τ and the cluster–specific realizations τ_i may be interpreted as effects of omitted covariates. Assuming normality one immediately gets a linear random effects model of the form (7.1.1), (7.1.2) with

$$\beta' = (\tau, \gamma'), \quad z_{it}' = (1, x_{it}'), \quad w_{it} = 1, \quad b_i = (\tau_i - \tau) \sim N(0, \sigma^2).$$

Note that the random–intercept model also takes into account intracluster correlation of the Gaussian outcomes. From (7.1.5) it is easily seen that the correlation between y_{it} and y_{is} is given by

$$\rho_{ts} = \rho = \frac{\sigma^2}{\sigma_\varepsilon^2 + \sigma^2}, \qquad t \neq s.$$

Random intercept models are also called *error components* or *variance components models* (see, e.g., Hsiao, 1986). The primary objective is to analyse the variance components σ_ε^2 and σ^2, which stand for variability within (resp. between) the clusters.

Random slopes

Random intercept models do not alleviate the restrictive assumption that the slope coefficients are equal for each observation. Varying slope coefficients arise in particular in longitudinal studies, where intercept and slope coefficients are specific to each time series. To take into account such parameter heterogeneity the random intercept model (7.1.6) can be extended, treating not only the intercept but all coefficients as random. The corresponding model has the form

$$y_{it} = \tau_i + x'_{it}\gamma_i + \varepsilon_{it}.$$

Note that in the longitudinal setting the ith cluster corresponds to a time series. Then effects of past responses are also allowed to vary from time series to time series.

Suppose now that the regression coefficients $\beta_i = (\tau_i, \gamma_i)$ vary independently across clusters according to a normal density with mean $E(\beta_i) = \beta$ and covariance matrix $\mathrm{cov}(\beta_i) = Q$ being positive definite,

$$\beta_i \sim N(\beta, Q).$$

The mean β can be interpreted as the population–averaged effect of the regressors z_{it}. The covariance matrix Q contains variance components indicating the parameter heterogeneity of the population as well as covariance components representing the correlation of single components of β_i. Rewriting the coefficients β_i as

$$\beta_i = \beta + b_i, \qquad b_i \sim N(0, Q),$$

and assuming mutual independence of the sequences $\{b_i\}$ and $\{\epsilon_{it}\}$, one obtains a linear random effects model with

$$z'_{it} = w'_{it} = (1, x'_{it}).$$

Models where all coefficients are assumed to vary randomly over clusters are also called *random coefficient regression models*, see, e.g., Hsiao (1986). Sometimes, however, assuming that some coefficients are cluster–specific is less realistic than the assumption that some coefficients are constant across clusters. If β_{i1} denotes the cluster–specific coefficients and β_{i2} the remaining coefficients are constant across clusters, the parameter vector β_i can be partitioned into $\beta_i = (\beta_{i1}, \beta_{i2})$ with $\beta_{i2} = \beta_2$ for all i. The design vector $z_{it} = (1, x_{it})$ also has to be rearranged according to the structure

$z_{it} = (z_{it1}, z_{it2})$. Then the probability model for β_i can be expressed by a multivariate normal density with singular covariance matrix,

$$\beta_i = \begin{bmatrix} \beta_{i1} \\ \beta_{i2} \end{bmatrix} \sim N\left(\begin{bmatrix} \beta_1 \\ \beta_2 \end{bmatrix}, \begin{bmatrix} Q & 0 \\ 0 & 0 \end{bmatrix} \right), \qquad (7.1.7)$$

where the submatrix Q is assumed to be positive definite. Rewriting the coefficients β_i as

$$\beta_i = \beta + a_i, \qquad \text{with } a_i' = (b_i', 0'), \qquad b_i \sim N(0, Q),$$

one gets again a random effects model with

$$z_{it}' = (z_{it1}', z_{it2}'), \qquad w_{it} = z_{it1}. \qquad (7.1.8)$$

Due to the mixing of "fixed" and "random" coefficients, models of this type are also called linear mixed models.

Multilevel models

The linear random effects models considered so far are based on a single level structure. In some applications, however, clustering occurs on more than one level and the clusters are hierarchically nested. For example, in a standard application in educational research there are classrooms with varying numbers of students, and each student has a pair of scores, one at the beginning and the other at the conclusion of a specific course. In this case we have a two–level structure: At the first level the clusters correspond to $j = 1, \dots, n$ classrooms, each having $i = 1, \dots, I_j$ students and at the second level the clusters are formed by students, each having $t = 1, \dots, T_i$ observations. Models that take into account unobservable classroom–specific effects and student–specific effects have hierarchically nested clusters. Models of this type have been considered by Aitkin and Longford (1986), among others.

7.1.2 Statistical inference

The primary objective in analyzing linear random effects models is to estimate the parameters β, σ_ε^2 and Q, and the random effects b_i. Estimation is often based on a frequentist approach, where β, σ_ε^2 and Q are treated as "fixed" parameters. As an alternative, Bayesian estimation methods can be applied, where β, σ_ε^2 and Q are treated as random variables with some prior distribution (see, e.g., Broemeling, 1985). In the following we only consider the "empirical Bayes" approach. For a better understanding we first treat the more hypothetical situation, where the variance–covariance components σ_ε^2 and Q are known. For simplicity only single–level models are considered. Extensions to multilevel models can be derived, see, for example, Goldstein (1986), Longford (1987), Goldstein (1989).

Known variance–covariance components

First let the covariance matrix V_i (i.e., σ_ε^2 and Q) be known. Then estimation of the parameter β is usually based on the marginal model defined by (7.1.4) and (7.1.5). For this model the maximum likelihood estimator (MLE) is equal to the weighted–least–squares solution

$$\hat{\beta} = \left(\sum_{i=1}^n Z_i' V_i^{-1} Z_i \right)^{-1} \sum_{i=1}^n Z_i' V_i^{-1} y_i. \qquad (7.1.9)$$

The MLE can also be shown to be a best linear unbiased estimator (BLUE), see, e.g., Harville (1977) and Rao and Kleffe (1988).

Estimation of the unobservable random effects b_i is based on the posterior density of b_i, given the data $Y = (y_1, \ldots, y_n)$. Due to the normality and linearity assumptions the posterior of b_i is also normal. Moreover, the posterior depends only on y_i since the stochastic terms ϵ_{it} and b_i, $t = 1, \ldots, T_i$, $i = 1, \ldots, n$, are assumed to be completely independent. Following Bayesian arguments the optimal point estimator is the posterior mean

$$\hat{b}_i = E(b_i \mid y_i) = QW_i' V_i^{-1} \left(y_i - Z_i \hat{\beta} \right) \qquad (7.1.10)$$

which also can be shown to be a BLUE in the present setting (see, e.g., Harville, 1976; Rao and Kleffe, 1988). Since the posterior is normal, posterior mean and posterior mode coincide. Therefore, the estimators $\hat{b}_i$ are also obtained by maximizing the log–posterior density with respect to $b_1, \ldots, b_n$.

Unknown variance–covariance components

The estimating equations (7.1.9) and (7.1.10) are based on known variance–covariance components. Let $\theta = (\sigma_\varepsilon^2, Q)$ denote the vector of variance and covariance parameters. If θ is unknown it has to be replaced by a consistent estimate $\hat{\theta}$. Harville (1977), among others, distinguishes maximum likelihood estimation (MLE) and restricted maximum likelihood estimation (RMLE). Some authors also suggest minimum norm quadratic unbiased estimation (MINQUE) or minimum variance quadratic unbiased estimation (MIVQUE), which, however, are less in use since such estimates may be negative (see, e.g., Hsiao, 1986; Rao and Kleffe, 1988).

The MLEs for θ are obtained together with those for β by maximizing the marginal log–likelihood based on the marginal model defined by (7.1.4) and (7.1.5) with respect to β and θ. This is equivalent to the minimization of

$$l(\beta, \theta) = \sum_{i=1}^n \left[\log |V_i| + (y_i - Z_i\beta)' V_i^{-1} (y_i - Z_i\beta) \right].$$

The criticism of MLE for θ is that this estimator does not take into account the loss in degrees of freedom resulting from the estimation of β.

The larger the dimension of β the larger is the bias of the MLE for θ. RM-LEs generally yield a smaller bias. The idea of an RMLE is to construct a likelihood that depends only on θ. Such a likelihood can be derived using a Bayesian formulation of the linear random effects model, where the parameters β are considered random variables having a vague or totally flat prior distribution, for example,

$$\beta \sim N(\beta^*, \Gamma), \quad \text{with} \quad \Gamma \to \infty \quad \text{or} \quad \Gamma^{-1} \to 0,$$

so that the prior density of β is just a constant. The choice of β^* is immaterial since the covariance matrix Γ becomes infinite. Maximizing the limiting (as $\Gamma^{-1} \to 0$) marginal log–likelihood yields the RMLE for θ (see Harville, 1976). The numerical calculation of MLEs and RMLEs for θ is much more complicated than the estimation of β since the likelihoods depend nonlinearly on θ. Thus, iterative methods are required. Harville (1977) considers Newton–Raphson and scoring algorithms, and Laird and Ware (1982) recommend several versions of the EM–algorithm. Since the EM–algorithm will also be used for generalized linear models with random coefficients, a version of this algorithm used for RMLE is sketched in the following.

Some details on its derivation are given later. The resulting EM–algorithm jointly estimates $\delta = (\beta, b_1, \ldots, b_n)$ and $\theta = (\sigma_\varepsilon^2, Q)$ as follows:

1. Choose starting values $\theta^{(0)} = (\sigma_{\varepsilon(0)}^2, Q^{(0)})$.

For $p = 0, 1, 2, \ldots$

2. Compute $\hat{\delta}^{(p)} = (\hat{\beta}^{(p)}, \hat{b}_1^{(p)}, \ldots, \hat{b}_n^{(p)})$ from (7.1.9) and (7.1.10) with variance–covariance components replaced by their current estimates $\theta^{(p)} = (\sigma_\varepsilon^{2\,(p)}, Q^{(p)})$, together with current residuals $e_i^{(p)} = y_i - Z_i\hat{\beta}^{(p)} - W_i\hat{b}_i^{(p)}$, $i = 1, \ldots, n$, and posterior covariance matrices $\text{cov}(b_i|y_i; \theta^{(p)})$, $\text{cov}(\varepsilon_i|y_i; \theta^{(p)})$.

3. EM–step: Compute $\theta^{(p+1)} = (\sigma_\varepsilon^{2\,(p+1)}, Q^{(p+1)})$ by

$$\hat{\sigma}_\varepsilon^{2\,(p+1)} = \frac{1}{T_1 + \ldots + T_n} \sum_{i=1}^n \left[(e_i^{(p)})' e_i^{(p)} + \text{tr cov}(\varepsilon_i|y_i; \theta^{(p)})\right] \quad (7.1.11)$$

$$Q^{(p+1)} = \frac{1}{n} \sum_{i=1}^n \left[b_i^{(p)}(b_i^{(p)})' + \text{cov}(b_i|y_i; \theta^{(p)})\right]. \quad (7.1.12)$$

The posterior covariance matrices can be obtained from the joint normal density defined by the model. For details and alternative versions see Laird and Ware (1982) and Jones (1993).

However, RMLEs should not be used blindly, since RMLEs may have larger mean square errors than MLEs as has been pointed out by Corbeil and Searle (1976). Further tools of statistical inference on β and θ are less well developed. An approximation for the distribution of $\hat{\beta}$ that takes into account the additional variability due to the estimation of θ has been proposed by Giesbrecht and Burns (1985). Approximative distributions of MLEs and RMLEs for θ are considered by Rao and Kleffe (1988).

Derivation of the EM–algorithm*

Indirect maximization of the marginal density by an EM–algorithm starts from the joint log–density of observable data $Y = (y_1, \ldots, y_n)$ and unobservable effects $\delta = (\beta, b_1, \ldots, b_n)$; see Appendix A3. Since a flat prior is assumed for β, the joint log–likelihood is

$$\log f(Y, \delta; \theta) = \log f(Y|\delta; \sigma_\varepsilon^2) + \log f(b_1, \ldots, b_n; Q),$$

where the first term is determined by the first stage (7.1.1) resp. (7.1.3) of the model, and the second term by the second stage (7.1.2), the prior for $b_1, \ldots, b_n$. From the model assumptions one obtains, up to constants,

$$S_1(\sigma_\varepsilon^2) = -\frac{1}{2} \sum_{i=1}^{n} T_i \log \sigma_\varepsilon^2 - \frac{1}{2\sigma_\varepsilon^2} \sum_{i=1}^{n} \varepsilon_i' \varepsilon_i$$

$$S_2(Q) = -\frac{n}{2} \log \det(Q) - \frac{1}{2} \sum_{i=1}^{n} b_i' Q^{-1} b_i$$

$$= -\frac{n}{2} \log \det(Q) - \frac{1}{2} \sum_{i=1}^{n} \operatorname{tr}(Q^{-1} b_i b_i')$$

for the first and second terms. The E–step yields

$$M(\theta|\theta^{(p)}) = E\{S_1(\sigma_\varepsilon^2)|y; \theta^{(p)}\} + E\{S_2(Q)|y; \theta^{(p)}\}$$

with

$$E\{S_1(\sigma_\varepsilon^2)|y; \theta^{(p)}\} = -\frac{1}{2} \sum_{i=1}^{n} T_i \log \sigma_\varepsilon^2$$

$$-\frac{1}{2\sigma_\varepsilon^2} \sum_{i=1}^{n} [(e_i^{(p)})' e_i^{(p)} + \operatorname{tr} \operatorname{cov}(\varepsilon_i|y_i; \theta^{(p)})] \quad (7.1.13)$$

$$E\{S_2(Q)|y; \theta^{(p)}\} = -\frac{n}{2} \log \det(Q)$$

$$-\frac{1}{2} \sum_{i=1}^{n} \operatorname{tr}[Q^{-1} b_i^{(p)} (b_i^{(p)})'] + \operatorname{cov}(b_i|y_i; \theta^{(p)}). \quad (7.1.14)$$

Differentiation with respect to σ_ε^2 and Q using matrix calculus setting derivatives to zero and solving for σ_ε^2 and Q yields (7.1.11) and (7.1.12).

TABLE 7.1. Bitterness of wine data (Randall, 1989)

Judge	LOW TEMPERATURE				HIGH TEMPERATURE			
	No Contact		Contact		No Contact		Contact	
	Bottle 1	Bottle 2	Bottle 1	Bottle 2	Bottle 1	Bottle 2	Bottle 1	Bottle 2
1	2	3	3	4	4	4	5	5
2	1	2	1	3	2	3	5	4
3	2	3	3	2	5	5	4	4
4	3	2	3	2	3	2	5	3
5	2	3	4	3	3	3	3	3
6	3	2	3	2	2	4	5	4
7	1	1	2	2	2	3	2	3
8	2	2	2	3	3	3	3	4
9	1	2	3	2	3	2	4	4

7.2 Random effects in generalized linear models

Let us consider two simple examples with categorical responses.

Example 7.1: Ohio children data (Example 6.4 continued)
As already mentioned at the beginning of this section in the study of Ohio children one has the dichotomous response presence/absence of respiratory infection. The children are measured repeatedly (ages 7, 8, 9, 10) and the covariate of interest is the smoking behavior of mothers.   □

Example 7.2: Bitterness of white wines
In a study on the bitterness of white wine (Randall, 1989) it is of interest whether treatments that can be controlled during pressing the grapes influence the bitterness of wines. The two factors considered are the temperature and the admission of contact with skin when pressing the grapes. Both factors are given in dichotomous form. For each factor combination two bottles of white wine were chosen randomly and the bitterness of each of the $t = 1, \ldots, 8$ bottles was classified on a 5–categorical ordinal scale (1= nonbitter, ..., 5=very bitter) by $i = 1, \ldots, 9$ professional judges. The 5–categorical responses y_{it} are given in Table 7.1. Since judges cannot be expected to have the same sensitivity to bitterness an effect of judges should be incorporated. This may be done by allowing the judges to have shifted thresholds in a cumulative model.   □

For non–normal data we cannot expect to get a closed form like $y_i = Z_i\beta + \varepsilon_i^*$ as we had in (7.1.4) for normal data. For the introduction of random effects models in this case it is useful to reconsider the linear random effects model as a two–stage model:

At the first stage the observations y_{it} are treated as conditionally independent and normally distributed random variables, given the effects b_i,

$$y_{it}|b_i \sim N(\mu_{it}, \sigma_\varepsilon^2), \tag{7.2.1}$$

where the *conditional* mean $\mu_{it} = E(y_{it}|b_i)$ is given by $\mu_{it} = z_{it}'\beta + w_{it}'b_i$.

At the second stage the cluster–specific effects b_i are assumed to be i.i.d. with $b_i \sim N(0, Q)$. If the covariates x_{it} are stochastic or the responses y_{it} depend on past observations $y_{it}^* = (y_{i,t-1}, ..., y_{i1})$, as in longitudinal studies, the model is to be understood conditionally, i.e., (7.2.1) is the conditional density of y_{it}, given x_{it}, y_{it}^*, b_i. However, to avoid unnecessary inflation of notation the possible dependence on x_{it} or y_{it}^* is supressed in the following.

For non–normal responses y_{it} like the binary response infection (Example 7.1) or the multicategorical response bitterness (Example 7.2), model (7.2.1) can be extended in the following way:

At the first stage it is assumed that the conditional density $f(y_{it}|b_i)$ is of the simple uni– or multivariate exponential family type with conditional mean

$$\mu_{it} = E(y_{it}|b_i) = h(\eta_{it}), \quad \text{with} \quad \eta_{it} = Z_{it}\beta + W_{it}b_i, \tag{7.2.2}$$

where h is one of the response functions in Sections 2.1, 3.2 or 3.3, and η_{it} is the linear predictor. The design matrix Z_{it} is a function of the covariates x_{it}, and possibly past responses y_{it}^* as considered in the previous chapters. Intercepts, covariates and/or past responses whose effects are assumed to vary across the $i = 1, .., n$ clusters are collected in the design matrix W_{it}. Though it is not neccessary, W_{it} often is a submatrix of Z_{it}.

The second stage of linear random effects models is retained: Cluster–specific effects b_i are assumed to be independent and (if b_i in (7.2.2) is not restricted) normally distributed with mean $E(b_i) = 0$ and unknown covariance matrix $\text{cov}(b_i) = Q$, where Q has to be positive definite. In principle, each parametric density f having mean $E(b_i) = 0$ and unknown parameters Q is admissible so that the second stage of the generalized linear random effects model is specified more generally by independent densities

$$p(b_i; Q), \quad i = 1, ..., n. \tag{7.2.3}$$

As an additional assumption conditional independence of observations within and between clusters is required, i.e.,

$$f(Y|B; \beta) = \prod_{i=1}^{n} f(y_i|b_i; \beta), \quad \text{with} \quad f(y_i|b_i; \beta) = \prod_{t=1}^{T_i} f(y_{it}|b_i; \beta), \tag{7.2.4}$$

where $Y = (y_1, ..., y_n)$ and $B = (b_1, ..., b_n)$ represent the whole set of responses and random effects. If covariates are stochastic or responses depend on past observations densities are to be understood conditional on these quantities. Marginally, where the conditioning on b_i is dropped, observations *within* clusters are dependent. Observations on *different* clusters are conditionally and marginally independent.

Example 1 (binary logistic model)
For univariate responses y_{it} the design matrices Z_{it} and W_{it} reduce to design vectors z_{it} and w_{it} that can be constructed along the lines of Section 7.1.1. Therefore, univariate generalized linear models imposing no restrictions on the predictor η are easily extended to include random effects. Let us consider a binary logistic model with intercepts varying according to a normal distribution,

$$\pi_{it} = P(y_{it} = 1|b_i) = \exp(\eta_{it})/(1 + \exp(\eta_{it})), \qquad (7.2.5)$$

with

$$\eta_{it} = z'_{it}\beta + b_i, \quad b_i \sim N(0, \sigma^2),$$

where the design vector z_{it} is assumed to contain the intercept term "1."

For Example 7.1 with random threshold the design for smoking mother and a child of seven years is given by

$$\eta_{it} = [1, 1, 1, 0, 0]\beta + [1]b_i$$

where β contains the fixed effects "constant," "smoking status," "age 7 years," "age 8 years," "age 9 years."   □

The mixed–effects logistic model (7.2.5) is cluster– or subject–specific in contrast to population–averaged models where the predictor has the simple form $\eta_{it} = z'_{it}\beta$. If the subject–specific model holds the covariate effects measured by the population–averaged model with $\eta_{it} = z'_{it}\beta$ are closer to zero than the covariate effects of the underlying subject–specific model (see Neuhaus, Kalbfleisch and Hauck, 1991). The binary model (7.2.5) has been considered by Anderson and Aitkin (1985) in analysing interviewer variability. More complex binary logistic models involving multinormally distributed random effects have been studied by Stiratelli, Laird and Ware (1984) for longitudinal data. Extensions to multilevel structures can be found in Preisler (1989) and Wong and Mason (1985). Log–linear Poisson models with a normally distributed intercept are treated by Hinde (1982) and Brillinger and Preisler (1983).

However, generalized linear random effects approaches are also suited for analysing clustered *multicategorical* data. Let $y_{it} = (y_{it1}, ..., y_{itk})$ denote the k–categorical response of the tth observation in cluster i, where $y_{itj} = 1$ if category j is observed and $y_{itj} = 0$ otherwise, $j = 1, ..., k$. Then multinomial random effects models for nominal or ordered responses are completely specified by conditional probabilities $\pi_{it} = (\pi_{it1}, ..., \pi_{itq}), q = k - 1$, which depend on population–specific effects and cluster–specific effects,

$$\pi_{it} = h(\eta_{it}), \quad \eta_{it} = Z_{it}\beta + W_{it}b_i.$$

The structure of the design matrix W_{it} depends on the assumed parameter heterogeneity.

Example 2 (multinomial logistic model)
A multinomial logistic random effects model that takes into account the effect of unobservable characteristics being cluster– and category–specific is given by

$$\pi_{itj} = \exp(\eta_{itj})/(1 + \sum_{m=1}^{q} \exp(\eta_{itm})),$$

with

$$\eta_{itj} = \alpha_{ij} + x'_{it}\gamma_j, \quad j = 1, ..., q.$$

Assuming that the effects $\alpha_i = (\alpha_{i1}, ..., \alpha_{iq})$ vary independently from one cluster to another according to a normal distribution with mean $E(\alpha_i) = \tilde{\alpha}$ and covariance matrix $\text{cov}(\alpha_i) = Q$,

$$\alpha_i \sim N(\tilde{\alpha}, Q),$$

the linear predictors can be rewritten as

$$\eta_{itj} = \tilde{\alpha}_j + b_{ij} + x'_{it}\gamma_j, \quad \text{where} \quad b_{ij} = \alpha_{ij} - \tilde{\alpha}_j.$$

Defining $\beta = (\tilde{\alpha}_1, \gamma_1, ..., \tilde{\alpha}_q, \gamma_q)$ and $b_i = (b_{i1}, ..., b_{iq}) \sim N(0, Q)$ one obtains a conditional multinomial model of the form (7.2.2) with

$$Z_{it} = \begin{bmatrix} 1 & x'_{it} & & & 0 \\ & & 1 & x'_{it} & \\ & & & \ddots & \\ 0 & & & & 1 & x'_{it} \end{bmatrix}, \quad W_{it} = \begin{bmatrix} 1 & & 0 \\ & \ddots & \\ 0 & & 1 \end{bmatrix}.$$

This model can also be developed by applying the random utility principle that was proposed in Section 3.3. Then the latent utility variable has to be based on a linear random effects model that is similar to (7.1.5). Inclusion of cluster–specific slopes and nested cluster–specific effects can be performed along the lines of Section 7.1.1. □

Example 3 (ordered response)
In the bitterness–of–wine example judges respond on a five–point scale that may be considered ordinal. Random effects versions of *ordinal* response models (see Sections 3.3 and 3.4) can be derived in the same way as before. However, one has to be cautious with cumulative models. Such

models are based on thresholds being constant for all observations. To allow for variation over clusters the thresholds may be parameterized by linear combinations of the observed covariates, as stated in Section 3.3.2. Sometimes, however, the heterogeneity of thresholds cannot be explained adequately by the observed covariates since there exist some unobservable cluster–specific characteristics. To take into account unobservable threshold heterogeneity cumulative approaches with randomly varying thresholds may be appropriate. If F denotes a known distribution function, e.g., the logistic, conditional response probabilities are given by

$$\pi_{it1} = F(\theta_{i1} + x'_{it}\gamma), \quad \pi_{itr} = F(\theta_{ir} + x'_{it}\gamma) - \pi_{it,r-1}, \quad r = 2, ..., q,$$

with cluster–specific thresholds being ordered,

$$-\infty = \theta_{i0} < \theta_{i1} < ... < \theta_{iq} < \theta_{im} = \infty, \qquad (7.2.6)$$

and covariate effects γ being constant over clusters. The simplest random effects model is given by cluster–specific shifting of thresholds where

$$\theta_{ir} = \theta_r + b_i, \qquad b_i \sim N(0, \sigma^2).$$

Then the linear predictor has the form

$$\eta_{itr} = \theta_r + b_i + x'_{it}\gamma.$$

The matrices in the linear form $\eta_{it} = Z_{it}\beta + W_{it}b_i$ are given by

$$Z_{it} = \begin{pmatrix} 1 & & & x'_{it} \\ & \ddots & & \vdots \\ & & 1 & x'_{it} \end{pmatrix}, \qquad W_{it} = \begin{pmatrix} 1 \\ \vdots \\ 1 \end{pmatrix}$$

where $\beta = (\theta_1, ..., \theta_q, \gamma)$.

An extended version assumes all the thresholds to be cluster–specific. The simplest assumption is the normal distribution $\theta_i = (\theta_{i1}, ..., \theta_{iq}) \sim N(\theta_i, Q)$. However, if the threshold means θ_r are not well separated and the variances are not small enough, the ordering (7.2.6) may be violated and numerical problems will occur in the estimation procedure. To overcome this problem the thresholds may be reparameterized as already stated in Section 3.3.3 using

$$\alpha_{i1} = \theta_{i1}, \quad \alpha_{ij} = \log(\theta_{ij} - \theta_{i,j-1}), \quad j = 2, ..., q.$$

The reparameterized thresholds $\alpha_i = (\alpha_{i1}, ..., \alpha_{iq})$ may vary unrestricted in $\mathbb{R}^q$ according to a normal distribution with mean $E(\alpha_i) = \tilde{\alpha}$ and covariance matrix $\text{cov}(\alpha_i) = Q$. The conditional response probabilities are now

$$\pi_{it1} = F(\alpha_{i1} + x'_{it}\gamma)$$

$$\pi_{itj} = F(\alpha_{i1} + \sum_{m=2}^{j} \exp(\alpha_{im}) + x'_{it}\gamma) - \pi_{it,j-1}, \quad j = 2, ..., q.$$

Rewriting the reparameterized thresholds as

$$\alpha_i = \tilde{\alpha} + b_i, \quad \text{with} \quad b_i \sim N(0, Q),$$

and defining $\beta = (\tilde{\alpha}, \gamma)$ yields a cumulative logistic random effects model, where the design matrices are given by

$$Z_{it} = \begin{bmatrix} 1 & & & 0 & x'_{it} \\ & \ddots & & & 0 \\ & & \ddots & & \vdots \\ 0 & & & 1 & 0 \end{bmatrix}, \quad W_{it} = \begin{bmatrix} 1 & & 0 \\ & \ddots & \\ 0 & & 1 \end{bmatrix}.$$

It should be noted that a simple shifting of thresholds $\theta_{ir} = \theta_r + b_i$ after reparameterization corresponds to the adding of a random effect to α_{i1} yielding $\alpha_{i1} = \alpha_1 + b_i$, $b_i \sim N(0, \sigma^2)$.

Cumulative random effects models allowing for varying slopes are obtained in the usual way, e.g., by using a linear random effects model as latent variable model (see, e.g., Harville and Mee, 1984; Jansen, 1990). □

7.3   Estimation based on posterior modes

In this section we consider a procedure for the simultaneous estimation of the parameters β, Q and the $i = 1, \ldots, n$ random effects b_i that is based on posterior modes, thus avoiding integration. The procedure corresponds to an EM–type algorithm, where posterior modes and curvatures are used instead of posterior means and covariances. Such an approach was already sketched in Section 2.3.2 for analyzing Bayes models. For binary logistic models with Gaussian random effects it has been used by Stiratelli, Laird and Ware (1984). See also Harville and Mee (1984) for the application to cumulative random effects models and Wong and Mason (1985), who adopted the approach to logistic regression models with nested random effects structures. The EM–type algorithm may be derived from an EM–algorithm that maximizes the marginal likelihood involving Q only. In linear random effects models the resulting estimators $\hat{Q}$ correspond to RMLEs (see Section 7.1.2). Estimation of the parameters β and the random effects b_i is embedded in the E–step of the EM–type algorithm by maximizing the joint posterior density with Q fixed at $\hat{Q}$.

Since the complete estimation procedure is based on a marginal likelihood, which only depends on the variance–covariance components Q, a vague or flat prior density with covariance matrix $\Gamma \to \infty$ is assigned to the parameters β (see Section 7.1.2). Collecting the population and cluster–specific effects into

$$\delta' = (\beta', b'_1, \ldots, b'_n)$$

the limiting (as $\Gamma \to \infty$) prior density of δ satisfies the proportionality

$$p(\delta; Q) \propto \prod_{i=1}^{n} p(b_i; Q).$$

7.3.1   Known variance–covariance components

For simplicity we first consider the estimation of δ for known variance–covariance components Q. Estimation of δ is based on its posterior density, given the data $Y = (y_1, .., y_n)$. Applying Bayes theorem and using the independence assumption (7.2.4) the posterior is given by

$$f(\delta|Y; Q) = \frac{\prod_{i=1}^{n} f(y_i|b_i, \beta) \prod_{i=1}^{n} p(b_i; Q)}{\int \prod_{i=1}^{n} f(y_i|b_i, \beta) p(b_i; Q) db_1 \cdot \ldots \cdot db_n d\beta}. \qquad (7.3.1)$$

An optimal point estimator for δ is the posterior mean that was used to estimate random effects for Gaussian data; see equation (7.1.10). However, due to the lack of analytically and computationally tractable random effects posterior densities, numerical integration or Monte Carlo methods are required in general. In the present setting, these methods imply an enormous numerical effort since the integral structure in (7.3.1) is nested. Therefore *posterior mode estimation* is used. Posterior modes and posterior curvatures are obtained by maximizing the logarithm of the posterior (7.3.1). Because of the proportionality

$$f(\delta|Y; Q) \propto \prod_{i=1}^{n} f(y_i|b_i, \beta) \prod_{i=1}^{n} p(b_i; Q)$$

maximization of (7.3.1) is equivalent to maximizing the log–posterior

$$\sum_{i=1}^{n} \log f(y_i|b_i, \beta) + \sum_{i=1}^{n} \log p(b_i; Q) \qquad (7.3.2)$$

with respect to δ. In general, computation has to be carried out iteratively, e.g., by Fisher scoring. As an example, we consider a normal random effects density $p(b_i; Q)$ with positive–definite covariance matrix Q so that (7.3.2) corresponds to

$$l(\delta) = \sum_{i=1}^{n} \log f(y_i|b_i, \beta) - \frac{1}{2} \sum_{i=1}^{n} b_i' Q^{-1} b_i \qquad (7.3.3)$$

after dropping terms that are constant with respect to b_i. Criterion (7.3.3) can be interpreted as a penalized log–likelihood for random effects deviations from the population mean zero. Similary as for spline smoothing in generalized linear models (Section 5.3), maximization of the penalized log–likelihood $l(\delta)$ is carried out by Fisher scoring. In each scoring step, efficient use may be made of the block structure of (expected negative) second derivatives of $l(\delta)$. Details are given in Section 7.3.3.

7.3.2   Unknown variance–covariance components

Generally the variance–covariance components Q are unknown. Estimation of Q is based on an EM–type algorithm, which can be deduced from an EM–algorithm that maximizes the marginal log–likelihood

$$l(Q) = \log \int \prod_{i=1}^{n} f(y_i|\beta, b_i) p(b_i; Q) db_1 \cdot \ldots \cdot db_n d\beta \qquad (7.3.4)$$

indirectly. In general, direct maximization of (7.3.4) is cumbersome due to the nested high–dimensional integral structure, which can be solved analytically only for special situations, e.g., linear random effects models. Indirect maximization of (7.3.4) by an EM–algorithm starts from the joint log–density

$$\log f(Y, \delta; Q) = \log f(Y|\delta) + \log p(\delta; Q)$$

of the observable data Y and the unobservable effects δ. For details of the EM–algorithm see Appendix A3. In the $(p+1)$th cycle of the algorithm the E–step consists of computing

$$M(Q|Q^{(p)}) = E\{\log f(Y, \delta; Q)|Y; Q^{(p)}\}$$

which denotes the conditional expectation of the complete data log–density, given the observable data Y and the estimate $Q^{(p)}$ from the previous cycle. Because now $Q^{(p)}$ from the previous cycle is known we have essentially the case of known variance–covariance components considered in Section 7.3.1. To avoid integrations necessary for computing conditional expectations of random effects (or quadratic forms of them) appearing in $\log p(\delta; Q)$, conditional expectations are replaced by conditional modes obtained from maximizing (7.3.3) for $Q = Q^{(p)}$. The resulting EM–type algorithm is described in detail in the following subsection.

7.3.3   Algorithmic details*

Fisher scoring for given variance–covariance components

Let us first consider Fisher–scoring maximization of the penalized log–likelihood $l(\delta)$ given by (7.3.3), with Q known or given. The components of the score function $s(\delta) = \partial l/\partial \delta = (s_\beta, s_1, \ldots, s_n)$ are then given by

$$s_\beta = \frac{\partial l(\delta)}{\partial \beta} = \sum_{i=1}^{n} \sum_{t=1}^{T_i} Z'_{it} D_{it}(\delta) \Sigma_{it}^{-1}(\delta)(y_{it} - \mu_{it}(\delta)),$$

$$s_i = \frac{\partial l(\delta)}{\partial b_i} = \sum_{t=1}^{T_i} W'_{it} D_{it}(\delta) \Sigma_{it}^{-1}(\delta)(y_{it} - \mu_{it}(\delta)) - Q^{-1} b_i, \quad i = 1, \ldots, n,$$

with $D_{it}(\delta) = \partial h(\eta_{it})/\partial \eta$, $\Sigma_{it}(\delta) = \text{cov}(y_{it}|\beta, b_i)$, and $\mu_{it}(\delta) = h(\eta_{it})$.

The expected conditional (Fisher) information matrix $F(\delta) = \text{cov } s(\delta)$ is partitioned into

$$
F(\delta) = \begin{bmatrix}
F_{\beta\beta} & F_{\beta 1} & F_{\beta 2} & \cdots & F_{\beta n} \\
F_{1\beta} & F_{11} & & & 0 \\
F_{2\beta} & & F_{22} & & \\
\vdots & & & \ddots & \\
F_{n\beta} & 0 & & & F_{nn}
\end{bmatrix}
$$

with

$$
F_{\beta\beta} = -E\left(\frac{\partial^2 l(\delta)}{\partial\beta\partial\beta'}\right) = \sum_{i=1}^{n}\sum_{t=1}^{T_i} Z'_{it} D_{it}(\delta)\Sigma_{it}^{-1}(\delta)D'_{it}(\delta)Z_{it},
$$

$$
F_{\beta i} = F'_{i\beta} = -E\left(\frac{\partial^2 l(\delta)}{\partial\beta\partial b'_i}\right) = \sum_{t=1}^{T_i} Z'_{it} D_{it}(\delta)\Sigma_{it}^{-1}(\delta)D'_{it}(\delta)W_{it},
$$

$$
F_{ii} = -E\left(\frac{\partial^2 l(\delta)}{\partial b_i\partial b'_i}\right) = \sum_{t=1}^{T_i} W'_{it} D_{it}(\delta)\Sigma_{it}^{-1}(\delta)D'_{it}(\delta)W_{it} + Q^{-1}.
$$

The posterior mode estimator $\hat{\delta}$ that satisfies the equation $s(\delta) = 0$ can be calculated by the Fisher–scoring algorithm

$$
\delta^{(k+1)} = \delta^{(k)} + F^{-1}\left(\delta^{(k)}\right)s\left(\delta^{(k)}\right), \tag{7.3.5}
$$

where k denotes an iteration index. However, the dimensionality of $F(\delta)$ may cause some problems. These can be avoided due to the partitioned structure of $F(\delta)$. Since the lower right part of $F(\delta)$ is block diagonal the algorithm (7.3.5) can be reexpressed more simply as

$$
F_{\beta\beta}^{(k)}\Delta\beta^{(k)} + \sum_{i=1}^{n} F_{\beta i}^{(k)}\Delta b_i^{(k)} = s_\beta^{(k)},
$$

$$
F_{i\beta}^{(k)}\Delta\beta^{(k)} + F_{ii}^{(k)}\Delta b_i^{(k)} = s_i^{(k)}, \quad i = 1, .., n,
$$

where $\Delta\beta^{(k)} = \beta^{(k+1)} - \beta^{(k)}$ and $\Delta b_i^{(k)} = b_i^{(k+1)} - b_i^{(k)}$. After some transformations the following algorithm is obtained, where each iteration step implies working off the data twice to obtain first the corrections

$$
\Delta\beta^{(k)} = \left\{F_{\beta\beta}^{(k)} - \sum_{i=1}^{n} F_{\beta i}^{(k)}(F_{ii}^{(k)})^{-1}F_{i\beta}^{(k)}\right\}^{-1}\left\{s_\beta^{(k)} - \sum_{i=1}^{n} F_{\beta i}^{(k)}(F_{ii}^{(k)})^{-1}s_i^{(k)}\right\}
$$

and then

$$
\Delta b_i^{(k)} = (F_{ii}^{(k)})^{-1}\left\{s_i^{(k)} - F_{i\beta}^{(k)}\Delta\beta^{(k)}\right\}, \quad i = 1, ..., n.
$$

If the cluster sizes T_i are large enough the resulting estimates $\hat{\delta} = (\hat{\beta}, \hat{b}_1, \ldots,$ $\hat{b}_n)$ become approximately normal,

$$\hat{\delta} \overset{a}{\sim} N(\delta, F^{-1}(\delta)),$$

under essentially the same conditions, which assure asymptotic normality of the MLE in GLMs. Then the posterior mode and the (expected) curvature $F^{-1}(\hat{\delta})$ of $l(\delta)$, evaluated at the mode, are good approximations to the posterior mean and covariance matrix. $F^{-1}(\delta)$ is obtained using standard formulas for inverting partitioned matrices (see, e.g., Magnus and Neudecker, 1988). The result is summarized as follows:

$$F^{-1}(\delta) = \begin{bmatrix} V_{\beta\beta} & V_{\beta 1} & V_{\beta 2} & \cdots & V_{\beta n} \\ V_{1\beta} & V_{11} & V_{12} & \cdots & V_{1n} \\ V_{2\beta} & V_{21} & V_{22} & \cdots & V_{2n} \\ \vdots & \vdots & & & \vdots \\ V_{n\beta} & V_{n1} & \cdots & \cdots & V_{nn} \end{bmatrix},$$

with

$$V_{\beta\beta} = (F_{\beta\beta} - \sum_{i=1}^{n} F_{\beta i} F_{ii}^{-1} F_{i\beta})^{-1}, \quad V_{\beta i} = V'_{i\beta} = -V_{\beta\beta} F_{\beta i} F_{ii}^{-1},$$

$$V_{ii} = F_{ii}^{-1} + F_{ii}^{-1} F_{i\beta} V_{\beta\beta} F_{\beta i} F_{ii}^{-1}, \quad V_{ij} = V'_{ji} = F_{ii}^{-1} F_{i\beta} V_{\beta\beta} F_{\beta j} F_{jj}^{-1}, \ i \neq j.$$

EM–type algorithm

In the M–step $M(Q|Q^{(p)})$ has to be maximized with respect to Q. Since the conditional density of the observable data, $f(Y|\delta)$, is independent from Q, the M–step reduces to maximizing

$$M(Q|Q^{(p)}) = E\{\log p(\delta; Q)|Y; Q^{(p)}\} = \int \log p(\delta; Q) f(\delta|Y; Q^{(p)}) d\delta,$$

where $f(\delta|Y; Q^{(p)})$ denotes the posterior (7.3.1), evaluated at $Q^{(p)}$. As an example consider again a normal random effects density, where the M–step simplifies to the update

$$Q^{(p+1)} = \frac{1}{n} \sum_{i=1}^{n} (\text{cov}(b_i|y_i; Q^{(p)}) + E(b_i|y_i; Q^{(p)}) E(b_i|y_i; Q^{(p)})').$$

This step is the same as for Gaussian observations (see equation (7.1.12)). Although the algorithm is simple, it is difficult to carry out the update exactly, since the posterior mean and covariance are only obtained by numerical or Monte Carlo integration in general (see Section 7.4.2). Therefore, the posterior means are approximated by the posterior modes $\hat{b}_i$ and the posterior covariances by the posterior curvatures $\hat{V}_{ii}$, which both are calculated by the Fisher–scoring algorithm (7.3.5).

The resulting EM–type algorithm with the Fisher–scoring algorithm embedded in each E–step jointly estimates δ and Q as follows:

1. Choose a starting value $Q^{(0)}$.

For $p = 0, 1, 2, \ldots$

2. Compute posterior mode estimates $\hat{\delta}^{(p)}$ and posterior curvatures $\hat{V}_i^{(p)}$ by the Fisher–scoring algorithm (7.3.5), with variance–covariance components replaced by their current estimates $Q^{(p)}$.

3. EM–step: Compute $Q^{(p+1)}$ by

$$Q^{(p+1)} = \frac{1}{n} \sum_{i=1}^{n} (\hat{V}_i^{(p)} + \hat{b}_i^{(p)} (\hat{b}_i^{(p)})').$$

Note that the EM–type algorithm may be viewed as an approximate EM–algorithm, where the posterior of b_i is approximated by a normal distribution. In the case of linear random effects models the EM–type algorithm corresponds to an exact EM algorithm since the posterior of b_i is normal so that posterior mode and mean coincide as do posterior covariance and curvature.

7.4   Estimation by integration techniques

In this section we consider maximum likelihood estimation of the "fixed" parameters β and Q and posterior mean estimation of the $b_1, \ldots, b_n$ in the generalized linear random effects model. Together, both methods represent a two–step estimation scheme, where first the "fixed" parameters are estimated. Then random effects are predicted on the basis of estimated "fixed" parameters. If the dimension of the random effects is high, the simultaneous scheme from Section 7.3 is of computational advantage since numerical integration is avoided in contrast to the sequential scheme. For simplicity in the following we assume that the nuisance parameter ϕ is known.

7.4.1   Maximum likelihood estimation of fixed parameters

As in linear random effects models the unknown fixed parameters β and Q can be estimated by maximizing the marginal log–likelihood

$$l(\beta, Q) = \sum_{i=1}^{n} \log L_i(\beta, Q), \quad \text{with} \quad L_i(\beta, Q) = \int f(y_i | b_i; \beta) p(b_i; Q) db_i$$

$$(7.4.1)$$

that is obtained after integrating out the random effects b_i from the conditional exponential family densities, where $f(y_i|b_i; \beta)$ is given by (7.2.4) and $p(b_i; Q)$ is the density of random effects with $\text{cov}(b_i) = Q$. Analytical solutions of the possibly high–dimensional integrals are only available for special cases, e.g., for linear random effects models of the form (7.1.2). Therefore, numerical or Monte Carlo integration techniques are required for a wide range of commonly used random effects models, e.g., binomial logit or log–linear Poisson models with normally distributed random effects.

If the random effects density $p(b_i; Q)$ is symmetric around the mean, application of such techniques is straightforward after reparameterizing the random effects in the conditional mean $E(y_i|b_i)$ by

$$b_i = Q^{1/2}a_i$$

where $Q^{1/2}$ denotes the left Cholesky factor, which is a lower triangular matrix, so that $Q = Q^{1/2}Q^{T/2}$, and a_i is a standardized random vector having mean zero and the identity matrix I as covariance matrix. To obtain a linear predictor in β as well as in the unknown variance–covariance components of $Q^{1/2}$ we apply some matrix algebra (see, e.g., Magnus and Neudecker, 1988, p.30) so that

$$\eta_{it} = Z_{it}\beta + W_{it}Q^{1/2}a_i$$

can be rewritten in the usual linear form

$$\eta_{it} = [Z_{it} \, , \, a_i' \otimes W_{it}] \begin{bmatrix} \beta \\ \theta \end{bmatrix}$$

where the operator $\otimes$ denotes the Kronecker product and the vector θ corresponds to the vectorization of $Q^{1/2}$, $\theta = \text{vec}(Q^{1/2})$. In the case of a scalar random effect a_i the Kronecker product simplifies to a_iW_{it}. For illustration consider a simple binary response model with two–dimensional random effect (b_{i1}, b_{i2}). Let $Q^{1/2} = (q_{ij})$ denote the root of Q and $\text{vec}(Q^{1/2}) = (q_{11}, q_{21}, q_{12}, q_{22})$ denote the corresponding vectorized form. The linear predictor is given by

$$\eta_{it} = z_{it}'\beta + (w_{it1}, w_{it2}) \, Q^{1/2} \begin{pmatrix} a_{i1} \\ a_{i2} \end{pmatrix}$$

where $(a_{i1}, a_{i2}) \sim N(0, I)$. For the second term simple computation shows

$$(w_{it1}, w_{it2}) \, Q^{1/2} \begin{pmatrix} a_{i1} \\ a_{i2} \end{pmatrix} = (a_{i1}, a_{i2}) \otimes (w_{it1}, w_{it2}) \, \text{vec}(Q^{1/2})$$

$$= a_{i1}w_{it1}q_{11} + a_{i1}w_{it2}q_{21} + a_{i2}w_{it1}q_{12} + a_{i2}w_{it2}q_{22}.$$

Since for the Cholesky factor $Q^{1/2}$ we have $q_{12} = 0$, we may omit q_{12} in $\text{vec}(Q^{1/2})$ and the corresponding column in $(a_{i1}, a_{i2}) \otimes (w_{it1}, w_{it2})$.

Defining the new parameter vector

$$\alpha = (\beta, \theta),$$

the reparameterized log–likelihood

$$l(\alpha) = \sum_{i=1}^{n} \log L_i(\alpha), \quad \text{with} \quad L_i(\alpha) = \int f(y_i|a_i; \alpha)g(a_i)da_i, \qquad (7.4.2)$$

is obtained, where g denotes a density with zero mean and the identity matrix I as covariance matrix. Specifying g and maximizing $l(\alpha)$ with respect to α yields MLEs for β and $Q^{1/2}$. Note that we do not assume $Q^{1/2}$ to be positive definite. As a consequence $Q^{1/2}$ is not unique. Uniqueness could be achieved by requiring the diagonal elements of $Q^{1/2}$ to be strictly positive. However, to avoid maximization under such constraints we do not restrict ourselves to positive–definite roots. Note also that $Q^{1/2}$ contains elements that are zero by definition. To obtain derivatives of $l(\alpha)$ we first differentiate with respect to the entire parameter vector θ and then delete entries that correspond to elements that are zero by definition in θ.

Two approaches for maximizing (7.4.1) or (7.4.2) are in common use: The *direct* approach applies Gauss–Hermite or Monte Carlo integration techniques directly to obtain numerical approximations of the marginal likelihood or, more exactly, of its score function. Iterative procedures are then used for obtaining the maximum likelihood estimates. The *indirect* approach for maximizing the marginal likelihood is an EM–algorithm, where conditional expectations in the E–step are computed by Gauss–Hermite or Monte Carlo integrations, and maximization in the M–step is carried out by Fisher scoring. This indirect maximization method seems to be numerically more stable and simpler to implement. However, it takes more computation time. Algorithmic details of both approaches are given in Section 7.4.3.

7.4.2   Posterior mean estimation of random effects

Predicted values of the random effects b_i are required to calculate fitted values of the conditional means μ_{it}. Since the cluster–specific effects b_i represent random variables, estimation is based on the posterior density of b_i given the observations $Y = (y_1, ..., y_n)$. Due to the independence assumptions the posterior of b_i depends only on y_i. Using Bayes' theorem the posterior

$$f(b_i|y_i; \beta, Q) = \frac{f(y_i|b_i; \beta)p(b_i; Q)}{\int f(y_i|b_i; \beta)p(b_i; Q)db_i}. \qquad (7.4.3)$$

is obtained. Since the parameters β and Q are not known they are replaced by some consistent estimators $\hat{\beta}$ and $\hat{Q}$ such that statistical inference on b_i is based on the empirical Bayes principle. The empirical Bayesian point estimator that is "best" in the mean square error is the posterior mean

$$b_i^m = E(b_i|y_i) = \int b_i f(b_i|y_i; \hat{\beta}, \hat{Q})db_i.$$

For confidence intervals the posterior covariance matrix

$$V_i^m = \text{cov}(b_i|y_i) = E(b_i b_i'|y_i) - b_i(b_i)'$$

is also required. To calculate the posterior mean b_i^m or the posterior covariance V_i^m one has to carry out integrations that have the common structure

$$S(q(b_i)) = \int q(b_i)f(y_i|b_i; \hat{\beta}, \hat{Q})p(b_i; \hat{Q})db_i, \qquad (7.4.4)$$

where $q(b_i)$ stands for 1, b_i, or $b_i b_i'$. Unfortunately, the integrals cannot be solved analytically for most situations. Therefore, some approximation technique is required, e.g., numerical or Monte Carlo integration. Details are given in Section 7.4.3.

7.4.3   Algorithmic details*

Direct maximization

Let us first consider estimation of α based on the reparameterized marginal likelihood (7.4.2). Direct maximization is based on solving the ML equations

$$s(\alpha) = \frac{\partial l(\alpha)}{\partial \alpha} = \sum_{i=1}^{n} \frac{\partial L_i(\alpha)/\partial \alpha}{L_i(\alpha)} = 0, \qquad (7.4.5)$$

where $s(\alpha)$ denotes the score function. The problem is that the marginal likelihood contribution $L_i(\alpha)$ contains a possibly high–dimensional integral that cannot be solved analytically in general.

If the mixing density g is normal, evaluation of the integral in (7.4.2) can be accomplished by *Gauss–Hermite quadrature*. This is feasible in practice for low–dimensional random effects. For simplicity we confine ourselves to scalar random effects so that the integral is one–dimensional and θ represents a single variance component. For multivariate random effects Monte Carlo integration will be considered.

In the case of a scalar random effect the *Gauss–Hermite approximation* (see Appendix A4) of the likelihood contribution $L_i(\alpha)$ is given by

$$L_i(\alpha) \approx L_i^{gh}(\alpha) = \sum_{j=1}^{m} v_j f(y_i|d_j; \alpha), \qquad (7.4.6)$$

where d_j denotes one of the $j = 1, .., m$ quadrature points, and v_j represents the weight associated with d_j. Using the identity

$$\frac{\partial f(y_i|d_j; \alpha)}{\partial \alpha} = f(y_i|d_j; \alpha)\frac{\partial \log f(y_i|d_j; \alpha)}{\partial \alpha},$$

the score approximation

$$s(\alpha) \approx s^{gh}(\alpha) = \sum_{i=1}^{n} \sum_{j=1}^{m} c_{ij}^{gh}(\alpha) \frac{\partial \log f(y_i|d_j; \alpha)}{\partial \alpha} \tag{7.4.7}$$

is obtained, where

$$c_{ij}^{gh}(\alpha) = \frac{v_j f(y_i|d_j; \alpha)}{\sum_{k=1}^{m} v_k f(y_i|d_k; \alpha)}, \quad \text{with} \quad \sum_{j=1}^{m} c_{ij}^{gh}(\alpha) = 1,$$

denote weight factors depending on the parameters α that have to be estimated. The derivative

$$\partial \log f(y_i|d_j, \alpha)/\partial \alpha' = (\partial \log f(y_i|d_j, \alpha)/\partial \beta' \,, \, \partial \log f(y_i|d_j, \alpha)/\partial \theta)$$

corresponds to the score function of the GLM

$$E(\tilde{y}_{itj}) = h(\eta_{itj}), \quad \eta_{itj} = Z_{it}\beta + d_j \cdot W_{it}\theta, \tag{7.4.8}$$

for observations $\tilde{y}_{itj}, t = 1, \ldots, T_i$, where $\tilde{y}_{itj} = y_{it}$. Therefore, the components of $\partial log f/\partial\alpha$ are given by

$$\frac{\partial \log f(y_i|d_j; \alpha)}{\partial \beta} = \sum_{t=1}^{T_i} Z_{it}' D_{it}(\alpha, d_j) \Sigma_{it}^{-1}(\alpha, d_j)(y_{it} - \mu_{it}(\alpha, d_j)), \tag{7.4.9}$$

$$\frac{\partial \log f(y_i|d_j; \alpha)}{\partial \theta} = \sum_{t=1}^{T_i} d_j W_{it}' D_{it}(\alpha, d_j) \Sigma_{it}^{-1}(\alpha, d_j)(y_{it} - \mu_{it}(\alpha, d_j)), \tag{7.4.10}$$

with $D_{it}(\alpha, d_j) = \partial h(\eta_{itj})/\partial \eta, \Sigma_{it}(\alpha, d_j) = \mathrm{cov}(y_{it}|d_j)$, and $\mu_{it}(\alpha, d_j) = h(\eta_{itj})$.

If the number m of quadrature points is large enough, approximation (7.4.7) becomes sufficiently accurate. Thus, as n and m tend to infinity the MLEs for α will be consistent and asymptotically normal under the usual regularity conditions. On the other hand, the number of quadrature points should be as small as possible to keep the numerical effort low.

For high–dimensional integrals Gaussian quadrature techniques are less appropriate since the numerical effort increases exponentially with the dimension of the integral. Monte Carlo techniques are then more appropriate since these depend only linearly on the dimension. The simplest Monte Carlo approximation of $L_i(\alpha)$ is given by

$$L_i(\alpha) \approx L_i^{mc}(\alpha) = \frac{1}{m} \sum_{j=1}^{m} f(y_i|d_{ij}; \alpha)$$

where the $j = 1, .., m$ random values d_{ij} are i.i.d. drawings from the mixing density g. Since g is assumed to be completely known these simulations are straightforward and the integrals are just replaced by empirical means evaluated from simulated values of the reparameterized cluster–specific effects a_i. Replacing $L_i(\alpha)$ by $L_i^{mc}(\alpha)$ in (7.4.5) yields

$$s(\alpha) \approx s^{mc}(\alpha) = \sum_{i=1}^{n} \sum_{j=1}^{m} c_{ij}^{mc}(\alpha) \frac{\partial \log f(y_i|d_{ij}; \alpha)}{\partial \alpha}, \qquad (7.4.11)$$

where the weights $c_{ij}^{mc}(\alpha)$ are given by

$$c_{ij}^{mc}(\alpha) = \frac{f(y_i|d_{ij}; \alpha)}{\sum_{k=1}^{m} f(y_i|d_{ik}; \alpha)}, \quad \text{with} \quad \sum_{j=1}^{m} c_{ij}^{mc}(\alpha) = 1.$$

The components of $\partial \log f / \partial \alpha$ are defined by (7.4.9) and (7.4.10). Only quadrature points d_j are replaced by simulated values d_{ij}. In the general case where a_i is a vector the quadrature points $d_j = (d_{j1}, \ldots, d_{js})$ are also vectors with quadrature points as components. Here $j = (j_1, \ldots, j_s)$ is a multiple index with $j_i \in \{1, \ldots, m\}$ for the case of m quadrature points in each component. Accordingly for Monte Carlo techniques d_{ij} are vector drawings from the mixing density g. Then in the linear predictor (7.4.8) $d_j W_{it}\theta$ is replaced by $d_j' \otimes W_{it}\theta$ or $d_{ij}' \otimes W_{it}\theta$, respectively, and in (7.4.10) $d_j W_{it}'$ is replaced by $(d_j' \otimes W_{it})'$ or $(d_{ij}' \otimes W_{it})'$, respectively.

The MLEs for α have to be computed by an iterative procedure such as Newton–Raphson or Fisher scoring. Both algorithms imply calculation of the observed or expected information matrix. Due to the dependence of the weights c_{ij}^{gh} resp. c_{ij}^{mc} on the parameters to be estimated, the analytical derivation of information matrices is very cumbersome. As an alternative one might calculate the observed information matrix by numerical differentiation of s^{gh} resp. s^{mc}. However, numerical inaccuracy often yields MLEs that depend heavily on starting values or do not converge.

Indirect maximization

For indirect maximization of the log–likelihood (7.4.2) we consider an EM–algorithm, that is described in its general form in Appendix A3. If $Y = (y_1, \ldots, y_n)$ denotes the incomplete data, which are observable, and $A = (a_1, \ldots, a_n)$ stands for the unobservable data, which represent the reparameterized random effects, the EM–algorithm is based on the complete data log–density

$$\log f(Y, A; \alpha) = \sum_{i=1}^{n} \log f(y_i|a_i; \alpha) + \sum_{i=1}^{n} \log g(a_i). \qquad (7.4.12)$$

In the E–step of the $(p+1)$th EM–cycle one has to determine

$$M(\alpha|\alpha^{(p)}) = E\{\log f(Y, A; \alpha)|Y; \alpha^{(p)}\} = \int \log(f(Y, A; \alpha)) f(A|Y; \alpha^{(p)}) dA$$

which represents the conditional expectation of (7.4.12), given the incomplete data Y and an estimate $\alpha^{(p)}$ from the previous EM–cycle. The density $f(A|Y; \alpha^{(p)})$ denotes the posterior

$$f(A|Y; \alpha^{(p)}) = \frac{\prod_{i=1}^{n} f(y_i|a_i; \alpha^{(p)}) \prod_{i=1}^{n} g(a_i)}{\prod_{i=1}^{n} \int f(y_i|a_i; \alpha^{(p)}) g(a_i) da_i} \qquad (7.4.13)$$

which is obtained after applying Bayes' theorem in connection with the independence assumption (7.2.4). Due to (7.4.12) and (7.4.13) the function $M(\alpha|\alpha^{(p)})$ simplifies to

$$M(\alpha|\alpha^{(p)}) = \sum_{i=1}^{n} k_i^{-1} \int [\log f(y_i|a_i; \alpha) + \log g(a_i)] f(y_i|a_i; \alpha^{(p)}) g(a_i) da_i,$$

where

$$k_i = \int f(y_i|a_i; \alpha^{(p)}) g(a_i) da_i$$

is independent from the parameters α and the reparameterized random effects a_i. In general, carrying out the E–step is problematic since the integrals cannot be solved analytically. Therefore, numerical or Monte Carlo integration is required. Using a simple Monte Carlo integration, as in (7.4.11), yields the approximation

$$M(\alpha|\alpha^{(p)}) \approx M^{mc}(\alpha|\alpha^{(p)}) = \sum_{i=1}^{n} \sum_{j=1}^{m} c_{ij}^{mc} [\log f(y_i|d_{ij}; \alpha) + \log g(d_{ij})],$$

$$\qquad (7.4.14)$$

where the $j = 1, \dots, m$ weight factors

$$c_{ij}^{mc} = \frac{f(y_i|d_{ij}; \alpha^{(p)})}{\sum_{k=1}^{m} f(y_i|d_{ik}; \alpha^{(p)})}, \quad \text{with} \quad \sum_{j=1}^{m} c_{ij}^{mc} = 1, \qquad (7.4.15)$$

are completely known and not dependent on the parameters α. As in (7.4.11) the vectors d_{ij} are i.i.d. drawings from the mixing density g.

For normal densities g the integrals can also be approximated by *Gauss–Hermite quadrature*; see Hinde (1982), Brillinger and Preisler (1983), Anderson and Hinde (1988) and Jansen (1990) for scalar random effects and Anderson and Aitkin (1985) and Im and Gianola (1988) for bivariate random effects. Using *Gauss–Hermite quadrature* yields the approximation

$$M(\alpha|\alpha^{(p)}) \approx M^{gh}(\alpha|\alpha^{(p)}) = \sum_{i=1}^{n} \sum_{j} c_{ij}^{gh} [\log f(y_i|d_j; \alpha) + \log g(d_j)]$$

with the weight factors

$$c_{ij}^{gh} = \frac{v_j f(y_i|d_j; \alpha^{(p)})}{\sum_{k} v_k f(y_i|d_k; \alpha^{(p)})}, \quad \sum_{j} c_{ij}^{gh} = 1,$$

where the sum that is substituted for the s–dimensional integral is over the multiple index $j = (j_1, \ldots, j_s), j_i \in \{1, \ldots, m\}, i = 1, \ldots, s$, or the multiple index $k = (k_1, \ldots k_s)$, respectively. Instead of drawings d_{ij} one has the transformed quadrature points $d_j = (d_{j_1}, \ldots, d_{j_s})$ and the weights $v_j = v_{j_1} \cdot \ldots \cdot v_{j_s}$. Again the weight factors c_{ij}^{gh} do not depend on the parameter α.

For m quadrature points in each dimension the sum is over m^s terms. Therefore the numerical effort of Gaussian quadrature techniques increases exponentially with the integral dimension. For high–dimensional integrals Monte–Carlo techniques should be preferred since the sum in (7.4.14) is only over m terms.

The M–step consists of maximizing $M(\alpha|\alpha^{(p)})$ with respect to α. Considering a Monte Carlo approximation of the form (7.4.14) this is equivalent to solving the equation

$$u(\alpha|\alpha^{(p)}) = \frac{\partial M^{mc}(\alpha|\alpha^{(p)})}{\partial \alpha} = \sum_{i=1}^{n} \sum_{j=1}^{m} c_{ij}^{mc} \frac{\partial \log f(y_i|d_{ij}; \alpha)}{\partial \alpha} = 0, \quad (7.4.16)$$

where the components of $\partial \log f / \partial \alpha = (\partial \log f / \partial \beta, \partial \log f / \partial \theta)$ correspond to (7.4.9) and (7.4.10) with the exception that the quadrature points d_j are replaced by the simulated values d_{ij}. Note the similarity of (7.4.16) and the direct maximization equation (7.4.11). Both equations differ only in the weight factors, which are now independent from the parameters to be estimated and, therefore, completely known. Additionally the derivative $\partial \log f / \partial \alpha$ is equivalent to the score function of the GLM

$$E(\tilde{y}_{itj}) = h(\eta_{itj}), \quad \eta_{itj} = Z_{it}\beta + d_{ij}' \otimes W_{it}\theta,$$

with observations $\tilde{y}_{it1} = \tilde{y}_{it2} = \ldots = \tilde{y}_{itm} = y_{it}$. Therefore $u(\alpha|\alpha^{(p)})$ can be interpreted as a weighted score function of such a GLM. Then, given the weights c_{ij}^{mc}, the solution of (7.4.16) corresponds to a weighted MLE, which can be computed by an iteratively weighted–least–squares or Fisher–scoring algorithm. For given weights c_{ij}^{mc} the expected conditional information matrix has the form

$$U(\alpha|\alpha^{(p)}) = -E\left(\frac{\partial^2 E^{mc}(\alpha|\alpha^{(p)})}{\partial \alpha \partial \alpha'}\right) = \sum_{i=1}^{n} \sum_{j=1}^{m} c_{ij}^{mc} F_{ij}(\alpha),$$

where $F_{ij}(\alpha)$ is partitioned into

$$F_{ij}(\alpha) = \begin{bmatrix} F_{ij}^{\beta\beta} & F_{ij}^{\beta\theta} \\ F_{ij}^{\theta\beta} & F_{ij}^{\theta\theta} \end{bmatrix}$$

with

$$F_{ij}^{\beta\beta} = \sum_{t=1}^{T_i} Z_{it}' D_{it} \Sigma_{it}^{-1} D_{it}' Z_{it},$$

$$F_{ij}^{\beta\theta} = (F_{ij}^{\theta\beta})' = \sum_{t=1}^{T_i} Z_{it}' D_{it} \Sigma_{it}^{-1} D_{it}' (d_{ij}' \otimes W_{it}),$$

$$F_{ij}^{\theta\theta} = \sum_{t=1}^{T_i} (d_{ij}' \otimes W_{it})' D_{it} \Sigma_{it}^{-1} D_{it}' (d_{ij}' \otimes W_{it}),$$

and $D_{it} = D_{it}(\alpha, d_{ij}) = \partial h(\eta_{itj})/\partial \eta$, $\Sigma_{it} = \Sigma_{it}(\alpha, d_{ij}) = \mathrm{cov}(y_{it}|d_{ij})$, $\mu_{it}(\alpha, d_{ij}) = h(\eta_{itj})$. Then the M–step consists of the iteration scheme

$$\alpha_{k+1} = \alpha_k + U^{-1}(\alpha_k|\alpha^{(p)})u(\alpha_k|\alpha^{(p)}), \qquad (7.4.17)$$

with starting value $\alpha_0 = \alpha^{(p)}$ and k as an iteration index.

The following is a sketch of the Monte Carlo version of the EM–algorithm (MCEM) with a Fisher–scoring algorithm embedded in each M–step for estimating the fixed parameters of a generalized linear random effects model:

1. Calculate initial values $\beta^{(0)}$ using an original GLM without random effects, initialize $\theta^{(0)}$ with an arbitrary starting value, e.g., $\theta^{(0)} = 0$, and set $\alpha^{(0)} = (\beta^{(0)}, \theta^{(0)})$.

For $p = 0, 1, 2, \ldots$

2. Approximate $M(\alpha|\alpha^{(p)})$ by $M^{mc}(\alpha|\alpha^{(p)})$, i.e., compute the weights (7.4.15) using the values $\alpha^{(p)}$.

3. Carry out the algorithm (7.4.17) to obtain updates $\alpha^{(p+1)}$.

As for every EM–algorithm convergence of the MCEM–algorithm is slow. Moreover, computing time depends on m, the number of simulations. Therefore, m should be chosen as small as possible. However, to make sure that the MLEs do not depend on m an adaptive procedure is recommended. That means the number of simulations should be increased successively with increasing EM–cycles as long as estimators based on different m's are not nearly identical. The same has been pointed out by several authors for a Gauss–Hermite version of the EM–algorithm; see e.g., Hinde (1982), Brillinger and Preisler (1983) and Jansen (1990).

An estimator for the asymptotic covariance of the MLE's $\hat{\alpha}$ is not provided by EM. However, it can be obtained in the same way as in the direct maximization routine, e.g., by numerical differentiation of (7.4.11) with respect to α and evaluation at $\hat{\alpha}$.

Posterior mean estimation

Next we discuss methods for carrying out the integrations necessary for posterior mean estimation of random effects in Section 7.4.2.

If the random effects are normally distributed, Gauss–Hermite quadrature may be applied. To make sure that the integration variables are orthogonal in the random effects density the random effects b_i are replaced by the Cholesky parameterization

$$b_i = \hat{Q}^{1/2}a_i,$$

which was proposed in the previous section. Applying the Gauss–Hermite quadrature as reported in Appendix A4 yields the approximation

$$S(q(b_i)) \approx \sum_j v_j q(\bar{b}_j) f(y_i|\bar{b}_j; \hat{\beta}, \hat{Q}),$$

where the sum is over the multiple index $j = (j_1, \ldots, j_s)$, $j_i \in \{1, \ldots, m\}$, $i = 1, \ldots s$, and

$$\hat{b}_j = (\hat{b}_{j_1}, ..., \hat{b}_{j_s}), \quad \text{with} \quad \hat{b}_{j_r} = \hat{Q}^{1/2}d_{j_r},$$

is a vector of the transformed quadrature points $d_{j_r}, r = 1, \ldots, s$, and

$$v_j = \prod_{r=1}^{s} v_{j_r}, \quad \text{with} \quad \sum_j v_j = 1,$$

is the product of the $r = 1, \ldots, s$ weights v_{j_r} each associated with the quadrature point d_{j_r}. To make sure that the resulting approximations of b_i^m resp. V_i^m do not depend on the number of quadrature points the calculations should be repeated with an increasing number of quadrature points, until approximations based on different m's are nearly equal.

If the random effects density is not normal, Gauss–Hermite quadrature may also be used. Then, as stated in Appendix A4, we have to match a normal density f_N with mean b_i^m and covariance matrix V_i^m to the posterior (7.4.3), so that

$$S(q(b_i)) = \int \frac{q(b_i)f(y_i|b_i; \hat{\beta})p(b_i; \hat{Q})}{f_N(b_i; b_i^m, V_i^m)} f_N(b_i; b_i^m, V_i^m)db_i.$$

To determine the unknown parameters b_i^m and V_i^m, Gauss–Hermite integration has to be applied iteratively with some starting values for b_i^m and V_i^m. A natural choice is the posterior mode $\hat{b}_i$ and the curvature $\hat{V}_i$, which can be computed along the lines of Section 7.3.

An attractive alternative to Gauss–Hermite quadrature is Monte Carlo integration, especially when the dimension of the integral to be approximated is high. The integrals (7.4.4) are then estimated by empirical means, which are obtained from replacing the unobservable random effects by simulated values that are i.i.d. drawings from the random effects density. However, if the cluster size T_i is small, a large number of drawings is required to obtain consistent estimates of b_i^m and V_i^m. More efficient integral estimates, which require fewer drawings, are obtained applying *importance* or *rejection sampling*. Both sampling techniques, which are described in their general form in Appendix A5, match a density g to the posterior (7.4.3), where g should be proportional to the posterior, at least approximately, and it should be easy to sample from g.

Zellner and Rossi (1984), for example, suggest *importance sampling* (compare Appendix A5), which is based on a normal "importance function" g with the posterior mode $\hat{b}_i$ as mean and the posterior curvature $\hat{V}_i$ as covariance matrix. They justify such a choice asymptotically as the cluster size T_i tends to infinity. Then the posterior becomes normal, so that posterior mode and mean and posterior curvature and covariance coincide asymptotically. The corresponding estimator for integral (7.4.4) is

$$\hat{S}(q(b_i)) = \frac{1}{m} \sum_{j=1}^{m} q(d_{ij}) w_{ij}, \quad \text{with} \quad w_{ij} = \frac{f(y_i|d_{ij}, \hat{\beta}) p(d_{ij}; \hat{Q})}{g(d_{ij}; \hat{b}_i, \hat{V}_i)},$$

where the simulated values $d_j = (d_{j1}, .., d_{js})$ are i.i.d. drawings from the normal density g and the resulting estimator for b_i^m has the form

$$b_i^m = \hat{S}(q(b_i))/\hat{S}(1) = \sum_{j=1}^{m} d_j c_{ij},$$

$$\text{with} \quad c_{ij} = w_{ij}/\sum_{k=1}^{m} w_{ik}, \quad \text{and} \quad \sum_{j=1}^{m} c_{ij} = 1.$$

The quality of this estimator heavily depends on the weights c_{ij}. If one or few of them are extreme relative to the others, the estimator $\hat{b}_i$ is dominated by a single or few values d_j. Such a phenomenon especially occurs when the cluster sizes T_i are small. Then the chosen importance function g is apparently less appropriate, and importance functions being multimodal are recommended (see, e.g., Zellner and Rossi, 1984).

Rejection sampling (see Appendix A5) has been used by Zeger and Karim (1991) in a Gibbs sampling approach to random effects models. They match a normal density g with mean $\hat{b}_i$ and covariance matrix $c_2 \hat{V}_i$ to the posterior (7.4.3), where the parameter $c_2 \geq 1$ blows up $\hat{V}_i$. The parameters $\hat{b}_i$ and $\hat{V}_i$ again denote the posterior mode and curvature, which can be estimated along the lines of Section 7.3.

The parameter c_2 and a further parameter c_1 have to be chosen so that the inequality

$$c_1 g(b_i; \hat{b}_i, c_2 \hat{V}_i) \geq f(b_i|y_i; \hat{\beta}, \hat{Q}) \tag{7.4.18}$$

holds for all values of b_i. Zeger and Karim (1991) recommend setting $c_2 = 2$ and choosing c_1 so that the ordinates at the common mode $\hat{b}_i$ of the density g and the posterior kernel are equal, e.g.,

$$c_1 = \frac{f(y_i|\hat{b}_i;\hat{\beta})p(\hat{b}_i;\hat{Q})}{g(b_i;\hat{b}_i,c_2\hat{V}_i)}.$$

If (7.4.18) is fullfilled, a simulated value d_k drawn from g behaves as if distributed according to the posterior (7.4.3) as long as d_k is not rejected. Let u_k denote a random number being [0,1]–distributed; then d_k is rejected if

$$\frac{f(y_i|d_k;\hat{\beta})p(d_k;\hat{Q})}{c_1 g(d_k;\hat{b}_i,c_2\hat{V}_i)} > u_k.$$

Having generated $j = 1,\ldots,m$ accepted drawings d_j the posterior mean and the posterior covariance are estimated by

$$b_i^m = \frac{1}{m}\sum_{j=1}^{m} d_j \qquad \text{and} \qquad V_i^m = \frac{1}{m}\sum_{j=1}^{m}(d_j - \hat{b}_i)(d_j - \hat{b}_i)'.$$

Moreover, the $j = 1,\ldots,m$ simulated values d_j can be used to construct the empirical posterior distribution. Zeger and Karim (1991) propose a fully Bayesian approach, by treating β, Q as random with appropriate priors and estimating them together with random effects by Gibbs sampling.

7.5   Examples

Example 7.3: Ohio children data (Example 7.1, continued)
Analysis is based on the random effects logit model

$$\log \frac{P(\text{infection})}{P(\text{no infection})} = b_i + \beta_0 + \beta_s x_s + \beta_{A1} x_{A1} + \beta_{A2} x_{A2} + \beta_{A3} x_{A3}$$

$$+ \beta_{SA1} x_S x_{A1} + \beta_{SA2} x_S x_{A2} + \beta_{SA3} x_S x_{A3}$$

where x_S stands for smoking status in effect coding ($x_S = 1$: smoking, $x_S = -1$: nonsmoking) and x_{A1}, x_{A2}, x_{A3} represents age in effect coding. The parameters $\beta_0, \beta_S, \beta_{A1}, \beta_{A2}, \beta_{A3}, \ldots$ are fixed effects and b_i stands for a random intercept, where $b_i \sim N(0,\sigma^2)$. Apart from the random intercept, the linear predictor is the same as considered in Chapter 6 (Example 6.4).

Table 7.2 shows the estimation results for the fixed effects model and several estimation procedures for the random effects model. Standard deviations are given in brackets. Following a proposal by Gourieroux and Montfort (1989) standard errors are based on the estimated Fisher matrix

TABLE 7.2. Random intercept logit model for Ohio children data (effect coding of smoking and age, standard deviations in parentheses)

	Fixed effects		Gauss–Hermite m=10		Monte Carlo m=10		EM–type
$\hat\beta_0$	-1.696	(0.062)	-2.797	(0.205)	-2.292	(0.276)	1.952
$\hat\beta_S$	0.136	(0.062)	0.189	(0.110)	0.219	(0.152)	0.140
$\hat\beta_{A1}$	0.059	(0.106)	0.088	(0.186)	0.077	(0.260)	0.068
$\hat\beta_{A2}$	0.156	(0.103)	0.245	(0.189)	0.213	(0.268)	0.186
$\hat\beta_{A3}$	0.066	(0.105)	0.101	(0.191)	0.088	(0.263)	0.078
$\hat\beta_{SA1}$	-0.115	(0.106)	-0.178	(0.186)	-0.155	(0.260)	-0.135
$\hat\beta_{SA2}$	0.069	(0.103)	0.114	(0.189)	0.099	(0.268)	0.085
$\hat\beta_{SA1}$	0.025	(0.105)	0.041	(0.191)	0.035	(0.263)	0.030
$\hat\sigma$	—	—	2.136	(0.203)	1.817	(0.238)	1.830

$\sum_i s_i(\hat\alpha) s_i(\hat\alpha)'$ where $s_i(\hat\alpha)$ is the contribution of the ith observation to the score function (approximated by Gauss–Hermite or Monte Carlo methods). The estimated effects for the random effects models are larger than for the fixed effect model. This is to be expected because there is a bias of estimates toward zero if the random component is falsely omitted. Comparison of the estimates for the random effects models shows that the Gauss–Hermite procedure yields slightly stronger effects (except for smoking status) and a higher value for the heterogeneity parameter. The estimated standard deviation of the random intercept is quite large; thus heterogeneity should not be ignored.

As far as the inference of smoking is concerned the same conclusions must be drawn as in Example 6.4. For the fixed effects model the smoking effect seems to be significant. However, the standard deviation is not trustworthy since independent observations are falsely assumed. In the random effects model the smoking effect is not significant but the estimate is positive signaling a tendency toward increased infection rates. The interpretation of the other effects is similar to Example 6.4.

However, it should be noted that the estimates for random effects models are different from the estimates based on the naive independence assumption. For marginal models the point estimates (in this example) are stable for differing correlation assumptions (see Table 6.8).   □

Example 7.4: Bitterness of white wines (Example 7.2, continued)
Each of the judges tastes eight wines varying with respect to "temperature" ($x_T = 1$: low, $x_T = 0$: high), "contact" ($x_C = 1$: yes, $x_C = 0$: no) and

TABLE 7.3. Estimation results of bitterness of wine data

	Fixed effects model	Random Effects		
		MCEM 10	GHEM 10	EM–type
α_1	-5.289 (0.0)	-6.479	-6.496	-6.315
α_2	0.974 (0.0)	1.195	1.177	1.150
α_3	0.739 (0.0)	0.977	0.956	0.931
α_4	0.426 (0.105)	0.631	0.616	0.596
β_T	2.373 (0.0)	2.994	2.947	2.867
β_c	1.571 (0.0)	1.925	1.901	1.844
σ^2	—	1.157	1.496	1.512
log–likelihood	-87.31	- 79.86	-82.064	—

"bottle" (first/second). Since "bottle" is not influential it is omitted in the analysis.

In the first step a fixed effects cumulative logistic model with thresholds $\alpha_1, \ldots, \alpha_4$ may be used:

$$P(y_{it} = 1) = F(\alpha_1 + \beta_T x_T + \beta_C x_C), \qquad (7.5.1)$$

$$P(y_{it} \le r) = F(\alpha_1 + \sum_{s=2}^{r} \exp(\alpha_s) + \beta_T x_T + \beta_C x_C), \quad r = 2, ..., 4.$$

The covariate effect β_T represents the influence of "low temperature" and β_C the influence of "contact with skin."

Table 7.3 gives the MLE's with corresponding p–values. The deviance has value 15.87 at 26 df. So, at first glance the model fit looks quite good. However, the goodness–of–fit test is based on the assumption of independent responses y_{it}. For the responses $y_{it}, t = 1, ..., 8$ of the ith judge such an assumption does not hold. A model that allows each judge to have a specific level for the bitterness of wine includes a random effect varying across judges. In parameterization (7.5.1) α_1 sets the level whereas $\alpha_2, \ldots, \alpha_r$ are added for higher categories. Thus the level for judge i is specified by the addition of a random effect to α_1 in the form

$$P(y_{it} = 1|\alpha_{1i}) = F(a_i + \alpha_1 + \beta_T x_T + \beta_C x_C) \qquad (7.5.2)$$

$$P(y_{it} \le r|\alpha_{1i}) = F(a_i + \alpha_1 + \sum_{s=2}^{r} \exp(\alpha_s) + \beta_T x_T + \beta_C x_C), r = 2, \ldots, 4$$

with

$$a_i \overset{iid}{\sim} N(0, \sigma^2).$$

We use a maximum likelihood procedure based on the EM–algorithm as proposed in Section 7.3 and a 10–point Gauss–Hermite variant of the EM–algorithm and a Monte Carlo version of the EM–algorithm with 10 simulations per response y_{it}. The results are given in Table 7.3. The MLE's of the cumulative logistic model (7.5.1) given in column 1 of Table 7.3 have been used as starting values for the thresholds $\alpha_1, ..., \alpha_4$ and for β_T and β_C. In comparison with the cumulative logistic model (7.5.1) the threshold α_1 and the threshold differences $\alpha_2, ..., \alpha_4$ and the covariate effects β_T and β_C have changed in the random effects model (7.5.2). As was to be expected β_T and β_C are larger than for the fixed effects model. Moreover, the relatively high estimates of the variance component σ^2 indicate heterogeneity across judges. The results are quite similar for Monte Carlo procedure, Gauss Hermite integration and the empirical Bayes estimate.

7.6   Marginal estimation approach to random effects models

Collecting the clustered responses to a multivariate response vector $y_i = (y_{i1}, \ldots, y_{iT_i})$ the linear random effects model (7.2.1) can be shown to be equivalent to the linear marginal model

$$y_i \sim N(\nu_i, V_i),$$

where the marginal mean $\nu_i = E(y_i)$ is easily obtained from (7.1.4) and the marginal covariance $V_i = \text{cov}(y_i)$ is given in (7.1.5). In contrast to the linear random effects approach, where the cluster–specific mean of y_{it} is modelled as a function of population–averaged and cluster–specific effects, the marginal approach models the marginal or population–averaged mean of y_{it} just as a function of population–averaged effects. The essential point is that the population–averaged effect of the covariates measured by the parameters β is the same in both linear random effects models and linear marginal models, so that the distinction between these two approaches is irrelevant. For nonlinear models, however, the distinction between random effects and marginal approaches is important. Marginal approaches to population–averaged models are considered in Sections 3.5.2 and 6.2.2.

Zeger, Liang and Albert (1988) suggest analysing generalized linear random effects models within that marginal framework based on the marginal mean and marginal covariance structure of the responses y_i. Let us assume that a generalized linear random effects model, given by (7.2.2), (7.2.3) and (7.2.4), holds. Then

$$\mu_i' = E(y_i|b_i) = (\mu_{i1}', \ldots, \mu_{iT_i}') \tag{7.6.1}$$

denotes the conditional mean of y_i, given b_i, where μ_{it} is given by (7.2.2),

and

$$\Sigma_i = \text{cov}(y_i|b_i) = \text{diag}(\Sigma_{i1}, \ldots, \Sigma_{iT_i}), \quad \text{with} \quad \Sigma_{it} = \phi V(\mu_{it}),$$

denotes the conditional covariance of y_i, where Σ_{it} corresponds to the co-variance function of a simple exponential family. Then the marginal mean of y_i is obtained by integrating out the random effects b_i from the conditional mean (7.6.1),

$$\nu_i = (\nu_{i1}, \ldots, \nu_{iT_i})' = E(y_i) = \int \mu_i p(b_i; Q) db_i, \qquad (7.6.2)$$

where $p(b_i; Q)$ denotes the mixing density of the random effects b_i. Only for linear models one gets the simple form $\nu = h(Z_{it}\beta) = Z_{it}\beta$. The marginal covariance matrix is given by

$$\begin{aligned} V_i = \text{cov}(y_i) &= \int \Sigma_i p(b_i; Q) db_i + \int (\mu_i - \nu_i)(\mu_i - \nu_i)' p(b_i; Q) db_i \\ &= E\text{cov}(y_i|b_i) + \text{cov}(\mu_i) \end{aligned} \qquad (7.6.3)$$

where $\Sigma_i = \text{diag}(\Sigma_{i1}, \ldots, \Sigma_{i,T_i})$. However, analytical solutions of the integrals are available only for special cases, e.g., linear random effects models. Therefore, Zeger, Liang and Albert (1988) suggest linearizing the response function h. Assuming that the random effects b_i are small a first–order Taylor series expansion of $\mu_{it} = h(Z_{it}\beta + W_{it}b_i)$ around $b_i = 0$ yields

$$\mu_{it} = h(Z_{it}\beta + W_{it}b_i) \approx h(Z_{it}\beta) + \frac{\partial h(Z_{it}\beta)}{\partial \eta'} W_{it}b_i. \qquad (7.6.4)$$

Taking expectations with respect to the mixing density p one immediately gets the marginal mean approximation

$$\nu_{it} \approx \tilde{\nu}_{it} = h(Z_{it}\beta). \qquad (7.6.5)$$

The true marginal mean ν_{it} is equivalent to the approximation $\tilde{\nu}_{it}$ if the response function h is the identity so that the relationship between the mean μ_{it} and β resp. b_i is linear. For nonidentical response functions the quality of the marginal mean approximation depends on the magnitude of the variance–covariance components: the larger the variance–covariance components Q, the larger is the discrepancy between the parameters β of the random effects approach (7.2.2) and those of the marginal approach (7.6.5).

For illustration consider a log–linear Poisson random effects model of the form

$$\mu_{it} = \exp(z_{it}'\beta + w_{it}'b_i), \quad b_i \sim N(0, Q).$$

From (7.6.2) the marginal mean

$$\nu_{it} = \exp(z_{it}'\beta)E\{\exp(w_{it}'b_i)\}$$

is obtained, where the expectation has to be taken with respect to the normal random effects density having mean $E(b_i) = 0$ and covariance matrix $\mathrm{cov}(b_i) = Q$. Since the random variables $\exp(w'_{it}b_i)$ are log–normal with mean $\exp\left(\frac{1}{2}w'_{it}Qw_{it}\right)$ the marginal mean

$$\nu_{it} = \exp\left(z'_{it}\beta + \frac{1}{2}w'_{it}Qw_{it}\right)$$

is obtained. Apparently the marginal mean approximation $\tilde{\nu}_{it}$ neglects the effect of the variance–covariance components Q. Zeger, Liang and Albert (1988) also give an expression for the true marginal mean of a binary logistic random effects model, which also depends on the variance–covariance components. Moreover, Neuhaus, Kalbfleisch and Hauck (1991) showed in a general context that random effects approaches with nonlinear response functions are incompatible with marginal approaches based on $\tilde{\nu}_{it}$, and, in contrast to linear models, parameters β have different interpretations.

However, for small or moderate variance–covariance components Q the marginal mean structure of a generalized linear random effects model is adequately approximated by (7.6.5). Moreover, a first–order approximation of the marginal covariance (7.6.3) is obtained, if the components μ_{it} of the conditional mean μ_i are replaced by approximation (7.6.5) so that

$$V_i \approx \tilde{V}_i = D_i Z_i Q Z'_i D_i + A_i, \qquad (7.6.6)$$

where $A_i = \mathrm{diag}(A_{i1}, \ldots, A_{iT_i})$ denotes a block–diagonal matrix, with $A_{it} = \phi V(\tilde{\nu}_{it})$ evaluated at the marginal mean approximation $\tilde{\nu}_{it}, D_i = \mathrm{diag}(D_{i1}, ..., D_{iT_i})$ is also block–diagonal, with $D_{it} = \partial h/\partial \eta^m_{it}, \eta^m_{it} = Z_{it}\beta$, and $Z_i = (Z_{i1}, \ldots, Z_{iT})$ is a "grand" design matrix. The quality of approximation (7.6.5) and estimation of β and Q by a GEE approach has been studied by Zeger, Liang and Albert (1988).

7.7 Further approaches

When cluster sizes do not depend on cluster an alternative approach to reduce the number of parameters is to use conditional likelihood methods based on sufficient statistics. This approach is shortly sketched in Chapter 6 and outlined in Conaway (1989, 1990) and Hamerle and Ronning (1992). Similar approaches that are also suited for ordered data and are not considered have been proposed by Agresti (1993a,b), Agresti (1997) and Agresti and Lang (1993). Zeger and Karim (1991) considered generalized linear random effects models in a Bayesian framework and used the Gibbs sampler to overcome computational problems. An alternative (population–averaged) approach to account for heterogeneity that has led to a considerable number of papers is based on conjugate prior distributions. Instead of considering random coefficients the parameters of the exponential family distribution are determined by conjugate prior distributions yielding beta-binomial, Poisson–gamma or Dirichlet–multinomial models according to the type of data considered, see, e.g., Williams (1982), Moore (1987), Wilson and Koehler (1991), Brown and Payne (1986), Hausman, Hall and Griliches (1984) and Tsutakawa (1988). Recently, Waclawiw and Liang (1993) introduced an empirical Bayes technique using estimating functions in the estimation of both the random effects and their variance. In Waclawiw and Liang (1994) a fully parametric bootstrap method for deriving empirical Bayes confidence intervals is proposed. Breslow and Clayton (1993) consider penalized quasi–likelihood methods for the mean parameters and pseudo–likelihood for the variances.

Schall (1991) and McGilchrist (1994) derive approximate maximum likelihood estimates based on linearization of the model under consideration, Kuk (1995) proposes a method of adjusting estimates to result in asymptotical unbiased and consistent estimates. Variants of Monte Carlo EM algorithms are given in McCulloch (1997), whereas Drum and McCullagh (1993) consider residual maximum likelihood estimation for certain logistic mixed models.

8

State space models

This chapter surveys state space modelling approaches for analyzing non-normal time series or longitudinal data. The data situation is the same as in Chapter 6, i.e., categorical, counted or nonnegative data are observed over time. Typical examples are categorized daily rainfall data, the number of monthly polio incidences, or daily measurements on sulfur dioxide. State space models, also termed dynamic models, relate time series observations or longitudinal data $\{y_t\}$ to unobserved "states" $\{\alpha_t\}$ by an observation model for y_t given α_t. The states, which may be, e.g., unobserved trend and seasonal components or time–varying covariate effects, are assumed to follow a stochastic transition model. Given the observations $\{y_t\}$, estimation of states ("filtering" and "smoothing") is a primary goal of inference.

For approximately normal data, linear state space models and the famous linear Kalman filter have found numerous applications in the analysis of time series (see, e.g., Harvey, 1989; West and Harrison, 1989, for recent treatments) and longitudinal data (e.g., Jones, 1993). Extensions to non–Gaussian time series started with robustifying linear dynamic models and filters (compare, e.g., Martin and Raftery, 1987, and the references therein). Work on exponential family state space models or dynamic generalized models began only more recently (West, Harrison and Migon, 1985). While the formulation of non–normal state space models in Section 8.2 is straightforward and in analogy to random effects models in the previous chapter, the filtering and smoothing problem, which corresponds to estimation of random effects, becomes harder. Full Bayesian analysis based on posterior distributions or conditional mean filtering and smoothing will generally require repeated multidimensional integrations. We report on work in this area, which is still in progress, in Section 8.3.2. As an alternative, which is computationally less demanding, we consider posterior mode filtering and smoothing in more detail (Section 8.3.1). This latter approach is extended to longitudinal data in Section 8.4. It should be noted that most of the material in this chapter is written for the case of equally spaced time points. For linear state space models, nonequally spaced time points are treated, e.g., in Jones (1993).

8.1   Linear state space models and the Kalman filter

As a basis, this section gives a short review of linear dynamic models. Comprehensive treatments can be found, e.g., in Anderson and Moore (1979), Sage and Melsa (1971), Schneider (1986), Harvey (1989) and West and Harrison (1989).

8.1.1   Linear state space models

In the standard state space form, uni– or multivariate observations y_t are related to unobserved state vectors α_t by a *linear observation equation*

$$y_t = Z_t \alpha_t + \varepsilon_t, \quad t = 1, 2, \ldots, \tag{8.1.1}$$

where Z_t is an *observation* or *design matrix* of appropriate dimension, and $\{\varepsilon_t\}$ is a white noise process, i.e., a sequence of mutually uncorrelated error variables with $E(\varepsilon_t) = 0$ and $\text{cov}(\varepsilon_t) = \Sigma_t$. For *univariate* observations the design matrix reduces to a *design vector* z_t' and the covariance matrix to the variance σ_t^2. The observation equation (8.1.1) is then in the form of a dynamic linear regression model with time–varying parameters α_t. The sequence of states is defined by a linear *transition equation*

$$\alpha_t = F_t \alpha_{t-1} + \xi_t, \quad t = 1, 2, \ldots, \tag{8.1.2}$$

where F_t is a *transition* matrix, $\{\xi_t\}$ is a white noise sequence with $E(\xi_t) = 0$, $\text{cov}(\xi_t) = Q_t$, and the initial state α_0 has $E(\alpha_0) = a_0$ and $\text{cov}(\alpha_0) = Q_0$. The *mean* and *covariance structure* of the model is fully specified by assuming that $\{\varepsilon_t\}$ and $\{\xi_t\}$ are mutually uncorrelated and uncorrelated with the initial state α_0.

The *joint* and *marginal distributions* of $\{y_t, \alpha_t\}$ are completely specified by distributional assumptions on the errors and the initial state. Since linear state space models in combination with the linear Kalman filter and smoother are most useful for analyzing approximately Gaussian data, we assume joint normality throughout the section so that

$$\varepsilon_t \sim N(0, \Sigma_t), \quad \xi_t \sim N(0, Q_t), \quad \alpha_0 \sim N(a_0, Q_0), \tag{8.1.3}$$

and $\{\varepsilon_t\}$, $\{\xi_t\}$, α_0 are mutually independent.

It should be remarked that the covariance matrices are allowed to be singular. Therefore partially exact observations and time–constant states are not excluded by the model. Together with these distributional assumptions, (8.1.1) and (8.1.2) correspond to two–stage linear random effects models in Section 7.1.

In the simplest and basic state space form, the system matrices Z_t, F_t, Σ_t, Q_t and a_0, Q_0 are assumed to be deterministic and known. In many applications, however, the covariance matrices Σ_t, Q_t, the initial values a_0, Q_0, and in some cases the transition matrices F_t are unknown wholly or contain unknown hyperparameters, say θ, so that

$$\Sigma_t = \Sigma_t(\theta), \quad Q_t = Q_t(\theta), \quad a_0 = a_0(\theta), \quad Q_0 = Q_0(\theta).$$

Moreover the design matrix may depend on covariates or past observations so that

$$Z_t = Z_t(x_t, y^*_{t-1}), \qquad \text{with } y^*_{t-1} = (y_{t-1}, \ldots, y_1).$$

The design matrix may be called predetermined since it is known when y_t is observed. Models where Z_t, and possibly other system matrices, depend on past observations are termed conditionally Gaussian models. Whereas unknown hyperparameters complicate filtering and smoothing considerably, conditional models pose no further problems if all results are interpreted conditionally.

Comparing with the random effects models of the previous chapter, the main difference is that the sequence $\{\alpha_t\}$ is no longer i.i.d., but a Markovian process that need not even be stationary.

State space models have their origin in systems theory and engineering, with famous applications in astronautics in the 1960's (see, e.g., Hutchinson, 1984). In this context, the observation equation (8.1.1) describes radar observations y_t, disturbed by noise, on the state (position, velocity, ...) of a spacecraft, and the transition equation is a linearized and discretized approximation to physical laws of motion in space. Given the measurements $y_1, \ldots, y_t$, on–line estimation of α_t ("filtering") and prediction is of primary interest. A main reason for propagating and further developing the state space approach in statistics, and in particular in time series analysis, was that a number of prominent models, e.g., autoregressive–moving–average models, structural time series and dynamic regression models, can be described and dealt with in a flexible and unifying way. A prominent application in biostatistics is the monitoring of patients and other biometric or ecological processes, see, e.g., Smith and West (1983), Gordon (1986), Van Deusen (1989) and the survey in Frühwirth–Schnatter (1991).

In the following we present some simple univariate structural time series models and show how they can be put in state space form. For comprehensive presentations, we refer the reader to Gersch and Kitagawa (1988), Harvey (1989), and, with more Bayesian flavour, West and Harrison (1989).

The basic idea is to interpret the decomposition

$$y_t = \tau_t + \gamma_t + \varepsilon_t, \quad \varepsilon_t \sim N(0, \sigma_\varepsilon^2) \tag{8.1.4}$$

of a time series into a trend component τ_t, a seasonal component γ_t and an irregular component ε_t as the observation equation of a state space model, and to define stochastic trend and seasonal components by recursive transition equations. Simple nonstationary trend models are first– or second–order *random walks* (sometimes shortened $RW(1)$ or $RW(2)$)

$$\tau_t = \tau_{t-1} + u_t \quad \text{resp.} \quad \tau_t = 2\tau_{t-1} - \tau_{t-2} + u_t, \quad u_t \sim N(0, \sigma_u^2) \tag{8.1.5}$$

and the *local linear trend model*

$$\begin{aligned} \tau_t &= \tau_{t-1} + \lambda_{t-1} + u_t \\ \lambda_t &= \lambda_{t-1} + v_t, \quad v_t \sim N(0, \sigma_v^2), \end{aligned} \tag{8.1.6}$$

with mutually independent white noise processes $\{u_t\}, \{v_t\}$. If no seasonal component γ_t is present in the model, then (8.1.4), (8.1.5) or (8.1.6) can be put in state space form by defining

$$\alpha_t = \tau_t = 1 \cdot \alpha_{t-1} + u_t, \quad y_t = 1 \cdot \alpha_t + \varepsilon_t \tag{8.1.7}$$

for the $RW(1)$ model and

$$\alpha_t = \begin{bmatrix} \tau_t \\ \tau_{t-1} \end{bmatrix} = \begin{bmatrix} 2 & -1 \\ 1 & 0 \end{bmatrix} \begin{bmatrix} \tau_{t-1} \\ \tau_{t-2} \end{bmatrix} + \begin{bmatrix} u_t \\ 0 \end{bmatrix}, \quad y_t = (1,0)\alpha_t + \varepsilon_t,$$

for the $RW(2)$ model. For the local linear trend model, one has

$$\alpha_t = \begin{bmatrix} \tau_t \\ \lambda_t \end{bmatrix} = \begin{bmatrix} 1 & 1 \\ 0 & 1 \end{bmatrix} \begin{bmatrix} \tau_{t-1} \\ \lambda_{t-1} \end{bmatrix} + \begin{bmatrix} u_t \\ v_t \end{bmatrix}, \quad y_t = (1,0)\alpha_t + \varepsilon_t. \tag{8.1.8}$$

A stochastic seasonal component for quarterly data can be specified in dummy variable form by (e.g., Harvey, 1989, p.40)

$$\gamma_t = -\gamma_{t-1} - \gamma_{t-2} - \gamma_{t-3} + w_t, \quad w_t \sim N(0, \sigma_w^2). \tag{8.1.9}$$

If the disturbance term w_t were zero, (8.1.9) reduces to the requirement that seasonal effects sum up to zero. By introducing w_t, the seasonal effects can be allowed to change over time.

Together with one of the trend models, e.g., an $RW(2)$ model, the transition and observation equation in state space form are

$$\alpha_t = \begin{bmatrix} \tau_t \\ \tau_{t-1} \\ \gamma_t \\ \gamma_{t-1} \\ \gamma_{t-2} \end{bmatrix} = \begin{bmatrix} 2 & -1 & | & 0 & 0 & 0 \\ 1 & 0 & | & 0 & 0 & 0 \\ 0 & 0 & | & -1 & -1 & -1 \\ 0 & 0 & | & 1 & 0 & 0 \\ 0 & 0 & | & 0 & 1 & 0 \end{bmatrix} \begin{bmatrix} \tau_{t-1} \\ \tau_{t-2} \\ \gamma_{t-1} \\ \gamma_{t-2} \\ \gamma_{t-3} \end{bmatrix} + \begin{bmatrix} u_t \\ 0 \\ w_t \\ 0 \\ 0 \end{bmatrix}$$

and

$$y_t = (1, 0, 1, 0, 0)\alpha_t + \varepsilon_t. \tag{8.1.10}$$

As an alternative to (8.1.9), one may suppose that the seasonal dummy variables follow a random walk (Harrison and Stevens, 1976). A different approach is to model stochastic seasonality in trigonometric form. If there are s seasons (in the year or another period), then a stochastic seasonal component is the sum

$$\gamma_t = \sum_{j=1}^{[s/2]} \gamma_{jt}$$

of $[s/2]$ cyclical components defined by

$$\begin{aligned}
\gamma_{jt} &= \gamma_{j,t-1} \cos \lambda_j + \tilde{\gamma}_{j,t-1} \sin \lambda_j + w_{jt} \\
\tilde{\gamma}_{jt} &= -\gamma_{j,t-1} \sin \lambda_j + \tilde{\gamma}_{j,t-1} \cos \lambda_j + \tilde{w}_{jt}, \quad j = 1, \ldots, [s/2],
\end{aligned}$$

with seasonal frequencies $\lambda_j = 2\pi j/s$ and mutually independent white noise processes $\{w_{jt}\}, \{\tilde{w}_{jt}\}$ with a common variance σ_w^2 (Harvey, 1989, pp.40–43). As with the dummy variable form (8.1.9), one obtains a deterministic seasonal component if the disturbances $w_{jt}, \tilde{w}_{jt}$ are set to zero. With the use of standard trigonometric identities, γ_t can then be written as a sum of trigonometric terms instead of the recursive form. The component $\tilde{\gamma}_{jt}$ is only needed for recursive definition of γ_{jt} and is not important for interpretation. Note that for even s the component $\gamma_{[s/2],t}$ boils down to

$$\gamma_{[s/2],t} = \gamma_{[s/2],t-1} \cos \lambda_{[s/2]} + w_{[s/2],t}.$$

For quarterly data ($s = 4, \lambda_1 = \pi/2, \sin \lambda_1 = 1, \cos \lambda_1 = 0$) and with a local linear trend model the state space form becomes

$$\alpha_t = \begin{bmatrix} \tau_t \\ \lambda_t \\ \gamma_{1t} \\ \tilde{\gamma}_{1t} \\ \gamma_{2t} \end{bmatrix} = \begin{bmatrix} 1 & 1 & 0 & 0 & 0 \\ 0 & 1 & 0 & 0 & 0 \\ 0 & 0 & 0 & 1 & 0 \\ 0 & 0 & -1 & 0 & 0 \\ 0 & 0 & 0 & 0 & -1 \end{bmatrix} \begin{bmatrix} \tau_{t-1} \\ \lambda_{t-1} \\ \gamma_{1,t-1} \\ \tilde{\gamma}_{1,t-1} \\ \gamma_{2,t-1} \end{bmatrix} + \begin{bmatrix} u_t \\ v_t \\ w_{1t} \\ \tilde{w}_{1t} \\ w_{2t} \end{bmatrix},$$

$$y_t = (1, 0, 1, 0, 1)\alpha_t + \varepsilon_t. \tag{8.1.11}$$

All these state space models are time–invariant, i.e., the system matrices do not depend on the time index t. Assuming mutual independence among the error processes, all covariance matrices Q are diagonal but mostly singular. For example, in (8.1.10) $Q = \text{diag}(\sigma_u^2, 0, \sigma_w^2, 0, 0)$, and in (8.1.11) $Q = \text{diag}(\sigma_u^2, 0, \sigma_w^2, \sigma_w^2, \sigma_w^2)$. The variances σ_u^2, σ_w^2 are, generally unknown, hyperparameters of the model. The transition equation is of block–diagonal structure, with blocks corresponding to trend and seasonal components. This block structure is typical for structural time series models, and it reflects the flexibility of the approach: adding or deleting components and blocks corresponding to each other. So other models for trend and season-ality and additional ones for daily effects, calendar effects, global stationary components, etc., can be included, see, e.g., Kitagawa and Gersch (1984), Gersch and Kitagawa (1988) and Harvey (1989).

In addition, covariates and past responses may be incorporated. If their effects are supposed to be time–invariant and if we delete the seasonal component for simplicity, this leads to

$$y_t = \tau_t + (x'_t, y_{t-1}, \ldots,)\beta + \varepsilon_t. \tag{8.1.12}$$

Time–invariance of β can be described by the artificial dynamic relation $\beta_t = \beta_{t-1}(=\beta)$. Together with, e.g., an $RW(2)$ model for τ_t, one obtains

$$\alpha_t = \begin{bmatrix} \tau_t \\ \tau_{t-1} \\ \beta_t \end{bmatrix} = \begin{bmatrix} 2 & -1 & 0 \\ 1 & 0 & 0 \\ 0 & 0 & I \end{bmatrix} \begin{bmatrix} \tau_{t-1} \\ \tau_{t-2} \\ \beta_{t-1} \end{bmatrix} + \begin{bmatrix} u_t \\ 0 \\ 0 \end{bmatrix}$$

$$y_t = (1, 0, x'_t, y_{t-1}, \ldots)\alpha_t + \varepsilon_t. \tag{8.1.13}$$

The model (8.1.12), (8.1.13) can be interpreted as a semiparametric re-gression model: While the influence of covariates is parameterized in the usual way, the intercept is a trend component specified and estimated *non-parametrically* by an $RW(2)$ model, which is the discrete–time analog of a continuous–time cubic spline for τ_t (compare Section 5.1). We come back to this analogy in the sequel. Furthermore, continuous–time splines can be treated within the state space framework: Wahba (1978) shows that the assumption of an integrated Wiener process for τ_t leads to the common spline methodology, and Wecker and Ansley (1983) and Kohn and Ansley (1987) use this link for spline smoothing by Kalman filter and smoother methods.

Going a step further, one may also admit time–varying covariate effects:

$$y_t = \tau_t + \gamma_t + (x'_t, y_{t-1}, \ldots)\beta_t + \varepsilon_t. \tag{8.1.14}$$

The simplest choice for $\{\beta_t\}$ is an $RW(1)$ model with mutually independent error components. Then the corresponding block in the transition equation is specified by

$$\beta_t = \beta_{t-1} + w_t, \qquad w_t \sim N(0, Q), \qquad Q \text{ diagonal}. \tag{8.1.15}$$

A dynamic regression model like (8.1.14) where all coefficients are time–varying may cause some uneasiness. It should be noted, however, that (8.1.15) allows but does not postulate varying parameters β_t if Q is ad-mitted to be singular and is estimated as a hyperparameter. Variance com-ponents in Q with a (estimated) value 0 will then correspond to effects constant in time. Yet one may question the sense of such models and pos-sible causes for time–varying effects. In economic contexts, for example, changing structural and technological conditions, individual behaviour and attitude may be relevant causes. Dynamic modelling allows investigation of the (in–)stability of regressions and can lead to better prediction. A second main cause can be misspecification of models, in particular in connection with surrogates, errors in variables and omitted variables. Thinking positive, time–varying parameters can "swallow" or indicate such effects. In a refined analysis the model can be revised, and dynamic models can be used in this way as a tool for exploratory data analyses.

We would also like to point out a certain similarity to (generalized) additive models: In these models nonlinear covariate effects are included additively in the linear predictor by an unspecified function $s_t(x_t)$, which is estimated nonparametrically (see Section 5.3). This function $s_t(x_t)$ corresponds to $x_t\beta_t$, with β_t time varying according to a random walk, in dynamic models.

8.1.2   Statistical inference

As soon as a model can be written in state space form this provides the key for employing unified methods of statistical inference. Given the observations $y_1, \ldots, y_T$, estimation of α_t is the primary goal. This is termed

- filtering for $t = T$

- smoothing for $t < T$

- prediction for $t > T$.

We first consider the case of known hyperparameters, i.e., system and covariance matrices, initial values, etc., are known or given. For the sequel it is notationally convenient to denote histories of responses, covariates and states up to t by

$$y_t^* = (y_1', \ldots, y_t')', \qquad x_t^* = (x_1', \ldots, x_t')', \qquad \alpha_t^* = (\alpha_0', \ldots, \alpha_t')',$$

where y_0^*, x_0^* are empty. Initial values $y_0, y_{-1}, \ldots$, needed in autoregressive models, are assumed to be part of the covariates.

Linear Kalman filtering and smoothing

Under the normality assumption, the optimal solution to the *filtering* problem is given by the *conditional* or *posterior mean*

$$a_{t|t} := E(\alpha_t | y_t^*, x_t^*)$$

of α_t given y_t^* and x_t^*. Since the model is linear and Gaussian, the *posterior distribution* of α_t is also Gaussian,

$$\alpha_t | y_t^*, x_t^* \sim N(a_{t|t}, V_{t|t}),$$

with *posterior covariance matrix*

$$V_{t|t} := E\left[(\alpha_t - a_{t|t})(\alpha_t - a_{t|t})'\right].$$

Similarly, the best one–step predictor for α_t, given observations y_{t-1}^* up to $t-1$ only, is

$$a_{t|t-1} := E(\alpha_t | y_{t-1}^*, x_{t-1}^*),$$

and the one–step prediction density is Gaussian,

$$\alpha_t | y_{t-1}^*, x_{t-1}^* \sim N(a_{t|t-1}, V_{t|t-1}),$$

with posterior covariance matrix

$$V_{t|t-1} := E\left[(\alpha_t - a_{t|t-1})(\alpha_t - a_{t|t-1})'\right].$$

The famous linear Kalman filter and smoother computes the posterior means and covariance matrices in an efficient recursive way. The usual derivations of the Kalman filter and smoother take advantage of the fact that the posterior distributions are normal. Proofs as in Anderson and Moore (1979) and Harvey (1989) repeatedly apply formulas for expectations and covariance matrices of linear transformations in combination with Bayes' theorem.

Linear Kalman filter:

Initialization:	$a_{0|0} = a_0$, $V_{0|0} = Q_0$

For $t = 1, \ldots, T$:

prediction step:	$a_{t|t-1} = F_t a_{t-1|t-1},$
	$V_{t|t-1} = F_t V_{t-1|t-1} F_t' + Q_t,$
correction step:	$a_{t|t} = a_{t|t-1} + K_t(y_t - Z_t a_{t|t-1})$
	$V_{t|t} = V_{t|t-1} - K_t Z_t V_{t|t-1}$
Kalman gain:	$K_t = V_{t|t-1} Z_t' \left[Z_t V_{t|t-1} Z_t' + \Sigma_t\right]^{-1}.$

Given the observations $y_1, \ldots, y_{t-1}$ the prediction step updates the last filter value $a_{t-1|t-1}$ to the one–step–prediction $a_{t|t-1}$ according to the linear transition equation (8.1.2). As soon as y_t is observed, $a_{t|t-1}$ is corrected additively by the one–step–prediction error, optimally weighted by the Kalman gain K_t, to obtain the new filtering estimate $a_{t|t}$.

Mathematically equivalent variants, which are in some situations of computational advantage, are information filters, square root filters, etc. (see, e.g., Schneider, 1986). Let us only take a look at the correction step. Applying the matrix inversion lemma (e.g., Anderson and Moore, 1979), it can be rewritten as

$$
\begin{aligned}
V_{t|t} &= \left(V_{t|t-1}^{-1} + Z_t' \Sigma_t^{-1} Z_t\right)^{-1} \\
a_{t|t} &= a_{t|t-1} + V_{t|t} Z_t' \Sigma_t^{-1} \left(y_t - Z_t a_{t|t-1}\right).
\end{aligned}
\tag{8.1.16}
$$

This exhibits another interpretation: Information increments $Z_t' \Sigma_t^{-1} Z_t$ are added to the currently available information $V_{t|t-1}^{-1}$ to obtain the updated information $V_{t|t}^{-1}$, and the weighting Kalman gain is closely related to it.

The *smoother* for α_t given all observations $y_T^* = (y_1, \ldots, y_T)$ and x_T^*

$$a_{t|T} := E\left(\alpha_t | y_T^*, x_T^*\right),$$

and again the posterior is normal,

$$\alpha_t | y_T^*, x_T^* \sim N(\alpha_{t|T}, \dot{V}_{t|T}),$$

with

$$V_{t|T} := E\left[(\alpha_t - a_{t|T})(\alpha_t - a_{t|T})'\right].$$

Smoothers are usually obtained in subsequent backward steps, proceeding from T to 1. The fixed interval smoother given later is the traditional form of smoothing, see, e.g., Anderson and Moore (1979). Recently, faster variants for smoothing have been developed by De Jong (1989), Kohn and Ansley (1989) and Koopman (1993).

The ("fixed–interval") *smoother* consists of backward recursions for $t = T, \ldots, 1$:

$$
\begin{aligned}
a_{t-1|T} &= a_{t-1|t-1} + B_t(a_{t|T} - a_{t|t-1}) \\
V_{t-1|T} &= V_{t-1|t-1} + B_t(V_{t|T} - V_{t|t-1})B_t'
\end{aligned}
$$

with

$$B_t = V_{t-1|t-1} F_t' V_{t|t-1}^{-1}.$$

In each step, the smoothing estimate $a_{t-1|T}$ is obtained from the filtering estimate $a_{t-1|t-1}$ by adding the appropriately weighted difference between the smoothing estimate $a_{t|T}$ of the previous step and the prediction estimate $a_{t|t-1}$.

In the following, we will sketch the lines of argument for a derivation of Kalman filtering and smoothing, which corresponds to the historically first derivation (Thiele, 1880), makes the relationship to "linear smoothing" methods (Section 5.1) clearer, and is of importance for posterior mode estimation in non–Gaussian state space models.

Kalman filtering and smoothing as posterior mode estimation*

The starting point is the joint conditional density $p(\alpha_0, \alpha_1, \ldots, \alpha_T | y_T^*, x_T^*)$, i.e., the filtering and smoothing problems are treated simultaneously. Since the posterior is normal, posterior means and posterior modes are equal and can therefore be obtained by maximizing the posterior density. Repeated application of Bayes' theorem, thereby making use of the model assumptions in (8.1.1), (8.1.2), and taking logarithms shows that this maximization is equivalent to minimization of the penalized least–squares criterion

$$
\begin{aligned}
PLS(\alpha) \;=\; & \sum_{t=1}^{T} (y_t - Z_t\alpha_t)'\Sigma_t^{-1}(y_t - Z_t\alpha_t) + (\alpha_0 - a_0)'Q_0^{-1}(\alpha_0 - a_0) \\
& + \sum_{t=1}^{T} (\alpha_t - F_t\alpha_{t-1})'Q_t^{-1}(\alpha_t - F_t\alpha_{t-1})
\end{aligned}
\tag{8.1.17}
$$

with respect to $\alpha = (\alpha_0, \alpha_1, ..., \alpha_T)$. For simplicity, we have assumed that the covariance matrices Σ_t, Q_t are nonsingular. One may, however, drop this restriction.

As an example, consider the trend model $y_t = \tau_t + \varepsilon_t$ with τ_t as an $RW(2)$ model. Setting $\lambda := \sigma_\epsilon^2/\sigma_u^2$ and choosing a diffuse prior for α_0, (8.1.17) reduces to the minimization of

$$
\sum_{t=1}^{T} (y_t - \tau_t)^2 + \lambda \sum_{t=1}^{T} (\tau_t - 2\tau_{t-1} + \tau_{t-2})^2.
\tag{8.1.18}
$$

This criterion is exactly the *penalized least–squares criterion* of Whittaker (1923) for "optimal smoothing" of a trend component. Minimizing (8.1.18) tries to hold the balance between fit of the data and smoothness of the trend, expressed by the quadratic penalty term on the right side of (8.1.18). The *smoothness parameter* λ weights the two competing goals *data fit* and *smoothness*. Incorporation of further components results in more general discrete time–smoothing methods (e.g., Schlicht, 1981; Hebbel and Heiler, 1987; Pauly, 1989). Passing over to continuous time smoothing leads to common continuous time splines (Reinsch, 1967; Wahba, 1978; Kohn and Ansley, 1987). In this context the following Bayesian interpretation of state space models seems quite natural: The transition model (8.1.2) defines a prior distribution for the sequence $\{\alpha_t\}$ of states, and therefore is sometimes termed smoothness–prior. One may, however, also forget this smoothness–prior and the transition model and start in a model–free manner right from the criterion (8.1.17). To rewrite (8.1.17) in matrix notation, we define $y_0 := a_0$, $Z_0 := I$ and introduce the "stacked" observation vector

$$
y' = (y_0', y_1', \ldots, y_T'),
$$

the block–diagonal "grand" design matrix

$$
Z = \mathrm{diag}\,(Z_0, Z_1, \ldots, Z_T)
$$

and the block–diagonal weight matrix

$$
W = \mathrm{diag}\,(Q_0^{-1}, \Sigma_1^{-1}, \ldots, \Sigma_T^{-1}).
$$

Then the penalized–least–squares criterion can be written as

$$
PLS(\alpha) = (y - Z\alpha)'W(y - Z\alpha) + \alpha'K\alpha.
$$

The "penalty matrix" K is symmetric and block–tridiagonal, with blocks easily obtained from (8.1.17):

$$
K = \begin{bmatrix}
K_{00} & K_{01} & & & & 0 \\
K_{10} & K_{11} & K_{12} & & & \\
& K_{21} & \ddots & & \ddots & \\
& & \ddots & & \ddots & K_{T-1,T} \\
0 & & & & K_{T,T-1} & K_{TT}
\end{bmatrix},
$$

with

$$
\begin{aligned}
K_{t-1,t} &= K'_{t,t-1}, \quad 1 \le t \le T, \\
K_{00} &= F'_1 Q_1^{-1} F_1, \\
K_{tt} &= Q_t^{-1} + F'_{t+1} Q_{t+1}^{-1} F_{t+1}, \quad 1 \le t \le T, \\
F_{T+1} &= 0, \\
K_{t-1,t} &= -F'_t Q_t^{-1}, \quad 1 \le t \le T.
\end{aligned}
$$

Setting the first derivatives $2(-Z'W(y - Z\alpha) + K\alpha)$ of $PLS(\alpha)$ to zero and solving for the maximizer $\hat{\alpha} = (a_{0|T}, \ldots, a_{T|T})$, we obtain

$$
\hat{\alpha} = (Z'WZ + K)^{-1} Z'Wy. \tag{8.1.19}
$$

Since the posterior distribution is normal, the posterior mode $\hat{\alpha}$ coincides with the posterior mean $(a_{0|T}, \ldots, a_{T|T})$, which is obtained from the Kalman filter and smoother. It computes $\hat{\alpha}$ without explicitly inverting $Z'WZ + K$, but making efficient use of its block–banded structure. The penalized least squares approach sketched earlier indicates the close relationship to non– and semiparametric smoothing by splines; see Chapter 5 and Hastie and Tibshirani (1990): Instead of nonparametrically smoothing unknown *covariate functions*, we now consider smoothing unknown states as a *function of (discrete) time*.

Unknown hyperparameters

We adopt a frequentist viewpoint and treat hyperparameters θ, such as initial values or covariance matrices, as unknown constants. Under the normality assumption the method of maximum likelihood is then a natural choice for estimation. Two variants are commonly in use. The *direct* method is based on the factorization

$$
L(y_1, \ldots, y_T; \theta) = \prod_{t=1}^{T} p(y_t | y_{t-1}^*; \theta)
$$

of the likelihood, i.e., the joint density of $y_1, \ldots, y_t$ as a function of θ, into the product of conditional densities $p(y_t | y_{t-1}^*; \theta)$ of y_t given the data $y_{t-1}^* =$

$(y_1, \ldots, y_{t-1})$ up to time $t - 1$. Under the normality assumption, the joint and the conditional densities are normal, and corresponding conditional first and second moments can be computed for given θ by the Kalman filter and one–step predictions

$$y_{t|t-1} = E(y_t | y_{t-1}^*) = Z_t a_{t|t-1}$$

and corresponding one–step prediction error covariance matrices

$$Z_t V_{t|t-1} Z_t' + \Sigma_t.$$

The likelihood can therefore be computed for any fixed θ and can be maximized by numerical algorithms. Since this direct method becomes less favorable in the non–Gaussian situation, we do not present details but refer to Harvey (1989, p.127).

Indirect maximization starts from the joint likelihood

$$L(y_1, \ldots, y_T, \alpha_0, \ldots, \alpha_T; \theta)$$

of time series observations and unobserved states and uses the EM–principle for ML estimation of θ. The resulting algorithm, which will be used in modified form in Section 8.3.1, is described later.

Other estimation procedures are: generalized least squares (Pauly, 1989), special methods for time–invariant or stationary models (Aoki, 1987; Harvey, 1989, p.191), and Bayes methods, where θ is treated as a stochastic parameter with some prior distribution. The latter generally require additional efforts such as the multiprocess Kalman filter, numerical integration or data augmentation techniques, see, e.g., the survey in Frühwirth–Schnatter (1991).

Further tools of statistical inference such as testing, model checking and model selection are meanwhile also well developed; see Harvey (1989, ch.5).

EM–algorithm for estimating hyperparameters*

Let us consider the case of a univariate time–invariant state space model with unknown vector of hyperparameters $\theta = (\sigma^2, Q, a_0, Q_0)$. The joint log–likelihood of the complete data, in the terminology of the EM–principle, is, apart from additive constants not containing elements θ, given by

$$
\begin{aligned}
l(y_1, &\ldots, y_T, \alpha_0, \ldots, \alpha_T; \theta) \\
= &-\frac{T}{2} \log \sigma^2 - \frac{1}{2\sigma^2} \sum_{t=1}^{T} (y_t - z_t' \alpha_t)^2 \\
&-\frac{T}{2} \log(\det Q) - \frac{1}{2} \sum_{t=1}^{T} (\alpha_t - F_t \alpha_{t-1})' Q^{-1} (\alpha_t - F_t \alpha_{t-1}) \\
&-\frac{1}{2} \log(\det Q_0) - \frac{1}{2} (\alpha_0 - a_0)' Q_0^{-1} (\alpha_0 - a_0).
\end{aligned}
\tag{8.1.20}
$$

In the pth cycle of the algorithm, the E-step consists in computing

$$M(\theta|\theta^{(p)}) := E\left[l(y_1, \ldots, y_T, \alpha_0, \ldots, \alpha_T; \theta)|y_1, \ldots, y_T, \theta^{(p)}\right], \qquad (8.1.21)$$

the conditional expectation of the log–likelihood given the observations and the current iterate $\theta^{(p)}$. The next iterate $\theta^{(p+1)}$ is obtained as the maximizer of $M(\theta|\theta^{(p)})$ with respect to θ. In the present situation, conditional expectations of quadratic forms, such as $E(\alpha_t\alpha_t'|y_1, \ldots, y_T; \theta^{(p)})$, appearing in (8.1.21) can be computed by running the Kalman filter and smoother fixed at $\theta^{(p)}$, and the maximization problem can be solved analytically. Detailed derivations proceed along similar lines as in Section 7.1.2. The resulting iterative algorithm (compare Los, 1984; Schneider, 1986; Kirchen, 1988; Harvey, 1989) jointly estimates $a_{t|T}, V_{t|T}$ and θ as follows:

1. Choose starting values $\sigma^2_{(0)}, Q^{(0)}, Q_0^{(0)}, a_0^{(0)}$. For $p = 0, 1, 2, \ldots$

2. Smoothing: Compute $a_{t|T}^{(p)}, V_{t|T}^{(p)}, t = 1, \ldots, T$, by linear Kalman filtering and smoothing, with unknown parameters replaced by their current estimates $\sigma^2_{(p)}, Q^{(p)}, Q_0^{(p)}, a_0^{(p)}$.

3. EM–step: Compute $\sigma^2_{(p+1)}, Q^{(p+1)}, Q_0^{(p+1)}, a_0^{(p+1)}$ by

$$a_0^{(p+1)} = a_{0|T}^{(p)}, \qquad Q_0^{(p+1)} = V_{0|T}^{(p)},$$

$$\sigma^2_{(p+1)} = \frac{1}{T}\sum_{t=1}^{T}\left[\left(y_t - z_t'\alpha_{t|T}^{(p)}\right)^2 + z_t'V_{t|T}^{(p)}z_t\right],$$

$$Q^{(p+1)} = \frac{1}{T}\sum_{t=1}^{T}\left[\left(a_{t|T}^{(p)} - F_t a_{t-1|T}^{(p)}\right)\left(a_{t|T}^{(p)} - F_t a_{t-1|T}^{(p)}\right)' + V_{t|T}^{(p)}\right.$$
$$\left. - F_t B_t^{(p)} V_{t|T}^{(p)'} - V_{t|T}^{(p)'} B_t^{(p)'} F_t' + F_t V_{t-1|T}^{(p)} F_t'\right]$$

with $B_t^{(p)}$ defined as in the smoothing steps.

8.2 Non–normal and nonlinear state space models

Models with nonlinear observation and transition equation and additive Gaussian errors have been used early in engineering applications (Jazwinski, 1970; Sage and Melsa, 1971; Anderson and Moore, 1979). Since the linear Kalman filter is no longer applicable, various approximative filters have been proposed (extended Kalman filter, second–order filter, Edgeworth–expansions, Gaussian sum filter, etc.), which work satisfactorily in many situations. Non–Gaussian errors explicitly appear for the first time in robustified linear state space models, assuming distributions with heavier tails and suggesting approximate conditional mean (ACM) filters, see, e.g., Martin (1979), West (1981) and Martin and Raftery (1987). Non–normality becomes even more obvious for time series of counts, qualitative or non–negative data. Figure 8.1 (Section 8.3.1) shows a time series of categorized daily rainfall data from the Tokyo area for the years 1983 and 1984 (Kitagawa, 1987). For each day it is recorded whether rainfall over 1 mm occurred or not. For each calendar day $t, t = 1, \ldots, 366$, $y_t = 0, 1$, or 2 means that there was no rain this day in both years, rain in one of the years, or rain in both years. Compared to metrical time series it becomes apparently more difficult to discover trends or seasonal effects for the probability of rainfall, and some kind of smoothing would surely be helpful. The next section introduces models for such kinds of non–normal time series.

8.2.1 Dynamic generalized linear models

Let us rewrite the Gaussian linear observation equation (8.1.1) as

$$y_t | \alpha_t, y_{t-1}^*, x_t^* \sim N(\eta_t = Z_t \alpha_t, \Sigma_t),$$

where the design matrix Z_t is a function of past responses and covariates. An obvious generalization for the exponential family framework is the following *observation model*.

The conditional density $p(y_t | \alpha_t, y_{t-1}^*, x_t^*)$ is of the simple (uni– or multivariate) exponential family type with (conditional) mean

$$E(y_t | \alpha_t, y_{t-1}^*, x_t^*) = \mu_t = h(\eta_t), \quad \eta_t = Z_t \alpha_t, \quad t = 1, 2, \ldots, \quad (8.2.1)$$

where h is one of the common response functions, η_t the linear predictor, and Z_t is a function of covariates and, possibly, past responses. Modelling of Z_t is performed along the lines in previous chapters.

Equation (8.2.1) together with the exponential family assumption replaces the observation equation in linear normal models. For parameter transitions, we retain a *linear Gaussian transition model*, as long as this is compatible with the observation model:

$$\alpha_t = F_t \alpha_{t-1} + \xi_t, \quad t = 1, 2, \ldots \quad (8.2.2)$$

The error process $\{\xi_t\}$ is Gaussian white noise, $\xi_t \sim N(0, Q_t)$, with ξ_t independent of y_{t-1}^*, x_t^*, and of $\alpha_0 \sim N(a_0, Q_0)$.

For univariate time series the linear predictor reduces to $\eta_t = z_t' \alpha_t$. We can define structural models with trend τ_t, seasonal component γ_t and covariates in complete analogy as in Section 8.1.1 by decomposing the linear predictor into

$$\eta_t = \tau_t + \gamma_t + x_t' \beta_t, \quad (8.2.3)$$

with transition models for τ_t and γ_t in state space forms like (8.1.6) to (8.1.11). For example, a *binary dynamic logit model* with a trend component, a covariate and past response is defined by

$$P\left(y_t=1|a_t\right) = \exp\left(\tau_t + \beta_{1t}x_t + \beta_{2t}y_{t-1}\right) / \left(1 + \exp\left(\tau_t + \beta_{1t}x_t + \beta_{2t}y_{t-1}\right)\right) \tag{8.2.4}$$

together with a transition model such as (8.1.13) or (8.1.15), for $\alpha_t = (\tau_t, \beta_t)$, and a *dynamic log–linear Poisson* model similarly by

$$\lambda_t = E(y_t|\alpha_t) = \exp(z_t'\alpha_t). \tag{8.2.5}$$

For univariate dynamic models, a somewhat different kind of modelling is proposed by West et al. (1985). Instead of an explicit transition equation they assume conjugate prior–posterior distributions for the natural parameter and impose a linear prediction equation for α_t, involving the discount concept to circumvent estimation of unknown error covariance matrices. The observation equation is regarded only as a "guide relationship" between the natural parameter and $z_t'\alpha_t$ to determine prior–posterior parameters by a "method of moments." There is no general way to extend their method of moments (or some approximation by mode and curvature) to multivariate responses. The problems connected with the conjugate prior–posterior approach can be avoided, at least for the filtering problem, for a far simpler class of models, where only the "grand mean" or "level" varies over time while covariate effects are kept constant (Harvey and Fernandes, 1988; Harvey, 1989, ch. 6.6; Smith and Miller, 1986). Dynamic generalized linear models are also described in Lindsey (1993).

The modelling approach in this section is in analogy to corresponding two–stage models for random effects in Chapter 7 and has been proposed, e.g., in Fahrmeir (1992a,b) and Fahrmeir and Kaufmann (1991).

Categorical time series

An interesting class of multivariate dynamic models is those for *time series of multicategorical or multinomial responses*. If k is the number of categories, responses y_t can be described by a vector $y_t = (y_{t1}, \ldots, y_{tq})$, with $q = k - 1$ components. If only one multicategorical observation is made for each t, then $y_{tj} = 1$ if category j has been observed, and $y_{tj} = 0$ otherwise, $j = 1, \ldots, q$. If there are n_t independent repeated responses at t, then y_t is multinomial with repetition number n_t, and y_{tj} is the absolute (or relative) frequency for category j. For known n_t, multinomial models are completely determined by corresponding (conditional) response probabilities $\pi_t = (\pi_{t1}, \ldots, \pi_{tq})$, specified by $\pi_t = h(Z_t\alpha_t)$. In this way one obtains dynamic versions of the multinomial models for unordered or ordered categories of Chapter 3. The basic idea is to model the components $\eta_{tj}, j = 1, \ldots, q$, of the multivariate predictor $\eta_t = Z_t\alpha_t$ in (8.2.1) in analogy to (8.2.3) .

A *dynamic multivariate logistic model* with trend and covariate effects can be specified by

$$\pi_{tj} = \frac{\exp(\eta_{tj})}{1 + \sum_{r=1}^{q} \exp(\eta_{tr})} \qquad (8.2.6)$$

with

$$\eta_{tj} = \tau_{tj} + x_t' \beta_{tj}, \qquad j = 1, \ldots, q. \qquad (8.2.7)$$

Model (8.2.6) also arises by applying the principle of random utility to underlying linear dynamic models with stochastic trends as in Section 3.3. It can be put in state space form (8.2.1), (8.2.2) along similar lines as in Section 8.1.1. As an example consider the case $q = 2$ and let $\tau_{tj}, j = 1, 2$ be defined by second–order random walks and the time–varying covariate effects by first–order random walks

$$\beta_{tj} = \beta_{t-1,j} + \xi_{tj}, \qquad \xi_{tj} \sim N(0, Q_j).$$

The degenerate case $Q_j = 0$ corresponds to covariate effects constant in time. Setting

$$\alpha_t' = (\tau_{t1}, \beta_{t1}', \tau_{t2}, \beta_{t2}')$$

and assuming mutual independence, the transition equation is

$$\begin{bmatrix} \tau_{t1} \\ \tau_{t-1,1} \\ \beta_{t1} \\ \tau_{t2} \\ \tau_{t-1,2} \\ \beta_{t2} \end{bmatrix} = \begin{bmatrix} 2 & -1 & | & 0 & | & 0 & 0 & | & 0 \\ 1 & 0 & | & 0 & | & 0 & 0 & | & 0 \\ 0 & 0 & | & I & | & 0 & 0 & | & 0 \\ 0 & 0 & | & 0 & | & 2 & -1 & | & 0 \\ 0 & 0 & | & 0 & | & 1 & 0 & | & 0 \\ 0 & 0 & | & 0 & | & 0 & 0 & | & I \end{bmatrix} \begin{bmatrix} \tau_{t-1,1} \\ \tau_{t-2,2} \\ \beta_{t-1,1} \\ \tau_{t-1,2} \\ \tau_{t-2,2} \\ \beta_{t-1,2} \end{bmatrix} + \begin{bmatrix} u_{t1} \\ 0 \\ \xi_{t1} \\ u_{t2} \\ 0 \\ \xi_{t2} \end{bmatrix}.$$

$$(8.2.8)$$

The observation model (8.1.1) is then obtained with

$$Z_t = \begin{bmatrix} 1 & 0 & x_t' & | & 0 & 0 & 0 \\ 0 & 0 & 0 & | & 1 & 0 & x_t' \end{bmatrix} \qquad (8.2.9)$$

and the response function of the multivariate logistic model according to (8.2.6). The transition matrix is block–diagonal with blocks corresponding to alternative–specific trends and covariate effects. Apparently other trend models or additional seasonal components that have been discussed in Section 8.1.1 can be included by changing or adding corresponding elements in the linear predictor and blocks in (8.2.8) in the same way as for univariate normal linear state space models.

The simplest models for ordered categories are *dynamic cumulative models*. They can be derived from a threshold mechanism for an underlying linear dynamic model as in Section 3.4. The resulting (conditional) response probabilities are

$$\pi_{tj} = F\left(\tau_{tj} + x_t' \beta_t\right) - F\left(\tau_{t,j-1} + x_t' \beta_t\right), \qquad j = 1, \ldots, q, \qquad (8.2.10)$$

with ordered threshold parameters

$$-\infty = \tau_{t0} < \tau_{t1} < \ldots < \tau_{tq} < \infty,$$

a global covariate effect β_t, and F a known distribution function, e.g., the logistic one. The order restriction guarantees that the probabilities in (8.2.10) are non–negative. If thresholds vary according to one of the stochastic trend and seasonal models, this ordering can be destroyed with positive probability. This is not a problem in practice as long as thresholds are clearly separated and variances of the errors are small. However, the problem can be overcome by the same reparameterization as for random effects models: Introducing the parameter vector $\tilde{\tau}_t = (\tilde{\tau}_{t1}, \ldots, \tilde{\tau}_{tq})'$, thresholds may be reparameterized by

$$\tau_{t1} = \tilde{\tau}_{t1}, \qquad \tau_{tr} = \tilde{\tau}_{t1} + \sum_{s=2}^{r} \exp(\tilde{\tau}_{ts}), \qquad r = 2, \ldots, q,$$

or equivalently by

$$\tilde{\tau}_{t1} = \tau_{t1}, \qquad \tilde{\tau}_{tr} = \log(\tau_{tr} - \tau_{t,r-1}). \tag{8.2.11}$$

Then $\tilde{\tau}_t$ may vary without restriction.

Dynamic cumulative models can be written in state space form along the previous lines. In the simplest case thresholds and covariate effects obey a first–order random walk or are partly constant in time. Then, in original parameterization,

$$\alpha_t' = (\tau_{t1}, \ldots, \tau_{tq}, \beta_t') = \alpha_{t-1}' + \xi_t',$$

$$Z_t = \begin{bmatrix} 1 & & & x_t' \\ & \ddots & & \vdots \\ & & 1 & x_t' \end{bmatrix},$$

and the response function is the common one. Thresholds with other trend and seasonal components can again be modelled as in Section 8.1.1 and are incorporated by appropriate modifications of α_t and Z_t.

Dynamic versions of other models for ordered categories discussed in Section 3.3, such as sequential models, can be designed with analogous reasoning.

To specify the models completely in terms of densities, additional basic assumptions on conditional densities of responses and covariates are required.

(A1) Conditional on α_t and (y_{t-1}^*, x_t^*), current observations y_t are independent of α_{t-1}^*, i.e.,

$$p(y_t | \alpha_t^*, y_{t-1}^*, x_t^*) = p(y_t | \alpha_t, y_{t-1}^*, x_t^*), \qquad t = 1, 2, \ldots$$

This conditional independence assumption is implied in linear models by the assumptions on the error structure.

(A2) Conditional on y_{t-1}^*, x_{t-1}^*, covariates x_t are independent of α_{t-1}^*, i.e.,

$$p(x_t|\alpha_{t-1}^*, y_{t-1}^*, x_{t-1}^*) = p(x_t|y_{t-1}^*, x_{t-1}^*), \qquad t = 1, 2, \ldots$$

Loosely speaking, (A2) means that the covariate process contains no information on the parameter process. It can be omitted for deterministic covariates.

The next assumption is implied by the transition model (8.2.2) and the assumption on the error sequence, but is restated for completeness.

(A3) The parameter process is Markovian, i.e.,

$$p(\alpha_t|\alpha_{t-1}^*, y_{t-1}^*, x_t^*) = p(\alpha_t|\alpha_{t-1}), \qquad t = 1, 2, \ldots$$

8.2.2   Nonlinear and nonexponential family models*

As for static GLM's one may drop the assumption that the conditional density $p(y_t|\alpha_t, y_{t-1}^*, x_t^*)$ depends on the conditioning variables in the form of a linear predictor $\eta_t = Z_t\alpha_t$. Retaining the assumption that the conditional density is of the exponential family type, the observation model is given by the general specification

$$\mu_t = h_t(\alpha_t, y_{t-1}^*, x_t^*) \tag{8.2.12}$$

for the conditional mean. As long as this is sensible and compatible with the model we may retain a linear Gaussian transition model, but we may also generalize it, for example, to a nonlinear Gaussian transition equation

$$\alpha_t = f_t(\alpha_{t-1}) + \xi_t . \tag{8.2.13}$$

This family of state space models includes the standard form of nonlinear Gaussian state space models (e.g., Anderson and Moore, 1979; Sage and Melsa, 1971) with observation equation

$$y_t = h_t(\alpha_t) + \varepsilon_t. \tag{8.2.14}$$

One may go a step further and drop the exponential family assumption for the observation model. All such models can be written in the form of the following *general state space model*.

The observation model is specified by (conditional) *observation densities*

$$p(y_t|\alpha_t, y_{t-1}^*, x_t^*), \qquad t = 1, 2, \ldots, \tag{8.2.15}$$

and the transition model by *transition densities*

$$p(\alpha_t|\alpha_{t-1}), \qquad t = 1, 2, \ldots \tag{8.2.16}$$

Both densities will generally depend on a vector θ of hyperparameters. To specify the model completely in terms of joint densities, we suppose that the basic assumptions (A1), (A2) and (A3) of Subsection 8.2.1 hold.

For example, in the case of nonlinear Gaussian models defined by observation and transition equations as in (8.2.13) and (8.2.14)

$$
\begin{aligned}
p(y_t|\alpha_t, y_{t-1}^*, x_t^*) &\sim N(h_t(\alpha_t), \sigma_t^2) \\
p(\alpha_t|\alpha_{t-1}) &\sim N(f_t(\alpha_{t-1}), Q_t).
\end{aligned}
$$

More generally, errors $\{\varepsilon_t\}$ and $\{\xi_t\}$ may be non-Gaussian white noise processes with densities $d(\varepsilon)$ and $e(\xi)$. Then observation and transition densities are given by

$$
\begin{aligned}
p(y_t|\alpha_t, y_{t-1}^*, x_t^*) &\sim d(y_t - h_t(\alpha_t)), &(8.2.17)\\
p(\alpha_t|\alpha_{t-1}) &\sim e(\alpha_t - f_t(\alpha_{t-1})). &(8.2.18)
\end{aligned}
$$

Choosing densities d and e with heavy tails, e.g., Cauchy, Student or mixtures of normals, one obtains robustified linear or nonlinear models.

8.3 Non–normal filtering and smoothing

Estimation is based on posterior densities like $p(\alpha_0, \alpha_1, \ldots, \alpha_T|y_T^*)$ or $p(\alpha_t|y_t^*)$ for smoothing, or $p(\alpha_t|y_{t-1}^*)$, $p(\alpha_t|y_t^*)$ for prediction and filtering. One may distinguish three approaches: (i) conjugate prior–posterior analyses, trying to solve necessary integrations in Bayes' theorem analytically, perhaps making additional approximations, (ii) full Bayes or at least posterior mean analyses based on numerical integration or Monte Carlo methods, (iii) posterior mode estimation, avoiding integration. Type (i) has already been briefly discussed in Section 8.2. More detailed expositions can be found in the literature cited there. It should be noted that in smoothing there is no merit in assuming a conjugate prior. For example, with a log–linear Poisson observation model, calculation of the smoothing density cannot be simplified, even if a gamma distribution is assumed for λ_t.

We will consider the other two estimation methods for dynamic generalized linear models with exponential family observation densities $p(y_t|\eta_t)$, $\eta_t = Z_t\alpha_t$ of the form (8.2.1) and Gaussian linear transition models of the form (8.2.2). The focus will be on dynamic Poisson, binomial and multinomial models, which will be used in the applications. It should be remarked however, that most of the material in this section can be extended to nonlinear and nonexponential family state space models.

8.3.1   Posterior mode estimation

We first consider estimation of $\alpha_T^* = (\alpha_0, \ldots, \alpha_T)$ for known or given hyperparameters such as initial values and covariance matrices Q_t in the parameter model (8.2.2).

Repeated application of Bayes' theorem yields

$$
p(\alpha_T^*|y_T^*, x_T^*) = \prod_{t=1}^{T} p(y_t|\alpha_t^*, y_{t-1}^*, x_t^*) \cdot \prod_{t=1}^{T} p(\alpha_t|\alpha_{t-1}^*, y_{t-1}^*, x_t^*)
$$
$$
\cdot \prod_{t=1}^{T} p(x_t|\alpha_{t-1}^*, y_{t-1}^*, x_{t-1}^*)/p(y_T^*, x_T^*) \cdot p(\alpha_0).
$$

Using assumptions (A1), (A2) and (A3) we obtain

$$
p(\alpha_T^*|y_T^*, x_T^*) \propto \prod_{t=1}^{T} p(y_t|\alpha_t, y_{t-1}^*, x_t^*) \prod_{t=1}^{T} p(\alpha_t|\alpha_{t-1}) \cdot p(\alpha_0).
$$

Maximization of the conditional density is thus equivalent to maximizing the log–posterior

$$
PL(\alpha_T^*) = \sum_{t=1}^{T} l_t(\alpha_t) - \frac{1}{2}(\alpha_0 - a_0)' Q_0^{-1}(\alpha_0 - a_0)
$$
$$
- \frac{1}{2} \sum_{t=1}^{T} (\alpha_t - F_t \alpha_{t-1})' Q_t^{-1}(\alpha_t - F_t \alpha_{t-1}),
$$
(8.3.1)

where $l_t(\alpha_t) = \log\, p(y_t|\, \eta_t{=}Z_t\alpha_t)$ are the log–densities of the observation model (8.2.1). The log–posterior (8.3.1) is a *penalized (log–) likelihood criterion*. Compared to the penalized–least–squares criterion (8.1.17), least–squares distances implied by the Gaussian observation model are replaced by Kullback–Leibler distances. Posterior mode smoothers

$$
\hat{\alpha} = (a_{0|T}, \ldots, a_{t|T}, \ldots, a_{T|T})
$$

are maximizers of (8.3.1).

As an example, consider a binary logit model (8.2.4), excluding x_t and y_{t-1}. Choosing an $RW(2)$–model with a diffuse prior for the initial value τ_0, (8.3.1) becomes

$$
PL(\tau_T^*) = \sum_{t=1}^{T} [y_t \log \pi_t(\tau_t) + (1 - y_t) \log(1 - \pi_t(\tau_t))]
$$
$$
- \frac{1}{2\sigma_\tau^2} \sum_{t=1}^{T} (\tau_t - 2\tau_{t-1} + \tau_{t-2})^2,
$$
(8.3.2)

with $\pi_t(\tau_t) = \exp(\tau_t)/[1 + \exp(\tau_t)]$. Comparison with (8.1.18) shows that essentially the sum of squares $\sum(y_t - \tau_t)^2$ is replaced by the sum of binomial log–likelihood contributions.

The penalized log–likelihood criterion (8.3.1) is a discrete–time analog to the criteria of Green and Yandell (1985), O'Sullivan, Yandell and Raynor (1986) and Green (1987) for spline smoothing in non- or semiparametric generalized linear models (compare with Section 5.3). This corresponds to the relationship between Kalman smoothing and spline smoothing in linear models, mentioned in Section 8.1

Numerical maximization of the penalized log–likelihood (8.3.1) can be achieved by various algorithms. Fahrmeir (1992a) suggests the generalized extended Kalman filter and smoother as an approximative posterior mode estimator in dynamic generalized linear models. Fahrmeir and Kaufmann (1991) develop iterative forward–backward Gauss–Newton (Fisher–scoring) algorithms. Gauss–Newton smoothers can also be obtained by iterative application of linear Kalman filtering and smoothing to a "working" model, similarly as Fisher scoring in static generalized linear models can be performed by iteratively weighted least squares applied to "working" observations. This is also described in more detail later.

Generalized extended Kalman filter and smoother*

This algorithm can be derived in a straightforward but lengthy way as an approximate posterior mode estimator by extending Sage and Melsa's (1971, p.447) arguments for maximum posterior estimation in nonlinear systems from conditionally Gaussian to exponential family observations. To avoid unnecessary repetition, we omit any details. Basically, the filter is derived as a gradient algorithm via the discrete maximum principle, replacing Gaussian log–likelihoods and derivatives by corresponding terms for non–Gaussian observations. In Taylor expansions, unknown parameters have to be replaced by currently available estimates. This also concerns observation covariance matrices $\Sigma_t(\alpha_t)$, in contrast to the Gaussian case, where Σ_t is assumed to be known. The same final result can also be obtained by using Hartigan's (1969) linear Bayes arguments or by linearizing the observation equations around the current estimates. For simplicity we assume a linear transition equation, but more general models could also be treated. Note that in the following, filter and smoother steps $a_{t|t}, V_{t|t}, a_{t|t-1}, V_{t|t-1}, a_{t|T}, V_{t|T}$ are numerical approximations to posterior modes and curvatures (inverses of negative second derivatives of corresponding log–posteriors). For linear Gaussian models they coincide with posterior means and covariance matrices.

Filter steps:
For $t = 1, 2, \ldots$

(1) Prediction

$$a_{t|t-1} = F_t\, a_{t-1|t-1}, \quad a_{0|0} = a_0$$
$$V_{t|t-1} = F_t\, V_{t-1|t-1} F_t' + Q_t, \quad V_{0|0} = Q_0\,.$$

(2) Correction (scoring form)

$$V_{t|t} = \left(V_{t|t-1}^{-1} + R_t \right)^{-1}$$
$$a_{t|t} = a_{t|t-1} + V_{t|t} r_t\,,$$

where $r_t = \partial l_t / \partial \alpha_t$ and $R_t = -E(\partial^2 l_t / \partial \alpha_t \partial \alpha_t')$ are the score function and (expected) information matrix contribution of observation y_t, however, evaluated at the prediction estimate $a_{t|t-1}$, i.e.,

$$r_t = Z_t' D_t \Sigma_t^{-1}(y_t - h(Z_t a_{t|t-1}))\,, \tag{8.3.3}$$
$$R_t = Z_t' D_t \Sigma_t^{-1} D_t' Z_t\,, \tag{8.3.4}$$

with the first derivative $D_t = \partial h / \partial \eta_t$ and the covariance matrix Σ_t of y_t evaluated at $a_{t|t-1}$.

An alternative form of the correction step can be obtained in this case by an application of the matrix inversion lemma:

(2)* Correction (Kalman gain form)

$$a_{t|t} = a_{t|t-1} + K_t(y_t - h_t(Z a_{t|t-1}))$$
$$V_{t|t} = (I - K_t D_t' Z_t)\, V_{t|t-1}$$

with the Kalman gain

$$K_t = V_{t|t-1} Z_t' D_t \left(D_t' Z_t V_{t|t-1} Z_t' D_t + \Sigma_t \right)^{-1}\,,$$

and D_t, Σ_t evaluated at $a_{t|t-1}$.

Both forms (2) and (2)* of the correction steps are mathematically equivalent, as can be shown by an application of the "matrix inversion lemma." The Kalman gain form (2)* is the more familiar form of extended Kalman filtering for nonlinear Gaussian state space models.

The correction steps in scoring form are more general since they also apply to nonexponential observation models. They can be interpreted as follows: The inverse $V_{t|t-1}^{-1}$ is the (estimated) information on α_t given y_{t-1}^*. The matrix R_t is the information on α_t contributed by the new observation y_t, and the sum $V_{t|t-1}^{-1} + R_t$ is the information on α_t given y_t^*. Inversion gives the (estimated) covariance matrix $V_{t|t}$. Thus the correction step (2) has just the form of a single Fisher–scoring step.

This observation suggests introducing an additional iteration loop in the correction steps, with $a_{t|t-1}$ as a starting value. Such additional local iterations may be useful if $a_{t|t}$ is comparably far from $a_{t|t-1}$. In the applications of this section, additional iterations do not lead to any relevant differences in the estimates for binomial models. They are useful, however, for time series of counts where a number of observations equal to or near zero is followed by large values as in the telephone data example. Alternatively the estimates can be improved by additional Fisher–scoring iterations; see below.

Smoother steps:
For $t = T, \ldots, 1$

$$
\begin{aligned}
a_{t-1|T} &= a_{t-1|t-1} + B_t(a_{t|T} - a_{t|t-1}) \\
V_{t-1|T} &= V_{t-1|t-1} + B_t(V_{t|T} - V_{t|t-1})B_t'
\end{aligned}
$$

where

$$
B_t = V_{t-1|t-1} F_t' V_{t|t-1}^{-1}. \tag{8.3.5}
$$

These smoother steps run through as in the linear case.

Gauss–Newton and Fisher–scoring filtering and smoothing*

A maximizer of the penalized log–likelihood $PL(\alpha_T^*)$ with generally better approximation quality can be found by Gauss–Newton or Fisher–scoring iterations. We will show that this can be achieved by applying linear Kalman filtering and smoothing to a "working model" in each Fisher–scoring iteration. A different, though mathematically equivalent, form of the algorithm is derived in Fahrmeir and Kaufmann (1991). To simplify notation, we will write α for $\alpha_T^* = (\alpha_0, \ldots, \alpha_T)$. Then the penalized log–likelihood criterion (8.3.1) can be written as

$$
PL(\alpha) = l(\alpha) - \frac{1}{2}\alpha' K \alpha,
$$

where

$$
l(\alpha) = \sum_{t=0}^{T} l_t(\alpha_t),
$$

$l_0(\alpha_0) := -(\alpha_0 - a_0)' Q_0^{-1}(\alpha_0 - a_0)/2$, and the penalty matrix K is the same as in Section 8.1.2. Similary as in that section, we define the "stacked" observation vector

$$
y' = (a_0', y_1', \ldots, y_T'),
$$

the vector of expectations

$$
\mu(\alpha)' = (\alpha_0', \mu_1'(\alpha_1), \ldots, \mu_T'(\alpha_T)),
$$

with $\mu_t(\alpha_t) = h(Z_t\alpha_t)$, the block–diagonal covariance matrix

$$\Sigma(\alpha) = \operatorname{diag}\left(Q_0, \Sigma_1(\alpha_1), \ldots, \Sigma_T(\alpha_T)\right),$$

the block–diagonal design matrix

$$Z = \operatorname{diag}\left(I, Z_1, \ldots, Z_T\right),$$

and the block–diagonal matrix

$$D(\alpha) = \operatorname{diag}(I, D_1(\alpha_1), \ldots, D_T(\alpha_T)),$$

where $D_t(\alpha_t) = \partial h(\eta_t)/\partial\eta$ is the first derivative of the response function $h(\eta)$ evaluated at $\eta_t = Z_t\alpha_t$. Then the first derivative of $PL(\alpha)$ is given by

$$u(\alpha) = \partial PL(\alpha)/\partial\alpha = Z'D(\alpha)\Sigma^{-1}(\alpha)(y - \mu(\alpha)) - K\alpha.$$

The expected information matrix is

$$U(\alpha) = -E(\partial^2 PL(\alpha)/\partial\alpha\partial\alpha') = Z'W(\alpha)Z + K$$

with the weight matrix $W(\alpha) = D(\alpha)\Sigma^{-1}(\alpha)D'(\alpha)$. The expressions for first and expected second derivatives of $l(\alpha)$ are obtained as in Sections 2.2.1 and 3.4.1. A Fisher–scoring step from the current iterate α^0, say, to the next iterate α^1 is then

$$(Z'W(\alpha^0)Z + K)(\alpha^1 - \alpha^0) = Z'D(\alpha^0)\Sigma^{-1}(\alpha^0)(y - \mu(\alpha^0)) - K\alpha^0.$$

This can be rewritten as

$$\alpha^1 = (Z'W(\alpha^0)Z + K)^{-1}Z'W(\alpha^0)\tilde{y}^0,$$

with "working" observation

$$\tilde{y}^0 = D^{-1}(\alpha^0)(y - \mu(\alpha^0)) + Z\alpha^0.$$

Comparing with (8.1.19) in Section 8.1.2, it can be seen that one iteration step can be performed by applying the linear Kalman smoother with "working" weight $W = W(\alpha^0)$ to the "working" observation $\tilde{y}^0$. A closely related algorithm was recently obtained by Durbin and Koopman (1993). It is recommended that the iterations be initialized with the generalized extended Kalman smoother of Fahrmeir (1992a). Iterations will often stop after very few steps.

Comparison with spline smoothing in Section 5.3 sheds further light on the close relationship between nonparametric approaches and state space modelling: If we use posterior mode smoothing or start directly from the penalized log–likelihood criterion, then this is a kind of discrete–time spline smoothing for trend, seasonality and covariate effects in dynamic generalized linear models. In contrast to generalized additive models (Section 5.3.2) no inner backfitting loop is required in the Fisher–scoring iterations.

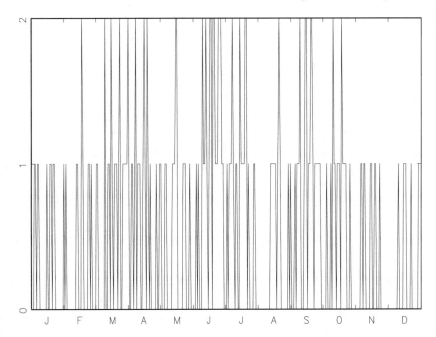

FIGURE 8.1. Number of occurrences of rainfall in the Tokyo area for each calendar day during 1983–1984.

Estimation of hyperparameters*

We only consider the situation of unknown initial values α_0, Q_0 and unknown error covariance matrix $Q_t = Q$ (independent of t). In analogy to the related situation in random effects models (Chapter 7), we suggest using an EM–type algorithm, which replaces posterior means and covariance matrices by posterior modes and curvatures obtained from one of the filtering and smoothing algorithms. The resulting EM–type algorithm is then formally identical with the EM–algorithm in Section 8.1.2 (omitting the estimation of σ^2). It has been studied in detail by Goss (1990) and is used in the following. However, extensions to more general situations and other approaches, e.g., cross–validation, still need to be studied.

Some applications

For illustration and comparison, we apply simple dynamic models and posterior mode filtering and smoothing to discrete–valued time series already analysed in the literature.

Example 8.1: Rainfall data (Example 5.5, continued)
Figure 8.1 displays the number of occurrences of rainfall in the Tokyo area for each calendar day during the years 1983–84. To obtain a smooth

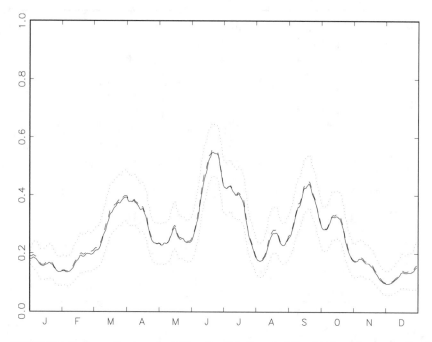

FIGURE 8.2. Smoothed probabilites $\hat{\pi}_t$ of daily rainfall, obtained by generalized Kalman (- - - -) and Gauss–Newton smoothing (———).

estimate of the probability π_t of occurrence of rainfall on calendar day $t, t = 1, ..., 366$, Kitagawa (1987) chose the following simple dynamic binomial logit model:

$$
\begin{aligned}
y_t &\sim \begin{cases} B(1, \pi_t), & t = 60 \quad \text{(February 29)} \\ B(2, \pi_t), & t \neq 60, \end{cases} \\
\pi_t &= h(\alpha_t) = \exp(\alpha_t)/(1 + \exp(\alpha_t)), \\
\alpha_{t+1} &= \alpha_t + \xi_t, \quad \xi_t \sim N(0, \sigma^2),
\end{aligned}
$$

so that $\pi_t = P$ (rain on day t) is reparameterized by α_t. Figure 8.2 shows corresponding smoothed estimates $\hat{\pi}_t = h(\hat{\alpha}_{t|366})$ together with pointwise confidence bands $(\hat{\pi}_t \pm \hat{\sigma}_t)$ based on generalized Kalman smoothing and the Gauss–Newton smoother, which was initialized by the Kalman smoother and stopped after two iterations. Starting values and the random walk variance were estimated by the EM–type algorithm as $\hat{a}_0 = -1.51, \hat{q}_0 = 0.0019, \hat{\sigma}^2 = 0.032$. In this example, generalized Kalman smoothing and Gauss–Newton smoothing lead to more or less the same pattern for the estimated probability of rainfall for calendar days. Whereas it is difficult to detect such a smooth pattern in Figure 8.1 by visual inspection, Figure 8.2 gives the impression of a distinct seasonal pattern: There are wet seasons in spring (March/April) and fall (September, October), June is

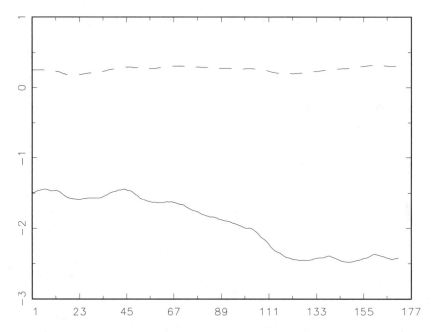

FIGURE 8.3. Smoothed trend (lower line) and advertising effect.

the month with the highest probability for rainfall, and winters are dry. Such a particular pattern may be compared to similar curves for former years, to see if there is any significant climatical change, or with curves for other geographical regions. Compared to Kitagawa's (1987) posterior mean smoother based on approximate numerical integration there are only minor departures for the initial time period. Compared to cubic spline smoothing (Example 5.3), the curve has a similar pattern as the curve of Figure 5.7. Use of a RW(2) model produces a posterior mode smoother that is almost indistinguishable from the cubic spline smoother of Figure 5.6. This is not astonishing due to the close relationship between the two approaches.   □

Example 8.2: Advertising data
West et al. (1985) analyzed binomial advertising data by a logit "guide relationship" together with a random walk model and subjectively chosen discount factors. The data are weekly counts y_t of the number of people, out of a sample of $n_t = 66$ for all t, who give a positive response to the advertisement of a chocolate bar. As a measure of advertisement influence, an "adstock coefficient" serves as a covariate x_t. The following questions might be of interest: Is there any general trend over time concerning positive response to advertising? How large is the effect of advertising, measured by the covariate? Is this effect stable over time or does it change? We reanalyze

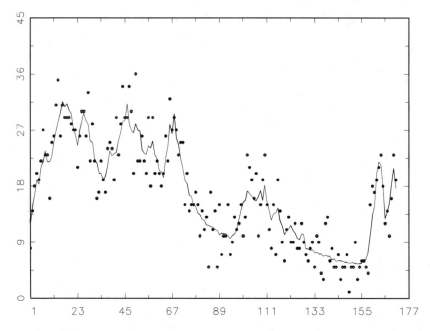

FIGURE 8.4. Advertising data and fitted values (———).

the data by the following closely related dynamic logit model, with $n_t = 66$:

$$\pi_t = h(\tau_t + x_t\beta_t), \quad \alpha_{t+1} = \alpha_t + \xi_t,$$

with $\alpha_t = (\tau_t, \beta_t)'$ and cov $\xi_t = \text{diag}\,(\sigma_0^2, \sigma_1^2)$. The posterior mode smoothers in Figure 8.3 show a slight decrease of the grand mean parameter whereas the positive advertising effect is more or less constant in time. This means that a certain amount of advertising, measured by x_t, has same positive, though not very large, effect regardless of whether we are at the beginning or end of the advertising campaign. However, there is a general negative and decreasing trend over time, so that additional advertisement efforts are necessary to keep the probability of positive response at least at a constant level. The variance components were estimated as $\hat{\sigma}_0^2 = 0.0025, \hat{\sigma}_1^2 = 0.0002$.

The data y_t together with fitted values $\hat{y}_t$ are again in good agreement with the results of West et al. (1985) (Figure 8.4). Our fitted values are somewhat nearer to the data at the beginning of the observation period. □

Example 8.3: Phone calls

The data, analysed again in West et al. (1985), consist of counts of phone calls, registrated within successive periods of 30 minutes, at the University of Warwick, from Monday, Sept.6, 1982, 0.00 to Sunday, Sept.12, 1982, 24.00. The data in Figure 8.5 show great variability, since counts are also

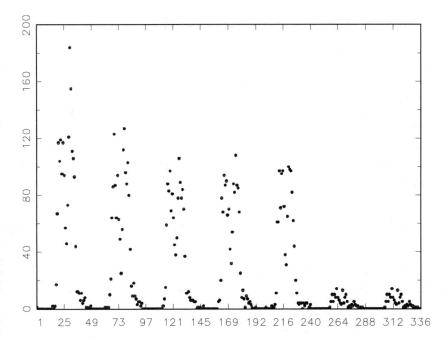

FIGURE 8.5. Number of phone calls for half–hour intervals.

registered during the nights and weekends. In particular, night hours often lead to longer periods of zero counts. For time series data of this kind, application of common linear state space methods will not be appropriate. Therefore, according to the suggestion of West et al., we analyze the data with a dynamic log–linear Poisson model (8.2.5), including a trend and seasonal component of the type in (8.1.11):

$$y_t \sim Po\left(\exp(z_t'\alpha_t)\right),$$

$$z_t' = (1, 1, 0, 1, 0, 1, 0, 1, 0, 1, 0), \quad \alpha_t' = (\tau_t, \gamma_{1t}, ..., \gamma_{10,t}),$$

$$\alpha_t = F\alpha_{t-1} + \xi_t, \quad \xi_t \sim N(0, Q)$$

with

$$F = \begin{bmatrix} 1 & & & 0 \\ & F_1 & & \\ & & \ddots & \\ 0 & & & F_5 \end{bmatrix},$$

$$F_i = \begin{bmatrix} \cos ip & \sin ip \\ -\sin ip & \cos ip \end{bmatrix}, \quad i = 1, ..., 5, \quad p = \pi/24,$$

and Q diagonal. The following variance component estimates corresponding to $\tau, \gamma_1, ..., \gamma_{10}$ were obtained after 64 iterations of the EM–type algorithm

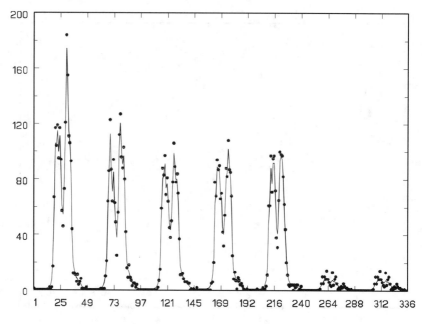

FIGURE 8.6. Observed and fitted (——) values.

with a relative stop criterion of $\varepsilon \leq 0.01 : \hat{q}_0 = 0.05, \hat{q}_1 = \hat{q}_2 = 0.003, \hat{q}_3 = ... = \hat{q}_8 = 0.0001, \hat{q}_q = \hat{q}_{10} = 0.00009$. Fitted values obtained by Gauss–Newton smoothing are shown in Figure 8.6. Peaks in the morning and afternoon hours and lows during lunch and night are clearly reproduced. As one would perhaps expect, Monday—the day after the weekend—has the highest rate of telephone calls. The daily pattern remains very similar for the whole week, but the level differs: Tuesday has a higher level than Wednesday, Thursday and Friday, while there are much fewer calls on Saturday and Sunday. (The last part of the graph is predicted counts for next Friday.)   □

8.3.2   Posterior mean estimation

This section discusses some approaches that have been suggested for calculating posterior densities like $p(\alpha_t|y_t^*)$ and $p(\alpha_t|y_{t-1}^*)$ for filtering and prediction, $p(\alpha_t|y_T^*)$ for smoothing, or at least first and second moments of these posterior densities. In the following,

$$a_{t|s}^m = E(\alpha_t|y_s^*) = \int \alpha_t p(\alpha_t|y_s^*)d\alpha_t, \quad s = 0, ..., T, \qquad (8.3.6)$$

denotes posterior means, i.e., predicted means for $s = t - 1$, filtered means for $s = t$, and smoothed means for $s = T$. Posterior covariances are defined in accordance with (8.3.6) by

$$
\begin{aligned}
V_{t|s}^m &= E[(\alpha_t - a_{t|s}^m)(\alpha_t - a_{t|s}^m)'|y_s^*] \\
&= \int \alpha_t \alpha_t' p(\alpha_t|y_s^*)d\alpha_t - a_{t|s}^m a_{t|s}^{m'}.
\end{aligned}
$$

There are two main approaches to calculate such posteriors: *Integration–based approaches* approximate analytically intractable integrals, which arise from repeated applications of Bayes' theorem, by numerical integration techniques or Monte Carlo integration. Direct application of numerical integration techniques may become computationally infeasible for higher dimensions of the state vector α since the numerical effort generally increases exponentially with dimension and over time. Therefore further approximations will generally be necessary. We discuss such methods in more detail at the end of this section.

The second approach is based on *Gibbs sampling*. This general Monte Carlo technique (see Appendix A5 for some further information) is particularly useful for problems in connection with Bayesian inference and can also be applied in our context. To estimate (marginal) posterior densities $p(\alpha_t|y_T^*)$, or first or second moments by Gibbs sampling, it is necessary to draw a large number of samples from the conditional posterior densities $p(\alpha_t|\alpha_s, s \neq t, y_T^*)$. Estimates for $p(\alpha_t|y_T^*)$ or its moments can then be computed by Monto Carlo techniques. How this can be done algorithmically is described later.

A Gibbs sampling approach*

A Gibbs sampling approach to dynamic models with normal mixture error structures has been proposed by Carlin, Polson and Stoffer (1992). Fahrmeir, Hennevogl and Klemme (1992) and Knorr–Held (1993) adopted their approach to dynamic generalized linear models with linear Gaussian transition models. However, the approach is also applicable to the more general class of nonlinear exponential family models. In the following, we only describe the procedure for the smoothing problem, but it carries over to prediction and filtering in a straightforward manner.

To obtain estimates for the marginal posterior densities $p(\alpha_t|y_T^*)$, or at least for the posterior moments $a_{t|T}^m$ and $V_{t|T}^m$, by Gibbs sampling, it is required that conditional posterior densities $p(\alpha_t|\alpha_{s \neq t}, y_T^*)$ of α_t given all other states $\alpha_s, s \neq t$, be available for sampling. That means it has to be possible to draw random numbers from these conditional densities, which are given by

$$
p(\alpha_t|\alpha_{s \neq t}, y_T^*) = \frac{p(\alpha_T^*, y_T^*)}{p(\alpha_{s \neq t}, y_T^*)} \tag{8.3.7}
$$

with $\alpha_T^* = (\alpha_0, ..., \alpha_T)$. Repeated application of Bayes' theorem under the assumptions (A1), (A2) and (A3) yields

$$p(\alpha_T^*, y_T^*) = p(\alpha_0) \prod_{t=1}^{T} p(\alpha_t | \alpha_{t-1}) \cdot \prod_{t=1}^{T} p(y_t | \alpha_t, y_{t-1}^*).$$

Proceeding in a similar way for the denominator in (8.3.7), one finally obtains

$$p(\alpha_t | \alpha_{s \neq t}, y_T^*) = \begin{cases} \dfrac{p(\alpha_{t+1} | \alpha_t) p(\alpha_t)}{p(\alpha_{t+1})}, & t = 0 \\[3mm] \dfrac{p(y_t | \alpha_t, y_{t-1}^*) p(\alpha_{t+1} | \alpha_t) p(\alpha_t | \alpha_{t-1})}{p(y_t | \alpha_{t-1}, y_{t-1}) p(\alpha_{t+1} | \alpha_{t-1})}, & t = 1, ..., T-1 \\[3mm] \dfrac{p(y_t | \alpha_t, y_{t-1}^*) p(\alpha_t | \alpha_{t-1})}{p(y_t | \alpha_{t-1}, y_{t-1}^*)}, & t = T. \end{cases}$$

$$(8.3.8)$$

Since the denominators in (8.3.8) do not depend on α_t, the following proportionality holds:

$$p(\alpha_t | \alpha_{s \neq t}, y_T^*) \propto \begin{cases} p(\alpha_{t+1} | \alpha_t) p(\alpha_t), & t = 0 \\ p(y_t | \alpha_t, y_{t-1}^*) p(\alpha_{t+1} | \alpha_t) p(\alpha_t | \alpha_{t-1}), & t = 1, ..., T-1 \\ p(y_t | \alpha_t, y_{t-1}^*) p(\alpha_t | \alpha_{t-1}), & t = T. \end{cases}$$

$$(8.3.9)$$

Carlin, Polson and Stoffer (1992) show that for linear Gaussian transition models of the form (8.2.2), the proportionality (8.3.9) specializes to

$$p(\alpha_t | \alpha_{s \neq t}, y_T^*) \propto \begin{cases} N(B_t b_t, B_t), & t = 0 \\ p(y_t | \alpha_t, y_{t-1}^*) N(B_t b_t, B_t), & t = 1, ..., T, \end{cases} \qquad (8.3.10)$$

with

$$B_t^{-1} = \begin{cases} Q_t^{-1} + F_{t+1}' Q_{t+1}^{-1} F_{t+1}, & t = 0, ..., T-1 \\ Q_t^{-1}, & t = T \end{cases}$$

$$b_t' = \begin{cases} \alpha_0' Q_0^{-1} + \alpha_{t+1}' Q_{t+1}^{-1} F_{t+1}, & t = 0 \\ \alpha_{t-1}' F_t' Q_t^{-1} + \alpha_{t+1}' Q_{t+1}^{-1} F_{t+1}, & t = 1, ..., T-1 \\ \alpha_{t-1}' F_t' Q_t^{-1}, & t = T \end{cases}$$

where $N(B_t b_t, B_t)$ stands for the normal density function having mean $B_t b_t$ and covariance B_t. Note that one has to assume nonsingularity of Q_0, Q_t.

In general, direct random drawings from the conditional densities $p(\alpha_t | \alpha_{s \neq t}, y_T^*)$ are not available. However, the densities from which we want to sample are dominated by two other density functions with closed form. Thus, rejection sampling (see Appendix A.5) can be used to obtain a random number from the density $p(\alpha_t | \alpha_{s \neq t}, y_T^*)$. In the context of rejection sampling a random number α_t is drawn from a density g and accepted with probability $f(\alpha_t)/(g(\alpha_t)M_t)$. The function $f(\alpha_t)$ has to be proportional to $p(\alpha_t|\alpha_{s\neq t}, y_T^*)$ and the constant M_t has to be chosen so that

$$M_t g(\alpha_t) \geq f(\alpha_t) \qquad \text{for all} \quad \alpha_t. \tag{8.3.11}$$

In view of (8.3.10) we set $f(\alpha_t) = p(y_t|\alpha_t, y_{t-1}^*)N(B_t b_t, B_t)$ and $g(\alpha_t) = N(B_t b_t, B_t)$. Then condition (8.3.11) corresponds to

$$M_t \geq p(y_t|\alpha_t, y_{t-1}^*) \qquad \text{for all} \quad \alpha_t \tag{8.3.12}$$

and the $N(B_t b_t, B_t)$-drawing d_t is accepted if

$$u \leq p(y_t|d_t, y_{t-1}^*)/M_t,$$

where u denotes a uniformly distributed random number.

Note that M_t should not be chosen too large, since the larger M_t the smaller the probability for accepting d_t and thus the efficiency of rejection sampling. As an example consider a binomial observation density $p(y_t|\alpha_t, y_{t-1}^*)$. Then the upper limit $M_t = 1$ would be satisfactory for condition (8.3.12).

With the conditional densities (8.3.8) and rejection sampling, the Gibbs sampling procedure runs as follows: Given a set of arbitrary starting values $(\alpha_t^{(0)})$, $t = 0, ..., T$ one has to draw $\alpha_0^{(1)}$ from the conditional density $p(\alpha_0|\alpha_1^{(0)}, ..., \alpha_T^{(0)}, y_T^*)$, then $\alpha_1^{(1)}$ from $p(\alpha_1|\alpha_0^{(1)}, \alpha_2^{(0)}, \alpha_3^{(0)}, ..., \alpha_T^{(0)}, y_T^*)$ and so on up to $\alpha_T^{(1)}$ from $p(\alpha_T|\alpha_0^{(1)}, \alpha_1^{(1)}, ..., \alpha_{T-1}^{(1)}, y_T^*)$ to complete one iteration. After a large number K of iterations that define one Gibbs run, the $(T+1)$–tuple $(\alpha_0^{(K)}, \alpha_1^{(K)}, ..., \alpha_T^{(K)})$ is obtained. Assuming the "mild conditions" of Geman and Geman (1984) hold, the joint density of this $(T+1)$–tuple converges at an exponential rate to the joint posterior density $p(\alpha_0, \alpha_1, ..., \alpha_T|y_T^*)$ as $K \to \infty$. Carrying out G Gibbs–runs yields $g = 1, ..., G$ i.i.d. $(T+1)$–tuples $(\alpha_0^{(K,g)}, \alpha_1^{(K,g)}, ..., \alpha_T^{(K,g)})$ from the joint posterior density $p(\alpha_0, \alpha_1, ..., \alpha_T|y_T^*)$. These can be used to estimate the marginal posterior density $p(\alpha_t|y_T^*)$ by

$$\hat{p}(\alpha_t|y_T^*) = \frac{1}{G}\sum_{g=1}^{G} p(\alpha_t|\alpha_{s\neq t}^{(K,g)}, y_T^*)$$

as long as the conditional density $p(\alpha_t|\alpha_{s\neq t}, y_T^*)$ is given in closed form. If no closed–form expression is available, the moments $a_{t|T}$ and $V_{t|T}$ of the marginal posterior density $p(\alpha_t|y_T^*)$ can be estimated in any case by simple Monte Carlo integrations yielding

$$\hat{a}_{t|T}^m = \frac{1}{G}\sum_{g=1}^{G} \alpha_t^{(K,g)}, \qquad \hat{V}_{t|T}^m = \frac{1}{G}\sum_{g=1}^{G}(\alpha_t^{(K,g)} - \hat{a}_{t|T}^m)(\alpha_t^{(K,g)} - \hat{a}_{t|T}^m)'.$$

Experience with dynamic binomial logit and log–linear Poisson models has shown that an appropriate choice of starting values $(\alpha_0^{(0)}, \alpha_1^{(0)}, ..., \alpha_T^{(0)})$ has a significant influence on the number K of iterations required for one Gibbs run. We recommend using posterior mode smoothing estimates, which are easily obtained by generalized extended Kalman filtering and smoothing (Section 8.3.1). Then, in contrast to arbitrarily chosen starting values, convergence of one Gibbs run takes only 20 to 40 iterations. Obviously the number of iterations also depends on the approximative normality of the observations y_t. For large counts y_t, for example, the number of iterations required until convergence is smaller than for small counts y_t involving many zeros and ones.

So far we have (tacitly) assumed that hyperparameters, in particular covariance matrices Q_t of the parameter model (8.2.2), are known or given. By treating hyperparameters as random variables with appropriate prior, they can be estimated together with states: One has to add further drawings from the posterior $p(Q|y_T^*)$ to the drawings (which are then to be understood for a given realization Q obtained by sampling from $p(Q|y_T^*)$). Details can be found in Knorr–Held (1993), where an inverse Wishart distribution is used for Q.

Integration–based approaches*

The first approach in this direction was presented by Kitagawa (1987). However, due to its complexity and numerical effort it is generally only applicable to dynamic generalized linear models with univariate response y_t and scalar states α_t. A similar approach has been proposed by West and Harrison (1989). They suggest the crude application of Gauss–Hermite quadrature to solve the analytically intractable integrals in the conditional moments $a_{t|s}^m$ and $V_{t|s}^m$. However, due to the recursive dependence of the integrals in posterior densities over time the numerical effort of their approach increases exponentially with time. Therefore, the method is mainly restricted to shorter time series.

A more practicable solution to the prediction and filter problem in dynamic generalized linear models with linear Gaussian transition models has been given by Schnatter (1992) and Frühwirth–Schnatter (1991). In contrast to Kitagawa (1987) and West and Harrison (1989), the prediction step is carried out only approximately so that the conditional densities $p(\alpha_t|y_{t-1}^*)$ are available in closed form. More precisely, the density $p(\alpha_t|y_{t-1}^*)$ is approximated by an $N(a_{t|t-1}, V_{t|t-1})$–density with mean

$$a_{t|t-1} = F_t a_{t-1|t-1}^m$$

and covariance matrix

$$V_{t|t-1} = F_t V_{t-1|t-1}^m F_t' + Q_t.$$

Such a normal approximation is sensible if the transition model is linear and Gaussian if the filtering density $p(\alpha_{t-1}|y_{t-1}^*)$ is approximately normal with mean $a_{t-1|t-1}^m$ and covariance matrix $V_{t-1|t-1}^m$. As a consequence the prediction step yields a normal approximation to the density $p(\alpha_t|y_{t-1}^*)$ which does not involve the use of numerical or Monte Carlo integration. The essential point, however, is that the correction step at time t is free from analytically intractable integrals that depend recursively on $\alpha_{t-1}, \alpha_{t-2}, ..., \alpha_0$. Thus, the numerical effort to calculate filtered moments $a_{t|t}$ and $V_{t|t}$ does no longer increase exponentially with time, but remains constant at each time point. This can be seen if the filtered moments $a_{t|t}^m$ and $V_{t|t}^m$ are approximated by

$$\tilde{a}_{t|t}^m = \frac{S(\alpha_t)}{S(1)} \quad \text{and} \quad \tilde{V}_{t|t}^m = \frac{S(\alpha_t \alpha_t')}{S(1)} - \tilde{a}_{t|t}^m \tilde{a}_{t|t}^{m'}, \qquad (8.3.13)$$

with

$$S(q(\alpha_t)) = \int q(\alpha_t) p(y_t|\alpha_t, y_{t-1}^*) N(a_{t|t-1}, V_{t|t-1}) d\alpha_t,$$

and $q(\alpha_t) = 1, \alpha_t$ or $\alpha_t \alpha_t'$. Note that $N(a_{t|t-1}, V_{t|t-1})$ stands for the normal density function, which approximates the prediction density $p(\alpha_t|y_{t-1}^*)$. Due to the closed form of this approximation, calculation of $S(q(\alpha_t))$ requires the solution of an integral depending only on the actual time point t. To solve it Gauss–Hermite quadrature may be applied as in Schnatter (1992) or Monte Carlo integration with importance sampling as in Hennevogl (1991). Both methods require filtered posterior modes $a_{t|t}$ and curvatures $V_{t|t}$ that can be obtained easily and fast by the generalized extended Kalman filter described in the previous section. For illustration we sketch the importance sampling approach where the integrals $S(q(\alpha_t))$ are estimated by empirical means obtained from simulated values of a so–called importance function (see Appendix A.5). This importance function should be, at least approximately, proportional to the posterior kernel $p(y_t|\alpha_t, y_{t-1}^*) p(\alpha_t|y_{t-1}^*)$. Assuming an approximate normal distribution having posterior mode $a_{t|t}$ and curvature $V_{t|t}$ as mean and covariance seems plausible. Then $S(q(\alpha_t))$ is estimated consistently by

$$\hat{S}(q(\alpha_t)) = \frac{1}{m} \sum_{j=1}^{m} q(d_{tj}) w_{tj}, \quad \text{with} \quad w_{tj} = \frac{p(y_t|d_{tj}, y_{t-1}^*) p(d_{tj}|y_{t-1}^*)}{N(a_{t|t}, V_{t|t})},$$

where $N(a_{t|t}, V_{t|t})$ stands for the normal density function having mean $a_{t|t}$ and covariance $V_{t|t}$, and the $j = 1, ..., m$ simulated values d_{tj} are i.i.d. $N(a_{t|t}, V_{t|t})$–drawings. To obtain estimators for the posterior mean $a_{t|t}^m$ and posterior covariance $V_{t|t}^m$, we only have to replace $S(q(\alpha_t))$ in (8.3.13) by the corresponding estimates $\hat{S}(q(\alpha_t))$.

Experience with artificial and real data (see, e.g., Hennevogl, 1991) has shown that Gauss–Hermite quadrature is more efficient than importance sampling as long as the dimension of the state vector α_t is moderate

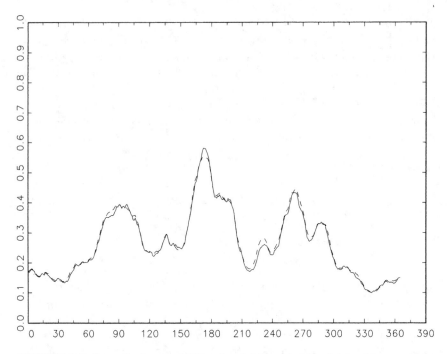

FIGURE 8.7. Smoothed probabilities $\hat{\pi}_t$ of daily rainfall, obtained by posterior mean (———) and posterior mode (- - - -) smoothing.

(up to three). For higher dimensions, importance sampling becomes more efficient. That means that importance sampling yields the same results as Gauss–Hermite quadrature, but the numerical effort is less than with Gauss–Hermite quadrature. Schnatter (1992) also gives a simulation–based comparison of approximative filtered posterior means obtained by Gauss–Hermite quadrature and approximative filtered posterior modes obtained by the generalized extended Kalman filter (see Section 8.3.1). In essence, the filtered sequences are often nearly identical, at least after a few filtering steps.

Example 8.4: Rainfall data (Example 8.1, continued)
Recall the rainfall data $y_t, t = 1, ..., 366$, which were analysed in Example 8.1 by a dynamic binomial logit model of the form

$$y_t \sim \begin{cases} B(1, \pi_t), & t = 60 \,(\text{February 29}) \\ B(2, \pi_t), & t \neq 60, \end{cases}$$

$$\pi_t = \exp(\alpha_t)/(1 + \exp(\alpha_t)),$$

together with a scalar random walk model $\alpha_{t+1} = \alpha_t + \xi_t$, to obtain smoothed posterior mode estimates of π_t by generalized extended Kalman filtering and smoothing. The unknown hyperparameters were estimated by an EM–type algorithm. Using these estimates, a Gibbs sampler based on $G = 50$ Gibbs runs was applied to the data. The resulting posterior mean smoother is displayed in Figure 8.7 together with the posterior mode smoother (dashed line). Apart from minor differences, both smoothers are nearly identical. The smoothed curve is also in close agreement with the smoothing estimates of Kitagawa (1987), who applied posterior mean smoothing by numerical integration to the data.   □

8.4 Longitudinal data

Let the longitudinal or panel data consist of observations

$$(y_{it}, x_{it}), \quad i = 1, \ldots, n, \quad t = 1, \ldots, T$$

for a population of n units observed across time. Fixed effects models for such data are the subject of Section 6.2. Models with random effects across units i (Chapter 7) are an alternative that is sensible in particular if T is small compared to n. The state space modelling approach to longitudinal data allows, in principle, introducing random effects across units and time by including them both in a "large" state vector. For normal data, models of this kind have been proposed and studied previously; see, e.g., Rosenberg (1973), Hsiao (1986) and in particular the recent book of Jones (1993). Corresponding work for non–normal data is more rudimentary. This section treats primarily the case where effects or "states" vary only across time, and the approach is useful in particular for longer observation periods and smaller cross sections.

8.4.1 State space modelling of longitudinal data

In the sequel it will be convenient to collect individual observations in "panel waves"

$$y_t' = (y_{1t}', \ldots, y_{nt}'), \quad x_t' = (x_{1t}', \ldots, x_{nt}'), \quad t = 1, \ldots, T,$$

and to denote "histories" up to t by $y_t^* = (y_1, \ldots, y_t)$, $x_t^* = (x_1, \ldots, x_t)$ as before. In view of the longitudinal data situation, it is natural to consider models for individual responses y_{it} conditional on the predetermined observations y_{t-1}^*, x_t^* of past responses and of past and current covariates, and on individual parameter vectors α_{it}. Generally, these parameter vectors may contain parameters that are constant over units ("cross–fixed") or time, and parameters that vary over units ("cross–varying") or time. Within the linear exponential family framework a corresponding *observation model* is given by:

The conditional densities $p\left(y_{it} | \alpha_{it}, y_{t-1}^*, x_t^*\right)$ are of the simple exponential family type with mean

$$E(y_{it}|\alpha_{it}, y_{t-1}^*, x_t^*) = \mu_{it} = h\left(\eta_{it}\right) \qquad (8.4.1)$$

and linear predictor $\eta_{it} = Z_{it}\alpha_{it}$, where the design matrix Z_{it} is a function of covariates and, possibly, past responses.

Collecting individual parameters into "panel wave" parameter vectors $\alpha_t = (\alpha_{1t}, \ldots, \alpha_{nt})$, the observation model is supplemented by a *transition model* $p(\alpha_t|\alpha_{t-1})$, which we assume to be linear and Gaussian for simplicity:

$$\alpha_t = F_t\alpha_{t-1} + v_t, \qquad t = 1, 2, \ldots, \qquad (8.4.2)$$

with the usual assumptions on $v_t \sim N(0, Q_t)$, $\alpha_0 \sim N(0, Q_0)$.

To specify the model in terms of joint distributions, we add a further assumption to the basic assumptions (A1), (A2) and (A3) of Section 8.2.1:

(A4) Given $\alpha_t, y_{t-1}^*, x_t^*$, individual responses y_{it} within y_t are conditionally independent, i.e.,

$$p(y_t|\alpha_t, y_{t-1}^*, x_t^*) = \prod_{i=1}^n p(y_{it}|\alpha_t, y_{t-1}^*, x_t^*), \qquad t = 1, 2, \ldots.$$

This conditional independence property corresponds to independence assumptions on units in purely cross–sectional situations. For longitudinal data (A4) allows for interaction among units via the common "history" y_{t-1}^*, x_t^*.

The general observation model covers some important submodels by appropriate specifications of the linear predictor η_{it} and, as a consequence, of α_{it}. Restricting discussion to univariate responses, let the linear predictor be decomposed into

$$\eta_{it} = \delta_{it} + x_{it}'\beta_{it},$$

so that $Z_{it} = (1, x_{it}')$, and $\alpha_{it} = (\delta_{it}, \beta_{it})$ with a random intercept δ_{it} and a random slope β_{it}. An important subclass is models with *additive intercepts*

$$\delta_{it} = \tau_t + \gamma_t + \delta_i, \qquad (8.4.3)$$

where τ_t, γ_t are trend and seasonal components as in the previous sections, and δ_i is cross–varying. Slopes may be constant, $\beta_{it} = \beta$, only time–varying $\beta_{it} = \beta_t$ or only cross–varying, $\beta_{it} = \beta_i$.

All models of this kind can be put in state space form by appropriate specifications of $\alpha_t = (\alpha_{1t}, \ldots, \alpha_{nt})$ and F_t, v_t in the transition equation (8.4.2) in similar ways as in the linear Gaussian case (see, e.g., Rosenberg, 1973; Jones, 1993). It seems that, in principle, the filtering and smoothing algorithms of the previous section can be applied to the sequence $\{y_t, x_t\}$ of panel waves to estimate the sequence $\{\alpha_t = (\alpha_{1t}, \ldots, \alpha_{nt})\}$. However, the dimension of α_t and of (8.4.2) is now $\dim(\alpha_{it})$ times n, the size of the cross section. Without further restrictions or simplification, the computational amount becomes infeasible even for moderate n. A first attempt to decompose the filtering problem into n parallel filtering algorithms α_{it}, $i = 1, \ldots, n$, has been made in Fahrmeir, Kaufmann and Morawitz (1989), but the approach is not completely satisfying yet and the problem deserves further study.

Therefore, we confine attention to the case without cross–varying parameters as in Fahrmeir and Goss (1992). Then the observation model specializes to

$$\mu_{it} = h(Z_{it}\alpha_t), \qquad (8.4.4)$$

since $\alpha_{it} = \alpha_t$ for all $i = 1, \ldots, n$. Model (8.4.4) together with (8.4.2) may be used to estimate "population–averaged" trends, seasonal components, etc. simultaneously with covariate effects. This modelling approach is useful for comparably small n and larger T.

8.4.2   Filtering and smoothing

As for time series one might consider posterior mean or posterior mode estimation. However, at the time of writing, we are not aware of posterior mean approaches to non–Gaussian longitudinal data. Posterior mode estimation can be extended to the longitudinal data situation by applying the time series approach to panel waves $\{y_t, x_t\}$ together with the conditional independence assumption (A4). The resulting penalized log–likelihood criterion for (8.4.2), (8.4.4) becomes (in the same way as in Section 8.3.1)

$$\begin{aligned} L^*(\alpha_T^*) &= \sum_{t=1}^{T}\sum_{i=1}^{n} l_{it}(\alpha_t) - \frac{1}{2}(\alpha_0 - a_0)' Q_0^{-1}(\alpha_0 - a_0) \\ &\quad - \frac{1}{2}\sum_{t=1}^{T}(\alpha_t - F_t\alpha_{t-1})' Q_t^{-1}(\alpha_t - F_t\alpha_{t-1}), \end{aligned} \qquad (8.4.5)$$

where $l_{it}(\alpha_t)$ is the (conditional) log–likelihood contribution of observation y_{it}. The log–likelihood contribution $l_t(\alpha_t)$ of "panel wave" t is the sum

$$l_t(\alpha_t) = \sum_{i=1}^{n} l_{it}(\alpha_t) \qquad (8.4.6)$$

of individual contributions, and score functions and information matrices are also sums of individual contributions:

$$r_t(\alpha_t) = \sum_{i=1}^{n} r_{it}(\alpha_t), \quad R_t(\alpha_t) = \sum_{i=1}^{n} R_{it}(\alpha_t), \qquad (8.4.7)$$

with r_{it}, R_{it} as in (8.3.3), (8.3.4) with additional index i. Hence, posterior mode filtering and smoothing algorithms will have similar structure as in the time series situation $n = 1$, with additional loops in view of (8.4.7). We consider only the computationally less demanding form of generalized extended Kalman filtering and smoothing for known hyperparameters. Unknown hyperparameters θ can be estimated by the same EM–type algorithm as time series.

Generalized Kalman filter and smoother for longitudinal data*

The filter correction steps are given in three alternative forms. The fully recursive standard form (3) is most convenient, but should be applied carefully for small sample size n (less than 10 in our experience). The information matrix form (3)' is mathematically equivalent to (3). For small n, (3) or (3)' should be replaced by (3)*.

(1) *Initialization.*

$$
\begin{aligned}
a_{0|0} &= a_0, \\
V_{0|0} &= Q_0.
\end{aligned}
$$

For $t = 1, \ldots, T$:

(2) *Filter prediction step.*

$$
\begin{aligned}
a_{t|t-1} &= F_t a_{t-1|t-1}, \\
V_{t|t-1} &= F_t V_{t-1|t-1} F_t' + Q_t.
\end{aligned}
$$

For $t > 1$, $a_{t-1|t-1}, V_{t-1|t-1}$ are the final estimates of the preceding correction step.

(3) *Filter correction steps.*
Initial values:

$$
\begin{aligned}
a_{0,t} &= a_{t|t-1}, \\
V_{0,t} &= V_{t|t-1}.
\end{aligned}
$$

For $i = 1, \ldots, n$

$$
\begin{aligned}
a_{it} &= a_{i-1,t} + K_{it}(y_{it} - \mu_{it}) \\
V_{it} &= (I - K_{it} D_{it}' Z_{it}) V_{i-1,t}
\end{aligned}
$$

where

$$
K_{it} = V_{i-1,t} Z_{it}' D_{it} \left[D_{it}' Z_{it} V_{i,t-1} Z_{it}' D_{it} + \Sigma_{it} \right]^{-1},
$$

and $\mu_{it}, \Sigma_{it}, D_{it}$ are evaluated at $a_{i-1,t}$. Final estimates: $a_{t|t}, V_{t|t}$. For $t = T, \ldots, 1$:

(4) *Backward smoothing steps.*

$$a_{t-1|T} = a_{t-1|t-1} + B_t \left(a_{t|T} - a_{t|t-1}\right)$$
$$V_{t-1|T} = V_{t-1|t-1} + B_t \left(V_{t|T} - V_{t|t-1}\right) B_t'$$

where

$$B_t = V_{t-1|t-1} F_t' V_{t|t-1}^{-1}.$$

Alternative filter correction steps:

(3)' *Recursive scoring correction steps.*
Initial values as in 3. For $i = 1, \ldots, n$

$$a_{it} = a_{i-1,t} + V_{it} r_{it}$$
$$V_{it} = \left(V_{i-1,t}^{-1} + R_{it}\right)^{-1},$$

with local score functions r_{it} and information matrices R_{it} evaluated at $a_{i-1,t}$.

(3)* *Global scoring correction step.*

$$a_{t|t} = a_{t|t-1} + V_{t|t} r_t$$
$$V_{t|t} = \left(V_{t|t-1}^{-1} + R_t\right)^{-1},$$

with

$$r_t = \sum_{i=1}^{n} r_{it}, \qquad R_t = \sum_{i=1}^{n} R_{it}$$

evaluated at $a_{t|t-1}$.

The additional loop in the correction step (3) results from a further application of this filter to the observations (y_{it}, x_{it}), $i = 1, \ldots, n$, within (y_t, x_t), observing that $\alpha_{it} = \alpha_t$ is constant for transitions from unit to unit and that observations are (conditionally) independent according to (A4). The correction steps (3)', replacing computation of the Kalman gains by evaluations of individual score function and information matrix contributions, are obtained by an application of the matrix inversion lemma. Written in this form, the correction steps can be interpreted as a recursive local scoring algorithm for computing the posterior mode estimate $a_{t|t}$ using $a_{t|t-1}, V_{t|t-1}$ as prior information. The global correction step (3)* is obtained if the scoring form (2) of the generalized extended Kalman filter of Section 8.3.1 is applied to (y_t, x_t). Observing (A4), global score functions r_t and information matrices R_t are the sums of the corresponding individual contributions.

For comparably large cross sections, the form (3) of the correction steps is computationally most convenient. It should be noted, however, that this

TABLE 8.1. Contingency table for 55 firms of branch "Steine und Erden"

P_t		+	+	+	=	=	=	-	-	-
P_{t-1}		+	=	-	+	=	-	+	=	-
D_t	O_t									
+	+	35	16	0	6	77	3	0	0	0
+	=	110	100	17	33	214	10	0	1	11
+	-	7	17	9	6	42	13	0	5	14
=	+	21	18	4	40	234	12	1	2	2
=	=	60	67	14	167	2632	162	0	70	74
=	-	4	9	8	12	311	91	0	28	49
-	+	1	2	0	6	31	9	1	8	16
-	=	5	12	1	21	523	113	7	221	352
-	-	2	4	3	5	244	90	5	213	472

form requires an (arbitrary) ordering of units in the cross section. Such an ordering may have an effect on the estimation for a smaller number of units or if the data set contains influential or outlying observations. Then the form (3)* is appropriate for small n, since it does not require an arbitrary ordering of units i as in the steps (3) or (3)'. The effect of such an ordering is negligible for larger n.

If the size n of the cross section is large enough, asymptotic theory for ML estimation in cross sections (e.g., Fahrmeir and Kaufmann, 1985) also applies to posterior mode estimation of α_t which is constant across units within (y_t, x_t). Then the posterior distribution of a_t is approximately normal under rather mild conditions. As a consequence, posterior modes $a_{t|T}$ provide reasonable approximations to posterior expectations of α_t.

Example 8.5: Business test (Example 6.3, continued).
We apply the method to monthly IFO business test data collected in the industrial branch "Steine und Erden", for the period of January 1980 to December 1990. Firms in this branch manufacture initial products for the building trade industry.

The response variable is formed by the production plans P_t. Its conditional distribution is assumed to depend on the covariates "orders in hand" O_t and 'expected business condition" D_t, and on the production plans P_{t-1} of the previous month. No interaction effects are included. This choice is motivated by previous results of König, Nerlove and Oudiz (1982) and Nerlove (1983, p.1273), applying log–linear probability models to a large panel of branches. Attempts to carry out analyses for subgroups or smaller industry branches separately, which should also be of scientific interest, often run into problems due to nonexistence of estimates. The reason is a particular property of this longitudinal data set: if in doubt, firms seem to

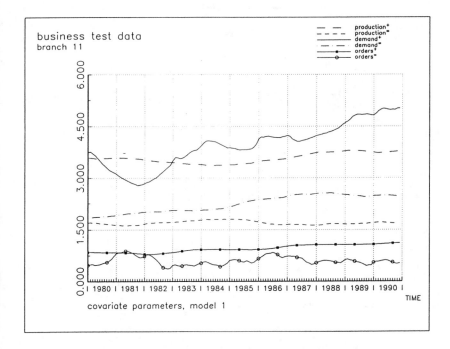

FIGURE 8.8. Covariate effects for model 1.

have a conservative tendency and prefer the "no change" ("=")–category. For multivariate categorical analyses, this results in a data pattern with a majority of entries in certain combinations while data are rather sparse in others. Table 8.1 displays such a pattern in the form of a contingency table for the variables P_t, P_{t-1}, D_t and O_t for the data set of 55 firms used in our analysis. As a consequence of such data sparseness, methods for analyzing time–varying effects by sequentially fitting models for cross sections (e.g., Stram et al., 1988) will often break down if applied to branches.

In the following each trichotomous ($k = 3$) variable is described by two ($q = 2$) dummy variables, with "−" as the reference category. Thus $(1,0)$, $(0,1)$ and $(0,0)$ stand for the responses +, = and −. The relevant dummies for "+" and "=" are shortened by $P_t^+, P_t^=$, etc. Then a cumulative logistic model with time–varying thresholds τ_{1t}, τ_{2t} and global covariate effects β_{1t} to β_{6t} is specified by

$$
\begin{aligned}
P(P_t = \text{`}+\text{'}) &= h(\tau_{1t} + \beta_{1t}P_{t-1}^+ + \beta_{2t}P_{t-1}^= + \beta_{3t}D_t^+ + \beta_{4t}D_t^= \\
&\quad + \beta_{5t}O_t^+ + \beta_{6t}O_t^=) \\
P(P_t = \text{`}+\text{'or`}=\text{'}) &= h(\tau_{2t} + \beta_{1t}P_{t-1}^+ + \beta_{2t}P_{t-1}^= + \beta_{3t}D_t^+ + \beta_{4t}D_t^= \\
&\quad + \beta_{5t}O_t^+ + \beta_{6t}O_t^=),
\end{aligned}
$$

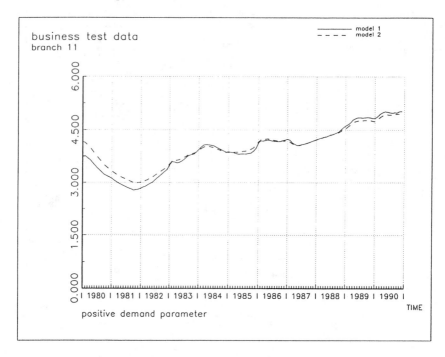

FIGURE 8.9. Time–varying effect of increasing demand for models 1 (———)
and 2 (- - - -).

where $P(P_t = \text{`}+\text{'})$ and $P(P_t = \text{`}+\text{' or `}=\text{'})$ stand for the probability
of increasing and nondecreasing production plans, and h is the logistic
distribution function. In a first analysis, parameter transitions of $\alpha_t =$
$(\tau_{1t}\tau_{2t}, \beta_{1t}, ..., \beta_{6t})$ were modelled by an eight–dimensional random walk of
first order, with ξ_0, Q_0 and Q unknown, Q_0 and Q diagonal. Smoothing
estimates of the covariate parameters for this "model 1" are displayed in
Figure 8.8. Apart from the D_t^+–parameter all effects are nearly constant
in time. An increase of production plans in the previous month (P_{t-1}^+) has
a high positive influence on current production plans, while the effect of
$P_{t-1}^=$ is still positive but distinctly smaller. Both effects are in agreement
with continuity in planning production. Compared to the effects of $D_t^+, D_t^=$,
which are both clearly positive on the average, the effects of increasing or
constant orders in hand ($O_t^+, O_t^=$) are still positive but surprisingly small.
This result, which is in agreement with previous findings, can be explained
as follows: The variable D serves as a substitute for expected demand. For
the purpose of short–range production planning, expected demand is more
relevant than current orders at hand, which are more relevant for current
production.

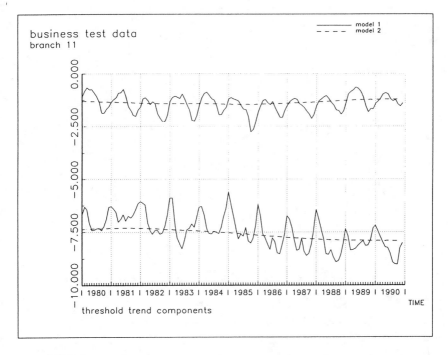

FIGURE 8.10. Trend parameters for models 1 (———) and 2 (- - - -).

Compared to the remaining effects, the parameter β_{3t} corresponding to the increase category D^+ of expected development of business has a remarkable temporal variation. Figure 8.9 exhibits a clear decline to a minimum at the beginning, and a distinct increase period coincides with the first months of the new German government in autumn 1982, ending with the elections to the German parliament in 1983. The growing positive effect of a positive state of business to the "increase" category of production plans indicates positive reactions of firms to the change of government. In Figure 8.10 both thresholds for model 1 exhibit seasonal variation corresponding to successive years. Threshold parameter τ_{1t} has peaks, mostly rather distinct, in December or January, and low values in the summer months. An explanation for this seasonal behaviour, which is not captured by covariate effects, may be the following: Firms in this specific branch manufacture initial products for the building industry. To be able to satisfy the increasing demand for their products in late winter/early spring, production plans are increased 2 to 3 months earlier. This is in agreement with the model, since higher values of τ_{1t} result in higher probabilities for increasing production plans, keeping covariate effects fixed. Similarly, decreasing values in spring and low values in summer reflect the tendency not to increase an already

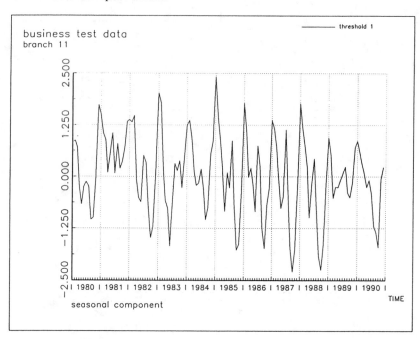

FIGURE 8.11. Seasonal effect of threshold 1, model 2.

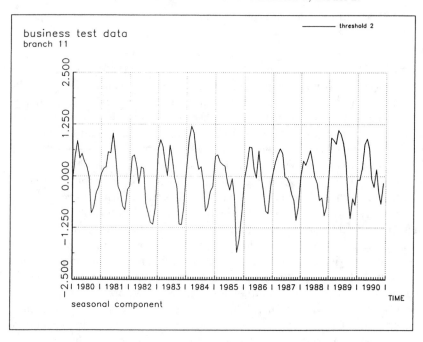

FIGURE 8.12. Seasonal effect of threshold 2, model 2.

comparably high level of production any further. The ups and downs of the second threshold parameter appear some months later. Interpretation is analogous and corresponds to seasonal ups and downs in the tendency of firms not to change their current production plans. To specify this seasonal effect more explicitly, a seasonal component in trigonometric form (Section 8.1.1) was included additionally (Model 2). Since seasonal variation is now modelled by these components, the trend parameters are now more or less constant in time (dashed line in Figure 8.10). Smoothing estimates of the seasonal components are shown in Figures 8.11 and 8.12 while Figure 8.9 (dashed line) shows the modified smoothed effect of the covariate D_t^+.

8.5   Further developments

Recent advances in Markov chain Monte Carlo (MCMC) simulation have led to a breakthrough in Bayesian inference in general, including full posterior analysis for non–Gaussian dynamic models. The Gibbs sampler is a useful yet special member of a whole family of techniques that are based on algorithms of the Metropolis–Hastings (MH)–type, compare Appendix A.5 for a short review.

Carter and Kohn (1994, 1996), Frühwirth–Schnatter (1994) and Shephard (1994) proposed modified Gibbs sampling schemes for estimating hyperparameters simultaneously with states in linear Gaussian state space models. The basic concepts of these samplers are also useful in conditionally Gaussian models, where error distributions are scale mixtures of normals. A popular choice are χ^2–mixture variables, leading to t–distributions for the errors and models that are robust against outliers in the observations errors and do not blur change points in the trend function or time–varying effects.

The design of efficient MCMC algorithms in non–Gaussian state space models is currently an intense research area. The algorithms of Shephard and Pitt (1995), Gamerman (1996) and Knorr–Held (1996) are based on blockmove MH algorithms. Knorr–Held (1996) and Fahrmeir and Knorr–Held (1997a) also deal with dynamic mixed models for longitudinal data that incorporate both time–varying parameters and random effects across units. A corresponding posterior mode approach has been developed by Biller (1997).

MCMC techniques for dynamic or state space models can also be adapted to semiparametric inference in generalized additive and varying coefficient models, see Fahrmeir and Knorr–Held (1997a) and Biller and Fahrmeir (1997).

9
Survival models

In recent years the analysis of survival time, lifetime or failure time data has received considerable attention. The methodology applies in medical trials where survival is of primary interest, and in reliability experiments where failure time is the duration of interest. We will mostly refer to survival time although in principle situations are considered where the time until the occurrence of some event is of interest.

There is a considerable number of excellent books on survival analysis. In particular, the case of continuous time is treated extensively in Lawless (1982), Kalbfleisch and Prentice (1980), Blossfeld, Hamerle and Mayer (1989) and Lancaster (1990). In the following models for continuous survival time, which are covered in these books, are only sketched. In particular models are considered that may be estimated in a way similar to generalized linear models.

In application, time is often measured as a discrete variable. For example, in studies on the duration of unemployment (see Example 9.1) in most cases time of unemployment is given in months. In the following a more extensive treatment is given for this case of grouped or discrete survival time.

9.1 Models for continuous time

9.1.1 Basic models

Survival time is considered a non–negative random variable T. For T continuous, let $f(t)$ denote the probability density function and $F(t)$ denote the corresponding distribution function given by

$$F(t) = P(T \leq t) = \int_0^t f(u)du.$$

The probability of an individual surviving until time t is given by the so–called *survivor function*

$$S(t) = P(T > t) = 1 - F(t)$$

which sometimes is also called *reliability function*. A basic concept in the analysis of survival time is the *hazard function* $\lambda(t)$, which is defined as the limit

$$\lambda(t) = \lim_{\Delta t \to 0} \frac{P(t \leq T < t + \Delta t | T \geq t)}{\Delta t}. \qquad (9.1.1)$$

The hazard function measures the instantaneous rate of death or failure at time t given the individual survives up until t. Sometimes it is also useful to consider the cumulative hazard function

$$\Lambda(t) = \int_0^t \lambda(u)du.$$

The connection between distribution function, density, hazard function and cumulative hazard function is given by the equations

$$\lambda(t) = \frac{f(t)}{S(t)}$$

$$S(t) = \exp\left(-\int_0^t \lambda(u)du\right) = \exp(-\Lambda(t))$$

$$f(t) = \lambda(t)\exp\left(-\int_0^t \lambda(u)du\right) = \lambda(t)\exp(-\Lambda(t))$$

(see, e.g., Lawless, 1982). The distribution of T is determined totally by one of these quantities. Basic models illustrate the relation between these basic concepts.

Exponential distribution

Assume a constant hazard function

$$\lambda(t) = \lambda, \quad t \geq 0$$

where $\lambda > 0$. Then the probability density function is given by

$$f(t) = \lambda e^{-\lambda t}$$

which is the density of an exponential distribution with parameter λ. The expectation is given by $1/\lambda$ and the variance by $1/\lambda^2$.

Weibull distribution

Let the hazard function be determined by

$$\lambda(t) = \lambda\alpha(\lambda t)^{\alpha-1}$$

where $\lambda > 0, \alpha > 0$ are parameters sometimes referred to as shape and scale parameters. Equivalently, one may consider the Weibull density function

$$f(t) = \lambda\alpha(\lambda t)^{\alpha-1}\exp(-(\lambda t)^{\alpha}) \qquad (9.1.2)$$

or the survivor function

$$S(t) = \exp(-(\lambda t)^{\alpha}).$$

Characterized by two parameters, the distribution is more flexible than the exponential model, which is included as the special case where $\alpha = 1$. The hazard function is increasing for $\alpha > 1$, decreasing for $\alpha < 1$ and constant for $\alpha = 1$. Expectation and variance are given by

$$E(T) = \Gamma\left(\frac{1+\alpha}{\alpha}\right)/\lambda,$$

$$\text{var}(T) = \left(\Gamma\left(\frac{\alpha+2}{\alpha}\right) - \Gamma\left(\frac{\alpha+1}{\alpha}\right)^2\right)/\lambda^2.$$

If T is Weibull distributed with parameters λ and α, the transformation $Y = \log T$ has the extreme value distribution with density function

$$f(y) = \frac{1}{\sigma}\exp[(y-u)/\sigma - \exp((y-u)/\sigma)] \qquad (9.1.3)$$

and distribution function

$$F(y) = 1 - \exp[-\exp((y-u)/\sigma)]$$

where $u = -\log\lambda$ and $\sigma = 1/\alpha$ are location and scale parameters. For $u = 0, \sigma = 1$ (9.1.3) is the standard (minimum) extreme value function.

9.1.2   Parametric regression models

Now let $x = (x_1, \ldots, x_p)$ denote a set of (time–independent) covariates that influence the lifetime T. Then one has to consider the distribution of T, given x, and thus population density $f(t)$, hazard function $\lambda(t)$ and survivor function $S(t)$ become density $f(t|x)$, hazard function $\lambda(t|x)$ and survivor function $S(t|x)$.

Location–scale models for log T

There are several approaches to the construction of regression models for lifetime data. *Location–scale models* for the log–lifetime $Y = \log T$ have the form

$$\log T = \mu(x) + \sigma \varepsilon \tag{9.1.4}$$

where σ is a constant scale parameter and ε is a noise variable independent of x. If ε follows the standard extreme value distribution (9.1.3) with $u = 0, a = 1$ then $Y = \log T$ has density function

$$f(y|x) = \frac{1}{\sigma} \exp \left[\frac{y - \mu(x)}{\sigma} - \exp \left(\frac{y - \mu(x)}{\sigma} \right) \right]$$

and the lifetime T is Weibull distributed with density function

$$f(t|x) = \frac{1}{\sigma} \exp \left[- \mu(x) \right] \left[t \exp(-\mu(x)) \right]^{1/\sigma - 1} \exp \left[- (t \exp(-\mu(x))^{1/\sigma} \right].$$
$$\tag{9.1.5}$$

That means the shape parameter λ in (9.1.2) is specified by $\lambda = \exp(-\mu(x))$ and the scale parameter α in (9.1.2) equals a constant $\alpha = 1/\sigma$. Only the shape parameter is influenced by the covariates, whereas the scale parameter is independent of x. Models where α is normally distributed or follows a log–gamma distribution are considered extensively in Lawless (1982).

Proportional hazards models

The Weibull distribution model with covariates given by (9.1.5) has the hazard function

$$\lambda(t|x) = \frac{\exp(-\mu(x))}{\sigma} \left[t \exp(-\mu(x)) \right]^{1/\sigma - 1}$$

or equivalently

$$\lambda(t|x) = \frac{t^{1/\sigma - 1}}{\sigma} \exp \left[\frac{-\mu(x)}{\sigma} \right]. \tag{9.1.6}$$

For two subpopulations or individuals characterized by x_1 and x_2 it is immediately seen that

$$\frac{\lambda(t|x_2)}{\lambda(t|x_1)} = \exp \left[\frac{\mu(x_2) - \mu(x_1)}{\sigma} \right].$$

That means that the ratio $\lambda(t|x_1)/\lambda(t|x_2)$ does not depend on time t. If the hazard for the first individual is twice the hazard for the second individual after a year, the ratio is the same after 2 years, 2.5 years, etc; the ratio of risks is the same at any time. Models that have this property are called *proportional hazards models.* The Weibull distribution model is a location–scale model of type (9.1.4) and a proportional hazards model. However, in general location–scale models for $\log T$ do not show proportional hazards. Assuming, for example, a normal distribution for ε in (9.1.4) yields hazards that are not proportional over time. A very general proportional hazards model due to Cox (1972) is given by

$$\lambda(t|x) = \lambda_0(t)\exp(x'\gamma) \qquad (9.1.7)$$

where the baseline hazard function $\lambda_0(t)$ is assumed to be the same for all observations but is not assumed known. In contrast to the Weibull model (9.1.6) no specific structure is assumed for the baseline hazard. Estimation of γ may be based on marginal likelihood (Kalbfleisch and Prentice, 1973) or the concept of partial likelihood (Cox, 1972, 1975; Tsiatis, 1981; Prentice and Self, 1983).

Linear transformation models and binary regression models

Linear transformation models have the general form

$$h(T) = x'\gamma + \varepsilon \qquad (9.1.8)$$

where h is an increasing continuous function and ε is a random error variable with distribution function F_ε. Obviously, location–scale models for the log–lifetime are special cases where $h = \log$. It is immediately seen that (9.1.8) is equivalent to

$$P(T \le t|x) = F_\varepsilon(h(t) - x'\gamma)$$

for all t. For the logistic distribution function $F_\varepsilon(z) = 1/(1+\exp(-z))$ one gets the *proportional odds model*

$$\log \frac{P(T \le t|x)}{P(T > t|x)} = h(t) - x'\gamma,$$

which for fixed t may be considered a binary response model with response

$$Y_t = \begin{cases} 1, & T \le t \\ 0, & T > t. \end{cases}$$

The connection between Cox's proportional hazards model and linear transformation models is seen if for F_ε the extreme value distribution $F_\varepsilon(z) = 1 - \exp(-\exp(z))$ is assumed. Then one has

$$\log(-\log P(T > t|x)) = h(t) - x'\gamma. \qquad (9.1.9)$$

Let $h(t)$ be defined by

$$h(t) = \log \int_0^t \lambda_0(s)ds$$

where $\lambda_0(t)$ is a function fulfilling $\lambda_0 \geq 0$. Then one gets for the hazard function

$$\lambda(t|x) = \lambda_0(t)\exp(-x'\gamma) \qquad (9.1.10)$$

where λ_0 is the baseline hazard function. Equivalently one has

$$S(t|x) = S_0(t)^{\exp(-x'\gamma)}$$

where $S_0(t) = \exp\left(-\int_0^t \lambda_0(s)ds\right)$ is the baseline survivor function. Obviously the model is equivalent to the Cox model (9.1.7); only the sign of the parameter vector γ has changed. Therefore the Cox model is a linear transformation model with unknown transformation h given the baseline hazard function $\lambda_0(s)$ is unspecified (see also Doksum and Gasko, 1990). In fact the class of models defined by (9.1.10) is invariant under the group of differentiable strictly monotone increasing transformations on t. If g is a transformation the hazard of $t' = g(t)$ given by $\lambda_0(g^{-1}(t'))\partial g^{-1}(t')/\partial t' \exp(-x'\gamma)$ is again of type (9.1.10) (see Kalbfleisch and Prentice, 1973).

9.1.3 Censoring

In survival analysis, most often only a portion of the observed times can be considered exact lifetimes. For the rest of the observations one knows only that the lifetime exceeds a certain value. This feature is referred to as censoring. More specifically a lifetime is called right–censored at t if it is known that the life time is greater than or equal t but the exact value is not known. There are several types of censoring due to the sampling situation.

Random censoring

Random censoring is a concept that is often assumed to hold in observation studies over time. It is assumed that each individual (unit) i in the study has a lifetime T_i and a censoring time C_i that are independent random variables. The observed time is given by $t_i = \min(T_i, C_i)$. It is often useful to introduce an indicator variable for censoring by

$$\delta_i = \left\{ \begin{array}{ll} 1, & T_i < C_i \\ 0, & T_i \geq C_i. \end{array} \right.$$

The data may now be represented by (t_i, δ_i). Let f_c, S_c denote the density function and survivor function for the censoring variable C. Then the likelihood for an uncensored observation $(t_i, \delta_i = 1)$ is given by

$$f_i(t_i)S_c(t_i)$$

as the product of the lifetime density at t_i and the probability for censoring time greater than t_i given by $P(C_i > t_i) = S_c(t_i)$. For a censored observation $(t_i, \delta_i = 0)$ the likelihood is given by

$$f_c(t_i)S_i(t_i)$$

as the product of the censoring density at t_i and the probability of lifetimes greater than t_i given by $P(T_i > t_i) = S_i(t_i)$. Combined into a single expression the likelihood for observation (t_i, δ_i) is given by

$$[f_i(t_i)S_c(t_i)]^{\delta_i} [f_c(t_i)S_i(t_i)]^{1-\delta_i} = [f_i(t_i)^{\delta_i} S_i(t_i)^{1-\delta_i}] [f_c(t_i)^{1-\delta_i} S_c(t_i)^{\delta_i}].$$

The likelihood for the sample $(t_i, \delta_i), i = 1, \ldots, n$, is given by

$$L = \prod_{i=1}^{n} [f_i(t_i)^{\delta_i} S_i(t_i)^{1-\delta_i}] [f_c(t_i)^{1-\delta_i} S_c(t_i)^{\delta_i}].$$

If the censoring time is not determined by parameters of interest, i.e., censoring is noninformative, the likelihood may be reduced to

$$L = \prod_{i=1}^{n} f(t_i)^{\delta_i} S(t_i)^{1-\delta_i}.$$

Type I censoring

Sometimes in life test experiments there is a fixed observation time. Exact lifetimes are only known for items that fail by this fixed time. All other observations are right censored. More generally each item may have a specific censoring time C_i that is considered fixed in advance. The likelihood for observation (t_i, δ_i) is given by

$$L_i = f_i(t_i)^{\delta_i} S_i(C_i)^{1-\delta_i}.$$

Since if $\delta_i = 0$ the observation $t_i = \min\{T_i, C_i\}$ has value $t_i = C_i$ the likelihood is equivalent to the reduced likelihood for random censoring. In fact, Type I censoring may be considered a special case of random censoring when degenerate censoring times are allowed.

Alternative censoring schemes like Type II censoring, where only a fixed proportion of observations is uncensored, and more general censoring schemes are considered in detail in Lawless (1982).

9.1.4   Estimation

If covariates are present the data are given by triples $(t_i, \delta_i, x_i), i \geq 1$, where t_i is the observed time, δ_i is the indicator variable and x_i denotes the vector of covariates of the ith observation. Aitkin and Clayton (1980) show how parametric survival models as the exponential model, the Weibull model and the extreme value model are easily estimated within the framework of generalized linear models. Consider the general proportional hazards model

$$\lambda(t|x) = \lambda_0(t) \exp(x'\gamma)$$

with survivor function

$$S(t|x) = \exp(-\Lambda_0(t) \exp(x'\gamma))$$

and density

$$f(t|x) = \lambda_0(t) \exp\left(x'\gamma - \Lambda_0(t) \exp(x'\gamma)\right)$$

where $\Lambda_0(t) = \int_0^t \lambda_0(u) du$. Assuming random censoring, which is noninformative, one gets the likelihood

$$
\begin{aligned}
L &= \prod_{i=1}^{n} f(t_i)^{\delta_i} S(t_i)^{1-\delta_i} = \\
&= \prod_{i=1}^{n} [\lambda_0(t_i) \exp(x_i'\gamma - \Lambda_0(t_i) \exp(x_i'\gamma))]^{\delta_i} \exp\left(-\Lambda_0(t_i) \exp(x_i'\gamma)\right)^{1-\delta_i} \\
&= \prod_{i=1}^{n} \mu_i^{\delta_i} e^{-\mu_i} \left(\frac{\lambda_0(t_i)}{\Lambda_0(t_i)}\right)^{\delta_i}
\end{aligned}
$$

where $\mu_i = \Lambda_0(t_i) \exp(x_i'\gamma)$. The second term depends only on the baseline hazard function and does not involve the parameter γ. The first term is equivalent to the kernel of the likelihood function of Poisson variates $\delta_i \sim P(\mu_i)$. The corresponding log–linear Poisson model, which has the same likelihood function, is given by

$$\log(\mu_i) = \log \Lambda_0(t_i) + x_i'\gamma$$

which is a linear model with a constant $\log \Lambda_0(t_i)$. For specific models considered in the following the log–likelihood may be maximized in the same way as for GLM's.

Exponential model

For the exponential model we have $\lambda(t) = \lambda_0 \exp(x'\gamma)$ with baseline hazard function $\lambda_0(t) = \lambda_0$ and $\Lambda_0(t) = \lambda_0 t$. The second term in the likelihood

$\lambda_0(t)/\Lambda_0(t) = 1/t$ does not depend on any further parameters. Thus maximization of the first term will do. This is equivalent to estimate γ for the Poisson model

$$\log(\mu_i) = \log(\lambda_0 t_i) + x_i'\gamma.$$

By taking $\log \lambda_0$ into the linear term one considers the model

$$\log(\mu_i) = \log t_i + x_i'\gamma. \tag{9.1.11}$$

Consequently the log–linear model may be fitted where $\log t_i$ is included in the regression model with a known coefficient of 1. In GLIM terminology such a variable is called offset (Aitkin et al., 1989).

Weibull model

For the Weibull model one has $\lambda_0(t) = t^{1/\sigma-1}/\sigma$ and $\Lambda_0(t) = t^{1/\sigma}$. Now, the second term in the likelihood $\lambda_0(t)/\Lambda_0(t) = 1/(t\sigma)$ depends on the unknown parameter σ. Instead of σ one may use $\alpha = 1/\sigma$. The log–likelihood now is given by

$$l = \sum_{i=1}^{n}(\delta_i \log(\mu_i) - \mu_i) - \sum_{i=1}^{n}\delta_i \log t_i + \delta \log \alpha$$

where $\delta = \sum_{i=1}^{n} \delta_i$ is the number of uncensored observations. The Poisson model corresponding to the first term is given by

$$\log(\mu_i) = \alpha \log t_i + x_i'\gamma.$$

From the likelihood equations

$$\begin{aligned}
\frac{\partial l}{\partial \gamma} &= \sum_i x_i(\delta_i - \mu_i) = 0 \\
\frac{\partial l}{\partial \alpha} &= \sum_i \log(t_i)(\delta_i - \mu_i) + \frac{\delta}{\alpha} = 0
\end{aligned}$$

one gets for the maximum likelihood estimate

$$\hat{\alpha} = \left(\frac{1}{\delta}\sum_i \log(t_i)(\mu_i - \delta_i)\right)^{-1}. \tag{9.1.12}$$

Aitkin and Clayton (1980) propose estimating γ and α iteratively by starting with $\hat{\alpha}^{(0)} = 1$, i.e., the exponential model. Fitting of model (9.1.11) yields estimates $\hat{\mu}_i^{(0)}$. Inserting $\mu_i^{(0)}$ in (9.1.12) yields an estimate $\tilde{\alpha}^{(0)}$. Now the Poisson model

$$\log(\mu_i) = \hat{\alpha}^{(1)} \log t_i + x\gamma$$

with offset $\hat{\alpha}^{(1)} \log t_i$ is fitted where $\hat{\alpha}^{(1)} = (\hat{\alpha}^{(0)} + \tilde{\alpha}^{(0)})/2$. This process is continued until convergence. According to Aitkin and Clayton (1980) the damping of the successive estimates of λ improves the convergence.

For the fitting of models based on the extreme value distributions see Aitkin and Clayton (1980).

9.2 Models for discrete time

Often time cannot be observed continuously; it is only known to lie between a pair of consecutive follow ups. Data of this kind are known as interval censored. Since many ties occur, these data cause problems when partial likelihood methods for continuous time models are used.

In the following let time be divided into k intervals $[a_0, a_1), [a_1, a_2), \ldots$ $\ldots, [a_{q-1}, a_q), [a_q, \infty)$ where $q = k - 1$. Often for the first interval $a_0 = 0$ may be assumed and a_q denotes the final follow up. Instead of observing continuous time the discrete time $T \in \{1, \ldots, k\}$ is observed where $T = t$ denotes failure within the interval $[a_{t-1}, a_t)$. The *discrete hazard function* is given by

$$\lambda(t|x) = P(T = t|T \geq t, x), \quad t = 1, \ldots, q, \qquad (9.2.1)$$

which is a conditional probability for the risk of failure in interval $[a_{t-1}, a_t)$ given the interval is reached.

The *discrete survivor function* for surviving interval $[a_{t-1}, a_t)$ is given by

$$S(t|x) = P(T > t|x) = \prod_{i=1}^{t}(1 - \lambda(i|x)). \qquad (9.2.2)$$

Alternatively one may consider the probability for reaching interval $[a_{t-1}, a_t)$ as a survivor function. With

$$\tilde{S}(t|x) = P(T \geq t|x) = \prod_{i=1}^{t-1}(1 - \lambda(i|x)) \qquad (9.2.3)$$

one gets $\tilde{S}(t|x) = S(t - 1|x)$. The unconditional probability for failure in interval $[a_{t-1}, a_t)$ is given by

$$P(T = t|x) = \lambda(t|x)\prod_{i=1}^{t-1}(1 - \lambda(i|x)) = \lambda(t|x)\tilde{S}(t|x). \qquad (9.2.4)$$

Assuming covariates that do not depend on time the data are given in the form (t_i, x_i, δ_i) where δ_i is the indicator variable for censoring given by

$$\delta_i = \begin{cases} 1, & \text{failure in interval } [a_{t_i-1}, a_{t_i}) \\ 0, & \text{censoring in interval } [a_{t_i-1}, a_{t_i}). \end{cases}$$

In cases where the intervals depend on the individuals, e.g., if individuals miss visits in a periodic follow up, the concept of discrete survival time must be somewhat modified (Finkelstein, 1986).

9.2.1   Life table estimates

A simple way to describe survival data for the total sample or for subpopulations (without reference to covariates) is by the life table. The method is described for discrete time but it is also a useful nonparametric estimate for continuous time observations after defining intervals. Let

d_r denote the number of observed lifetimes in interval $[a_{r-1}, a_r)$ (deaths)

w_r denote the numbers of censored observations in interval $[a_{r-1}, a_r)$ (withdrawals).

The number of observations at risk in the rth interval is given by

$$n_r = n_{r-1} - d_{r-1} - w_{r-1} \tag{9.2.5}$$

where with $d_0 = w_0 = 0$ we have $n_1 = n$. Without censoring, the natural estimate for $\lambda(t|x)$ is given by

$$\hat{\lambda}_t = \frac{d_t}{n_t}. \tag{9.2.6}$$

For $w_t > 0$ the so–called standard life table estimate takes the withdrawals into account by

$$\hat{\lambda}_t = \frac{d_t}{n_t - w_t/2}. \tag{9.2.7}$$

The latter estimate considers withdrawals being under risk for half the interval. Under censoring, the first estimate (9.2.6) is appropriate if all withdrawals are assumed to occur at the end of the interval $[a_{t-1}, a_t)$. If withdrawals are assumed to occur right at the beginning of the interval $[a_{t-1}, a_t)$ the appropriate choice is

$$\hat{\lambda}_t = \frac{d_t}{n_t - w_t}. \tag{9.2.8}$$

The standard life table estimate is a compromise between (9.2.6) and (9.2.8). Based on (9.2.2) the probability for surviving beyond a_t may be estimated by

$$\hat{S}(t) = \prod_{i=1}^{t}(1 - \hat{\lambda}_i). \tag{9.2.9}$$

Consequently the estimated probability for failure in interval $[a_{t-1}, a_t)$ is given by

$$\hat{P}(T = t) = \hat{\lambda}_t \prod_{i=1}^{t-1}(1 - \hat{\lambda}_i).$$

Without censoring ($w_t = 0$) the number of deaths is multinomially distributed with $(d_1, \ldots, d_q) \sim M(n, (\pi_1, \ldots, \pi_q))$ where $\pi_t = P(T = t)$. Using (9.2.5) and (9.2.9) yields the simple estimate

$$\hat{S}(t) = \frac{n - d_1 - \ldots - d_t}{n}$$

which is the number of individuals surviving beyond a_t divided by the sample size. Since $n - d_1 - \ldots - d_t \sim B(n, S(t))$ expectation and variance are given by

$$
\begin{aligned}
E(\hat{S}(t)) &= S(t) \\
\mathrm{var}(\hat{S}(t)) &= S(t)(1 - S(t))/n
\end{aligned}
$$

and the covariance for $t_1 < t_2$ is given by

$$\mathrm{cov}(\hat{S}(t_1), \hat{S}(t_2)) = \frac{(1 - S(t_1))S(t_2)}{n}.$$

For $\hat{\lambda}_t$ one gets

$$
\begin{aligned}
E(\hat{\lambda}_t) &= \lambda(t) \\
\mathrm{var}(\hat{\lambda}_t) &= \lambda(t)(1 - \lambda(t))E(1/n_t)
\end{aligned}
$$

where it is assumed that $n_t > 0$. In the censoring case ($w_t > 0$) we have a multinomial distribution $(d_1, w_1, \ldots, d_q, w_q) \sim M(n, (\pi_1^d, \pi_1^w, \ldots, \pi_q^d, \pi_q^w))$. Considering continuous lifetime T and censoring time C the probabilities are given by

$$\pi_r^d = P(T \in [a_{r-1}, a_r), T \le C), \quad \pi_r^w = P(C \in [a_{r-1}, a_r), C < T).$$

Since the frequencies $(d_1/n, w_1/n, \ldots, w_q/n)$ are asymptotically normally distributed the standard life table estimate $\hat{\lambda}_t = d_t/(n_t - w_t/2)$ is also asymptotically normal with expectation

$$\lambda_t^* = \frac{\pi_t^d}{\pi_t^0 + \pi_t^w/2}$$

where $\pi_t^0 = E(n_t/n)$. However, only in the case without withdrawals $\lambda_t^* = \lambda(t)$. Thus the standard life table estimate is not a consistent estimate. Lawless (1982) derives for the asymptotic variance for the random censorship model

$$\widehat{\text{var}}(\hat{\lambda}_t) = \frac{1}{n}(\lambda_t^* - \lambda_t^{*2}) \frac{\pi_t^0 - \pi_t^w/4}{(\pi_t^0 - \pi_t^w/2)(\pi_t^0 - \pi_t^w/2)}.$$

For the covariance $\text{cov}(\hat{\lambda}_{t_1}, \hat{\lambda}_{t_2}) = 0$ holds asymptotically. Since $(n_t - w_t/2)/n$ converges to $\pi_t^0 - \pi_t^w/2$ the usual estimate

$$\widehat{\text{var}}(\hat{\lambda}_t) = \frac{\hat{\lambda}_t - \hat{\lambda}_t^2}{n_t - w_t/2}$$

will overestimate $\text{var}(\hat{\lambda}_t)$ if λ_t and λ_t^* are not too different. For $\hat{S}(t)$ Lawless (1982) derives for large sample sizes

$$\text{var}(\hat{S}(t)) = S^*(t)^2 \sum_{i=1}^{t} \frac{\text{var}(1 - \hat{\lambda}_i)}{(1 - \lambda_i^*)^2}$$

where $S^*(t) = \prod_{i=1}^{t}(1 - \lambda_i^*)$. Approximating $\text{var}(1 - \hat{\lambda}_t)$ by $\hat{\lambda}_t(1 - \hat{\lambda}_t)/(n_t - w_t/2)$ and $S^*(t), \lambda_t^*$ by $\hat{S}(t), \hat{\lambda}_t$ yields Greenwood's (1926) often–used formula

$$\text{var}(\hat{S}(t)) = \hat{S}(t)^2 \sum_{i=1}^{t} \frac{\hat{\lambda}_i}{(1 - \hat{\lambda}_i)(n_t - w_i/2)}$$

as an approximation.

Example 9.1: Duration of unemployment
The data set comprises 1669 unemployed persons who are observed from January 1983 until December 1988 in the socioeconomic panel in Germany (Hanefeld, 1987). Time is measured in months. As absorbing state only the employment in a full–time job is considered. All other causes for the ending of unemployment (employment in a part–time job or going back to school) are considered censoring. Figure 9.1 shows the estimated hazard rates based on (9.2.8). Although showing the typical picture of unemployment data (short increase, slow decrease) the estimate is quite irregular. In particular, when the local sample size is small (for long–time unemployment) the curve is quite jagged. Figure 9.2 shows the corresponding survivor function based on the life table estimate. The rather smooth survivor function gives an estimate for the probability of being still unemployed after t months.   □

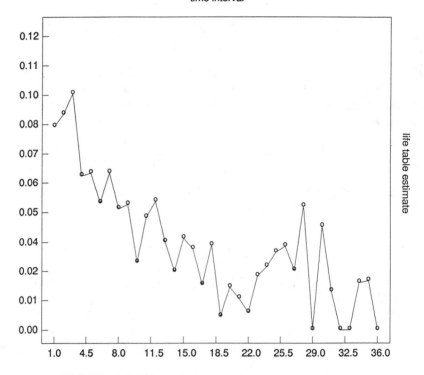

FIGURE 9.1. Life table estimate for unemployment data.

9.2.2 Parametric regression models

In this section we consider the case where discrete lifetime T_i depends on a vector of covariates x_i.

The grouped proportional hazards model

The proportional hazards or Cox model (9.1.7) for continuous time is given by

$$\lambda_c(t|x) = \lambda_0(t) \exp(x'\gamma) \qquad (9.2.10)$$

where $\lambda_c(t|x)$ stands for the continuous hazard function. If time T is considered a discrete random variable where $T = t$ denotes failure within the interval $[a_{t-1}, a_t)$ the assumption of (9.2.10) yields the grouped proportional hazards model

$$\lambda(t|x) = 1 - \exp(-\exp(\gamma_t + x'\gamma)) \qquad (9.2.11)$$

time interval

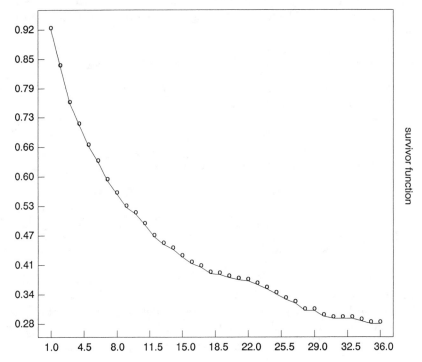

FIGURE 9.2. Estimated survivor function for unemployment data.

where the parameters

$$\gamma_t = \log(\exp(\theta_t) - \exp(\theta_{t-1})) \quad \text{with} \quad \theta_t = \log \int_0^{a_t} \lambda_0(u)du$$

are derived from the baseline hazard function $\lambda_0(u)$ (see Kalbfleisch and Prentice, 1973, 1980). It should be noted that the parameter vector γ is unchanged by the transition to the discrete version. That means, as far as the influence of covariates x is concerned, the discrete model allows the same analysis as the proportional hazards model. Alternative formulations of (9.2.11) are given by

$$\log(-\log(1 - \lambda(t|x))) = \gamma_t + x'\gamma \qquad (9.2.12)$$

and

$$\log(-\log(P(T > t|x))) = \theta_t + x'\gamma. \qquad (9.2.13)$$

Since the hazard rate is given by $\lambda(t|x) = P(T = t|T \geq t, x)$, it is seen that the grouped proportional hazards model (9.2.11) is a sequential model with distribution function $F(u) = 1 - \exp(-\exp(u))$ (see 3.3.16) in Chapter 3).

In general, grouping implies a loss of information. The extent of the loss of information is considered, e.g., by Gould and Lawless (1988) for a special model.

A generalized version: the model of Aranda–Ordaz

Instead of considering the multiplicative model (9.2.10) one may start from an additive form for continuous time by assuming

$$\lambda_c(t|x) = \lambda_0(t) + x'\gamma.$$

Then for the discrete time T one can derive

$$-\log(1 - \lambda(t|x)) = \rho_t - \rho_{t-1} + (a_t - a_{t-1})x'\gamma$$

where $\rho_t = \int_0^{a_t} \lambda_0(u)du$. If intervals are equidistant, i.e., $\Delta = a_t - a_{t-1}$, the discrete model has the form

$$-\log(1 - \lambda(t|x)) = \delta_t + x'\gamma \tag{9.2.14}$$

where Δ is absorbed into the parameter vector γ and $\delta_t = \rho_t - \rho_{t-1}$. Aranda–Ordaz (1983) proposed a general model family that includes the grouped Cox model (9.2.11) as well as the grouped version of the additive model (9.2.14). The model family is given by

$$
\begin{aligned}
\log(-\log(1 - \lambda(t|x))) &= \gamma_t + x'\gamma \quad \text{for} \quad \alpha = 0 \\
\left[\{-\log(1 - \lambda(t|x))\}^\alpha - 1\right]/\alpha &= \gamma_t + x'\gamma \quad \text{for} \quad \alpha \neq 0.
\end{aligned}
\tag{9.2.15}
$$

For $\alpha = 0$ one gets the grouped Cox model, for $\alpha = 1$ one gets model (9.2.14) with $\delta_t = 1 + \gamma_t$. Thus (9.2.15) includes both cases. The model may also be written in the form

$$\lambda(t|x) = F_\alpha(\gamma_t + x'\gamma)$$

where F_α is the distribution function

$$
F_\alpha(u) = \begin{cases}
1 - \exp\left(-(1 + \alpha u)^{1/\alpha}\right), & \text{for } u \in [-1/\alpha, \infty), \\
0, & \text{else,}
\end{cases}
$$

which depends on the additional parameter α. In this form it becomes clear that the grouped proportional harzards model is the limiting case $\alpha \to 0$. However, it also becomes clear that the range of parameters γ is restricted here. In order to avoid problems the linear predictor should fulfill $\gamma_t + x'\gamma > -1/\alpha$.

The logistic model

An alternative model, which has been considered by Thompson (1977), is
the logistic model for the discrete hazard

$$\lambda(t|x) = \frac{\exp(\gamma_t + x'\gamma)}{1 + \exp(\gamma_t + x'\gamma)}.$$

The model differs only slightly from the discrete logistic model given by
Cox (1972). Thompson (1977) also shows that the model is very similar to
the proportional hazards model if the grouping intervals become short.

Sequential model and parameterization of the baseline hazard

The common structure of the discrete survival models of the previous sections is that of the sequential model in Section 3.3.4. The discrete hazard
has the form

$$\lambda(t|x) = F(\gamma_{0t} + x'\gamma)$$

where F is a distribution function that for the model of Aranda–Ordaz depends on an additional parameter. In ordinal regression models the number
of response categories is usually very small. However, for survival models
the number of time intervals may be very large, e.g., if the data are given
in months. Thus, the number of parameters $\gamma_{01}, \ldots, \gamma_{0q}$ that represent the
baseline hazard may be dangerously high. An approximation of the base-
line hazard function used, e.g., by Mantel and Hankey (1978) is given by a
polynomial of degree s

$$\gamma_{0t} = \sum_{i=0}^{s} \alpha_i t^i. \tag{9.2.16}$$

An alternative approach used by Efron (1988) for the case without covariates is based on simple regression splines. He considers a cubic–linear spline
of the form

$$\gamma_{0t} = \alpha_0 + \alpha_1 t + \alpha_2 (t - t_c)_-^2 + \alpha_3 (t - t_c)_-^3$$

where $(t - t_c)_- = \min\{(t - t_c), 0\}$. Here t_c is a cut–off point chosen from the
data; the baseline hazard function is cubic before t_c and linear after t_c. The
reason for the approach is simple: for most survival data there are many
data available at the beginning and thus a more complicated structure may
be fitted, for higher values of t the data become sparse and a simple (linear)
structure is fitted. A more general model allows the parameter vector γ to
depend on time. In

$$\lambda(t|x) = F(\gamma_{0t} + x'\gamma_t) \tag{9.2.17}$$

the parameter γ_t varies over time. Of course by considering $x' = (x_1', x_2')$ the second term $x'\gamma_t$ may be split up in $x_1'\tilde{\gamma} + x_2'\tilde{\gamma}_t$ where the weight on the first subvector x_1' is time–independent (global) and the weight on subvector x_2' is time dependent. In the context of sequential models (see Section 3.3.4) time–dependent weighting is called category–specific where categories now refer to discrete time. For ML estimation the number of time intervals and the number of covariates with time–varying weights must be small compared to the number of observations. Alternative estimation procedures that handle the case of time–varying parameters more parsimoniously are considered in Section 9.4.2.

9.2.3   Maximum likelihood estimation

In the following the model is considered in the simple form

$$\lambda(t|x_i) = F(z_{it}'\beta) \tag{9.2.18}$$

where, e.g., model (9.2.17) is given by $z_{it} = (0, \ldots, 1, x_i, \ldots, 0)$ and $\beta = (\gamma_{01}, \gamma_1, \ldots, \gamma_{0q}, \gamma_q)$. The data are given by (t_i, δ_i, x_i), $i = 1, \ldots, n$, where discrete time $t_i = \min\{T_i, C_i\}$ is the minimum of survival time T_i and censoring time C_i. The indicator variable δ_i is determined by

$$\delta_i = \left\{ \begin{array}{ll} 1, & T_i < C_i \\ 0, & T_i \geq C_i. \end{array} \right.$$

Assuming independence of C_i and T_i (random censoring) the probability of observing $(t_i, \delta_i = 1)$ is given by

$$P(T_i = t_i, \delta_i = 1) = P(T_i = t_i)P(C_i > t_i). \tag{9.2.19}$$

The probability of censoring at time t_i is given by

$$P(C_i = t_i, \delta_i = 0) = P(T_i \geq t_i)P(C_i = t_i). \tag{9.2.20}$$

In (9.2.19) and (9.2.20) it is assumed that a failure in interval $[a_{t_i-1}, a_{t_i})$ implies a censoring time beyond a_{t_i-1} and censoring in interval $[a_{t_i-1}, a_{t_i})$ implies survival beyond a_{t_i-1}. Thus implicitly censoring is assumed to occur at the beginning of the interval. If censoring is assumed to occur at the end of the interval, (9.2.19) and (9.2.20) have to be substituted by

$$\begin{array}{rcl} P(T_i = t_i, \delta_i = 1) & = & P(T_i = t_i)P(C_i \geq t_i) \\ P(C_i = t_i, \delta_i = 0) & = & P(C_i = t_i)P(T_i \geq t_i). \end{array} \tag{9.2.21}$$

Combining (9.2.19) and (9.2.20) yields the likelihood contribution of observation (t_i, δ_i)

$$L_i = P(T_i = t_i)^{\delta_i} P(T_i \geq t_i)^{1-\delta_i} P(C_i > t_i)^{\delta_i} P(C_i = t_i)^{1-\delta_i}.$$

If the factor $c_i = P(C_i > t_i)^{\delta_i} P(C_i = t_i)^{1-\delta_i}$ that represents the censoring contributions does not depend on the parameters determining the survival time (noninformative in the sense of Kalbfleisch and Prentice, 1980), the likelihood contribution reduces to

$$L_i = c_i P(T_i = t_i)^{\delta_i} P(T_i \geq t_i)^{1-\delta_i}.$$

Including covariates and using the definition of the discrete hazard function one gets

$$L_i = c_i \lambda(t_i|x_i)^{\delta_i} \prod_{j=1}^{t_i-1} (1 - \lambda(j|x_i)). \qquad (9.2.22)$$

It is useful to consider (9.2.22) in a different form. For $\delta_i = 0$ (9.2.22) may be written by

$$L_i \propto \prod_{j=1}^{t_i-1} \lambda(j|x_i)^{y_{ij}} (1 - \lambda(j|x_i))^{1-y_{ij}}$$

where $y_i = (y_{i1}, \ldots, y_{i,t_i-1}) = (0, \ldots, 0)$. For $\delta_i = 1$ (9.2.22) may be written by

$$L_i \propto \prod_{j=1}^{t_i} \lambda(j|x_i)^{y_{ij}} (1 - \lambda(j|x_i))^{1-y_{ij}}$$

where $y_i = (y_{i1}, \ldots, y_{it_i}) = (0, \ldots, 0, 1)$. Here y_{ij} stands for the transition from interval $[a_{j-1}, a_j)$ to $[a_j, a_{j+1})$ given by

$$y_{ij} = \begin{cases} 1, & \text{individual fails in } [a_{j-1}, a_j) \\ 0, & \text{individual survives in } [a_{j-1}, a_j), \end{cases}$$

$j = 1, \ldots, t_i$. That means the total log–likelihood for model $\lambda(t|x_i) = F(z'_{it}\beta)$ given by

$$l \propto \sum_{i=1}^{n} \sum_{j=1}^{t_i-(1-\delta_i)} (y_{ij} \log \lambda(j|x_i) + (1 - y_{ij}) \log(1 - \lambda(j|x_i)))$$

is identical to the log–likelihood of the $\sum_i (t_i - 1 + \delta_i)$ observations $y_{11}, \ldots,$ $y_{1,t_1-(1-\delta_1)}, y_{21}, \ldots, y_{n,t_n-(1-\delta_n)}$ from the binary response model $P(y_{ij} = 1|x_i) = F(z'_{ij}\beta)$. Thus ML estimates may be calculated in the same way as for generalized linear models. The vector of binary responses and the design matrix are given by

$$y = \begin{bmatrix} y_{11} \\ \vdots \\ y_{1,t_1-(1-\delta_1)} \\ y_{21} \\ \vdots \\ y_{n,t_n-(1-\delta_n)} \end{bmatrix}, \quad Z = \begin{bmatrix} z'_{11} \\ \vdots \\ z'_{1,t_1-(1-\delta_1)} \\ z'_{21} \\ \vdots \\ z'_{n,t_n-(1-\delta_n)} \end{bmatrix}.$$

It should be noted that z_{it} in matrix Z for the simplest case does not depend on t. Alternatively, the observations may be reordered, yielding the likelihood function

$$l = \sum_{t=1}^{q} \sum_{i \in R_t} (y_{it} \log \lambda(t|x_i) + (1 - y_{it}) \log(1 - \lambda(t|x_i))) \qquad (9.2.23)$$

where $R_t = \{i : t \leq t_i - (1 - \delta_i)\}$ is the risk set, i.e., the set of individuals who are at risk in interval $[a_{t-1}, a_t)$.

If censoring is assumed to occur at the end of the interval, (9.2.20) has to be substituted by (9.2.21). Then the likelihood has the form

$$\begin{aligned} L_i &= c_i P(T_i = t_i)^{\delta_i} P(T_i > t_i)^{1-\delta_i} \\ &= c_i \lambda(t_i|x_i)^{\delta_i} \prod_{j=1}^{t_i-1} (1 - \lambda(j|x_i))^{\delta_i} \prod_{j=1}^{t_i} (1 - \lambda(j|x_i))^{1-\delta_i} \\ &= c_i \lambda(t_i|x_i)^{\delta_i} (1 - \lambda(t_i|x_i))^{1-\delta_i} \prod_{j=1}^{t_i-1} (1 - \lambda(j|x_i)) \\ &= c_i \prod_{j=1}^{t_i} \lambda(j|x_i)^{y_{ij}} (1 - \lambda(j|x_i))^{1-y_{ij}} \end{aligned}$$

where

$$y'_i = (y_{i1}, \ldots, y_{it_i}) = \begin{cases} (0,\ldots,0), & \delta_i = 0 \\ (0,\ldots,1), & \delta_i = 1. \end{cases}$$

The log–likelihood is now given by

$$\begin{aligned} l &= \sum_{i=1}^{n} \sum_{j=1}^{t_i} (y_{ij} \log(\lambda(j|x_i)) + (1 - y_{ij}) \log(1 - \lambda(j|x_i))) \\ &= \sum_{t=1}^{q} \sum_{i \in R'_t} (y_{it} \log(\lambda(t|x_i)) + (1 - y_{it}) \log(1 - \lambda(t|x_i))) \end{aligned}$$

where the risk set $R'_t = \{i|t \le t_i\}$ in interval $[a_{t-1}, a_t)$ now includes the individuals who are censored in this interval. In fact the individuals are under risk in $[a_{t-1}, a_t)$ since censoring is considered to happen at the end of the interval. Under this assumption response vector and design matrix have to be modified accordingly, now yielding $\sum_i t_i$ observations for the pseudo model $P(y_{ij} = 1|x_i) = F(z'_{ij}\beta)$.

Example 9.2: Duration of unemployment (Example 9.1 continued)
For the sample of unemployed persons considered in Example 9.1 several covariates are observed. Interesting covariates that are considered in the following are

Sex (dichotomous, 1: male, 2: female)

Education (four categories, 1: low, 2: medium, 3: high, 4: university)

Nationality (dichotomous, 1: German, 2: other)

Illness (dichotomous, 1: yes, 2: no)

Previous employment level (five categories, 1: full time, 2: part time, 3: job training, 4: school/university, 5: otherwise)

Country (ten categories, 1: Schleswig–Holstein, 2: Hamburg, 3: Niedersachsen, 4: Bremen, 5: Nordrhein–Westfalen, 6: Hessen, 7: Rheinland–Pfalz/Saarland, 8: Baden–Württemberg, 9: Bayern, 10: Berlin)

Age (four categories, 1: below 30, 2: 31 to 40, 3: 41 to 50, 4: above 50)

Begin of unemployment (four categories, 1: January to March, 2: April to June, 3: July to September, 4: October to December)

Table 9.1 shows the estimated values for the logistic model and the grouped proportional hazards model. The covariates are given in effect coding. All of the variables except educational level have very small p–values in at least some of the categories. For example, within the country effect most of the countries show no deviation from the baseline level of zero effect; only the countries 1,8,9 show a strong deviation. In particular, unemployed persons from country 9 have improved chances of ending their unemployment early. Both models, the logistic model and the grouped proportional hazards model yield almost the same parameter estimates. Also the fit of the model is almost identical.   □

9.2.4   Time–varying covariates

Maximum likelihood estimation has been considered for the model $\lambda(t|x_i) = F(z'_{it}\beta)$ where z_{it} is built from the covariates x_i. For the models considered in previous sections it is implicitly assumed that z_{it} depends on time t merely by incorporating the baseline hazard. However, more generally z_{it} may be a vector that incorporates covariates varying over time.

TABLE 9.1. Duration of unemployment

	Grouped proportional hazards model	p–values	Logistic model	p–values
1	-3.391	0.000	-3.358	0.000
POLY(1)	0.008	0.821	0.002	0.952
POLY(2)	-0.005	0.123	-0.004	0.164
POLY(3)	0.000	0.135	0.000	0.169
SEX	0.382	0.000	0.403	0.000
EDUCATION(1)	-0.139	0.364	-0.149	0.357
EDUCATION(2)	-0.072	0.317	-0.075	0.324
EDUCATION(3)	0.178	0.025	0.190	0.024
ILLNESS	-0.105	0.030	-0.111	0.029
NATIONALITY	0.233	0.000	0.244	0.000
LEVEL(1)	0.397	0.000	0.408	0.000
LEVEL(2)	-0.373	0.026	-0.380	0.027
LEVEL(3)	0.461	0.000	0.484	0.000
LEVEL(4)	0.032	0.801	0.027	0.841
COUNTRY(1)	-0.238	0.235	-0.257	0.220
COUNTRY(2)	-0.556	0.055	-0.580	0.052
COUNTRY(3)	-0.039	0.717	-0.043	0.702
COUNTRY(4)	-0.320	0.211	-0.333	0.211
COUNTRY(5)	-0.018	0.829	-0.024	0.791
COUNTRY(6)	0.059	0.629	0.063	0.622
COUNTRY(7)	0.135	0.304	0.143	0.308
COUNTRY(8)	0.309	0.001	0.327	0.001
COUNTRY(9)	0.406	0.000	0.425	0.000
AGE(1)	0.626	0.000	0.652	0.000
AGE(2)	0.352	0.000	0.361	0.000
AGE(3)	0.227	0.011	0.226	0.014
MONTH(1)	0.004	0.936	0.002	0.961
MONTH(2)	-0.144	0.045	-0.155	0.041
MONTH(3)	-0.060	0.352	-0.061	0.372

Let $x_{i1}, \ldots, x_{it}$ denote the sequence of observations of covariates for the ith unit until time t where x_{it} is a vector observed at the beginning of interval $[a_{t-1}, a_t)$ or is fixed at discrete time t. In interval $[a_{t-1}, a_t)$ the "history" of covariates

$$x_i'(t) = (x_{i1}', \ldots, x_{it}')$$

may influence the hazard rate in the model

$$\lambda(t|x_i(t)) = P(T = t|T \geq t, x_i(t)) = F(z_{it}'\beta) \qquad (9.2.24)$$

where z_{it} is composed from $x_i(t)$. There are many possibilities of specifying the vector z_{it}. A simple way where only the vector observed at time t is of influence is given by

$$z_{it}'\beta = \gamma_{0t} + x_{it}'\gamma$$

where $z_{it}' = (0, \ldots, 1, \ldots, 0, x_{it}'), \beta' = (\gamma_{01}, \ldots, \gamma_{0q}, \gamma')$. If the parameter varies over time one may specify

$$z_{it}'\beta = \gamma_{0t} + x_{it}'\gamma_t$$

where $z_{it}' = (0, \ldots, 1, \ldots, 0, 0, \ldots, x_{it}', \ldots, 0), \beta' = (\gamma_{01}, \ldots, \gamma_{0q}, \gamma_1', \ldots, \gamma_q')$. Of course time lags may be included by

$$z_{it}'\beta = \gamma_{0t} + x_{it}'\gamma_0 + \ldots + x_{i,t-r}'\gamma_{-r}$$

where

$$
\begin{aligned}
z_{it}' &= (0, \ldots, 1, \ldots, 0, x_{it}', x_{i,t-1}', \ldots, x_{i,t-r}'), \\
\beta' &= (\gamma_{01}, \ldots, \gamma_{0q}, \gamma_0', \gamma_{-1}', \ldots, \gamma_{-r}').
\end{aligned}
$$

In z_{it} characteristics of the interval (e.g., the length of the interval) may be included by

$$z_{it}'\beta = \gamma_{0t} + (a_t - a_{t-1})\gamma \qquad (9.2.25)$$

where $z_{it}' = (0, \ldots, 1, \ldots, 0, a_t - a_{t-1}), \beta' = (\gamma_{01}, \ldots, \gamma_{0q}, \gamma)$.

For time–dependent covariates one may distinguish between two types: *external* and *internal* covariates (Kalbfleisch and Prentice, 1980). External covariates are not directly involved with failure. The components of z_{it} that refer to the coding of the interval (i.e., the baseline hazard) as well as the difference $a_t - a_{t-1}$ in (9.2.25) are fixed in advance; they are not determined by the failure mechanism. If $x_{i1}, \ldots, x_{it}$ is the output of a stochastic process it may be considered external if the condition

$$P(x_{i,t+1}, \ldots, x_{iq}|x_i(t), y_i(t)) = P(x_{i,t+1}, \ldots, x_{iq}|x_i(t)), \qquad (9.2.26)$$

$$t = 1, \ldots, q,$$

holds where y_{ij} denotes failure in interval $[a_{j-1}, a_j)$ and $y_i(t) = (y_{i1}, \ldots$ $\ldots, y_{it})$. Equation (9.2.26) means that the path of the covariate process is not influenced by failure. A consequence of (9.2.26) is that

$$P(y_{it}|x_i(t), y_i(t-1)) = P(y_{it}|x_i(q), y_i(t-1))$$

and therefore conditioning may be done on the whole path $x_i(q)$ by

$$
\begin{aligned}
\lambda(t|x_i(t)) &= P(y_{it} = 1|x_i(t), y_{i1} = 0, \ldots, y_{i,t-1} = 0) = \\
&= P(y_{it} = 1|x_i(q), y_{i1} = 0, \ldots, y_{i,t-1} = 0) = \\
&= \lambda(t|x_i(q)).
\end{aligned}
$$

For external covariates under mild restrictions the likelihood (9.2.23) may still be used by substituting time–dependent covariates $x_i(t)$ for the time–independent covariate x_i. For a derivation, see the subsection "Maximum likelihood estimation*."

Internal covariates*

Internal covariates carry with their observed path information on failure such as characteristics of an individual that can be observed only as long as the individual is in the study and alive. Now the hazard function only incorporates the path until time t

$$\lambda(t|x(t)) = P(y_{it} = 1|x(t), y_{i1} = 0, \ldots, y_{i,t-1} = 0).$$

Condition (9.2.26) cannot be assumed to hold since covariates $x_{i,t+1}, \ldots, x_{iq}$ may no longer have any meaning if a patient dies in interval $[a_{t-1}, a_t)$. In particular for the latter type of covariates the simple connection between survivor function and hazard function given at the beginning of Section 9.2 no longer holds. One may consider a type of "survivor function" by defining

$$S(t|x(t)) = P(T > t|x(t))$$

where $T > t$ denotes the sequence $y_{i1} = 0, \ldots, y_{it} = 0$. Alternatively one may write

$$S(t|x(t)) = \prod_{s=1}^{t} P(T > s|T \geq s, x(t)).$$

However, since the factor $P(T > s|T \geq s, x(t))$ includes the history $x(t)$ it is not equal to $1 - \lambda(s|x(s))$, which only includes the history until time s.
Consider the general formula

$$P(A_s|B_s \cap C_s) = P(C_s|A_s \cap B_s)P(A_s|B_s)/P(C_s|B_s)$$

with $A_s = \{T > s\}, B_s = \{T \geq s, x(s)\}, C_s = \{x_{s+1}, \ldots, x_t\}$. Then one gets

$$P(T > s | T \geq s, x(t)) = P(T > s | T \geq s, x(s)) q_s = (1 - \lambda(s|x(s))) q_s$$

where $q_s = P(x_{s+1}, \ldots, x_t | T > s, x(s)) / P(x_{s+1}, \ldots, x_t | T \geq s, x(s))$. For external covariates $q_s = 1$ holds and the survivor function is again given by

$$S(t|x(t)) = S(t|x(q)) = \prod_{s=1}^{t} (1 - \lambda(s|x(s))).$$

Maximum likelihood estimation*

The data are given by $(t_i, \delta_i, x_i(t_i)), i = 1, \ldots, n$ where t_i is again the minimum $t_i = \min\{T_i, C_i\}$. The probability for the event $\{t_i, \delta_i = 1, x_i(t_i)\}$ is given by

$$P(t_i, \delta_i = 1, x_i(t_i)) = P(T_i = t_i, C_i > t_i, x_i(t_i))$$

$$= P(T_i = t_i, C_i > t_i, x_i(t_i) | H_{i,t_i-1}) \prod_{s=1}^{t_i-1} P(T_i > s, C_i > s, x_i(s) | H_{i,s-1})$$

where $H_{i,s} = \{T_i > s, C_i > s, x_i(s)\}, s = 1, \ldots, t_i - 1, H_{i0} = \{T_i > 0, C_i > 0\}$ is a sequence of "histories" fulfilling $H_{i,s+1} \subset H_{i,s}$.

By using the simple formula $P(A \cap B | C) = P(A | B \cap C) P(B | C)$ this may be rewritten into

$$P(t_i, \delta_i = 1, x_i(t_i))$$
$$= P(T_i = t_i | C_i > t_i, x_i(t_i), H_{i,t_i-1}) P(x_i(t_i), C_i > t_i | H_{i,t_i-1})$$
$$\cdot \prod_{s=1}^{t_i-1} P(T_i > s | C_i > s, x_i(s), H_{i,s-1}) P(x_i(s), C_i > s | H_{i,s-1}).$$

Assuming

$$P(T_i = s | T_i > s - 1, C_i > s, x_i(s)) = P(T_i = s | T_i > s - 1, x_i(s)) \quad (9.2.27)$$

which holds for T_i, C_i independent one gets

$$P(T_i = s | C_i > s, x_i(s), H_{i,s-1}) = P(T_i = s | T_i > s - 1, C_i > s, x_i(s))$$
$$= P(T_i = s | T_i > s - 1, x_i(s))$$
$$= \lambda(s | x_i(s))$$

and therefore

$$P\big(t_i, \delta_i = 1, x_i(t_i)\big) \;=\; \lambda\big(t_i|x_i(t_i)\big) \prod_{s=1}^{t_i-1} \Big[1 - \lambda\big(s|x_i(s)\big)\Big] \tag{9.2.28}$$
$$\cdot \prod_{s=1}^{t_i} P\big(x_i(s), C_i > s|H_{i,s-1}\big).$$

If censoring takes place at the beginning of the interval the interesting probability is that for the event $\{t_i, \delta_i = 0, x_i(t_i-1)\}$. The same probability has to be computed if data are discrete time points, because then $t_i, \delta_i = 0$ implies that the past covariates can be observed only until time $t_i - 1$. Assuming (9.2.26) holds it is given by

$$
\begin{aligned}
P(t_i, \delta_i = 0, x_i(t_i - 1)) \;&=\; P(T_i \geq t_i, C_i = t_i, x_i(t_i - 1)) \\
&=\; P(C_i = t_i|H_{i,t_i-1}, x_i(t_i - 1))P(x_i(t_i - 1)|H_{i,t_i-1}) \\
&\quad \cdot \prod_{s=1}^{t_i-1} P(T_i > s|C_i > s, x_i(s), H_{i,s-1})P(C_i > s, x_i(s)|H_{i,s-1}) \\
&=\; P(C_i = t_i|H_{i,t_i-1}) \\
&\quad \cdot \prod_{s=1}^{t_i-1} (1 - \lambda(s|x_i(s)))P(C_i > s, x_i(s)|H_{i,s-1}). \tag{9.2.29}
\end{aligned}
$$

Equations (9.2.28) and (9.2.29) yield

$$P(t_i, \delta_i, x_i(t_i)^{\delta_i}, x_i(t_i - 1)^{1-\delta_i}) = \lambda(t_i|x_i(t_i))^{\delta_i} \prod_{s=1}^{t_i-1} (1 - \lambda(s|x_i(s)))Q_i$$

where

$$
\begin{aligned}
Q_i \;&=\; P(C_i = t_i|H_{i,t_i-1})^{1-\delta_i} P(C_i > t_i, x_i(t_i)|H_{i,t_i-1})^{\delta_i} \\
&\quad \cdot \prod_{s=1}^{t_i-1} P(C_i > s, x_i(s)|H_{i,s-1}). \tag{9.2.30}
\end{aligned}
$$

If the factor Q_i is noninformative, i.e., it does not depend on the parameters determining survival time, then the likelihood is the same as (9.2.23), which was derived for time–independent covariates.

For n observations $(t_i, x_i(t_i), \delta_i)$ one gets for the log–likelihood

$$l \propto \sum_{i=1}^{n} \sum_{r=1}^{t_i-(1-\delta_i)} \Big[y_{ir} \log \lambda(r|x_i(r)) + (1 - y_{ir}) \log(1 - \lambda(r|x_i(r))) \Big]$$

or equivalently

$$l \propto \sum_{t=1}^{q} \sum_{i \in R_t} \Big[y_{it} \log \lambda(t|x_i(t_i)) + (1 - y_{it}) \log(1 - \lambda(t|x_i(t))) \Big] \tag{9.2.31}$$

where $R_t = \{i : t < t_i - (1 - \delta_i)\}$ is the risk set.

ML estimates may be estimated within the framework of generalized linear models with the response and design matrix as given in Section 9.2.3. The only difference is that z_{it} now includes time–varying components. For the consideration of time–dependent covariates, see also Hamerle and Tutz (1989).

9.3 Discrete models for multiple modes of failure

In the preceding sections methods have been considered for observations of failure or censoring where it is assumed that there is only one type of failure event. Often one may distinguish between several distinct types of terminating events. For example, in a medical treatment study the events may stand for several causes of death. In studies on unemployment duration one may distinguish between full–time and part–time jobs that end the unemployment duration. In survival analysis, models for this type of data are often referred to as competing risks models. We will use this name although in the case of unemployment data competing chances would be more appropriate. Most of the literature for competing risks considers the case of continuous time (e.g., Kalbfleisch and Prentice, 1980).

9.3.1 Basic models

Let $R \in \{1, \ldots, m\}$ denote the distinct events of failure or causes. Considering discrete time $T \in \{1, \ldots, q + 1\}$ the *cause–specific hazard function* resulting from cause or risk r is given by

$$\lambda_r(t|x) = P(T = t, R = r|T \geq t, x)$$

where x is a vector of time–independent covariates. The *overall hazard function* for failure regardless of cause is given by

$$\lambda(t|x) = \sum_{r=1}^{m} \lambda_r(t|x) = P(T = t|T \geq t, x).$$

Survivor function and unconditional probability of a terminating event are given as in the simple case of one terminating event by

$$S(t|x) = P(T > t|x) = \prod_{i=1}^{t}(1 - \lambda(i|x))$$

and

$$P(T = t|x) = \lambda(t|x)S(t - 1|x).$$

For an individual reaching interval $[a_{t-1}, a_t)$ the conditional response probabilities are given by

$$\lambda_1(t|x), \ldots, \lambda_m(t|x), 1 - \lambda(t|x)$$

where $1 - \lambda(t|x)$ is the probability for survival. Modelling of these events may be based on the approach for multicategory responses outlined in Chapter 3. A candidate for unordered events is the multinomial logit model

$$\lambda_r(t|x) = \frac{\exp(\gamma_{0tr} + x'\gamma_r)}{1 + \sum_{i=1}^{m} \exp(\gamma_{0ti} + x'\gamma_i)}, \qquad (9.3.1)$$

$r = 1, \ldots, m, t = 1, \ldots, q$. In model (9.3.1) the parameters $\gamma_{01j}, \ldots, \gamma_{0qj}$ represent the cause–specific baseline hazard function and γ_r is the cause–specific weight. In the same way as for the single–event case considered in Section 9.2 the baseline hazard function may be simplified by using a polynomial approximation and the weight γ_r may depend on time. Then the influence term $\gamma_{0tr} + x'\gamma_r$ is substituted by the more general term $\eta_r = z'_{tr}\gamma$. The general form of (9.3.1) is given by

$$\lambda_r(t|x) = h_r(Z_t\beta) \qquad (9.3.2)$$

where h_r is the local link function for responses in interval $[a_{t-1}, a_t)$ and Z_t is a design matrix composed of x and depending on time t. The simple model (9.3.1) has the logit response function

$$h_r(\eta_1, \ldots, \eta_m) = \frac{\exp(\eta_r)}{1 + \sum_{i=1}^{m} \exp(\eta_i)}$$

and design matrix

$$Z_t = \begin{bmatrix} 0 & 1 & & & 0 & x' & & \\ & & 1 & & & & x' & \\ & & & \ddots & & & & \ddots \\ 0 & & & 1 & 0 & & & x' \end{bmatrix} \qquad (9.3.3)$$

where the parameter vector is given by

$$\beta' = (\gamma_{011}, \ldots, \gamma_{01m}, \gamma_{021}, \ldots, \gamma_{0qm}, \gamma'_1, \ldots, \gamma'_m).$$

For the modelling of the conditional response in interval $[a_{t-1}, a_t)$ alternative models from Chapter 3 may be more appropriate. If the events are ordered, ordinal models like sequential or cumulative models may yield a simpler structure with fewer parameters.

A quite different approach to parametric models is based on models that have been derived for the continuous case. A proportional hazards model in which the cause–specific hazard function at continuous time t depends on x is given by

$$\lambda_r(t) = \lambda_{0r}(t)\exp(x'\gamma_r), \qquad (9.3.4)$$

$r = 1, \ldots, m$, (e.g., Kalbfleisch and Prentice, 1980). Derivation of the discrete model for observations in intervals yields parametric models at least in special cases. If the baseline hazard function does not depend on cause r the model has the simpler form

$$\lambda_r(t) = \lambda_0(t)\exp(x'\gamma_r).$$

For this model the discrete hazard function where $T = t$ denotes failure in interval $[a_{t-1}, a_t)$ may be derived by

$$\lambda_r(t|x) = \frac{\exp(\gamma_{0t} + x'\gamma_r)}{\sum_{j=1}^{m}\exp(\gamma_{0t} + x'\gamma_j)}\left\{1 - \exp\left(-\sum_{j=1}^{m}\exp(\gamma_{0t} + x'\gamma_j)\right)\right\} \quad (9.3.5)$$

where the baseline hazard function is absorbed into the parameters $\gamma_{0t} = \log\left(\int_{a_{t-1}}^{a_t}\lambda_0(t)\right)$. For the derivation see Hamerle and Tutz (1989). Model (9.3.5) has the general form (9.3.2) where the response function is given by

$$h_r(\eta_1, \ldots, \eta_m) = \frac{\exp(\eta_r)}{\sum_{j=1}^{m}\exp(\eta_j)}\left\{1 - \exp\left(-\sum_{j=1}^{m}\exp(\eta_j)\right)\right\}$$

and the design matrix has the form (9.3.3). If the covariates are stochastic processes $x_{i1}, \ldots, x_{it}$ cause–specific and global hazard function have the form

$$\lambda_r(t|x_i(t)) = P(T_i = t, R_i = r|T_i \geq t, x_i(t))$$

$$\lambda(t|x_i(t)) = \sum_{r=1}^{m}\lambda_r(t|x_i(t))$$

where $x_i(t) = (x_{i1}, \ldots, x_{it})$ is the sequence of observations until time t. The model for the hazard function has form (9.3.2) where the design matrix Z_t is a function of t and $x(t)$.

9.3.2   Maximum likelihood estimation

The data are given by $(t_i, r_i, \delta_i, x_i)$ where $r_i \in \{1, \ldots, m\}$ indicates the terminating event. We consider the case of random censoring with $t_i = \min\{T_i, C_i\}$, and censoring at the beginning of the interval. First the simpler case of time–independent covariates x is treated. The likelihood contribution of the ith observation for model (9.3.2) is given by

$$L_i = P(T_i = t_i, R_i = r_i)^{\delta_i}P(T_i \geq t_i)^{1-\delta_i}P(C_i > t_i)^{\delta_i}P(C_i = t_i)^{1-\delta_i}.$$

For noninformative censoring it may be reduced to

$$
\begin{aligned}
L_i &= P(T_i = t_i, R_i = r_i | x_i)^{\delta_i} P(T_i \geq t_i | x_i)^{1-\delta_i} \\
&= \lambda_{r_i}(t_i | x_i)^{\delta_i} P(T_i \geq t_i | x_i) \\
&= \lambda_{r_i}(t_i | x_i)^{\delta_i} \prod_{t=1}^{t_i-1} \left(1 - \sum_{r=1}^{m} \lambda_r(t | x_i)\right).
\end{aligned}
\tag{9.3.6}
$$

An alternative form of the likelihood is based on dummy variables given by

$$
y_{itr} = \begin{cases} 1, & \text{failure of type } r \text{ in interval } [a_{t-1}, a_t) \\ 0, & \text{no failure in interval } [a_{t-1}, a_t) \end{cases}
$$

$r = 1, \ldots, m$. Given an individual reaches interval $[a_{t-1}, a_t)$ the response is multinomial with $y_{it} = (y_{it1}, \ldots, y_{itm}) \sim M(1, \lambda_1(t | x_i), \ldots, \lambda_m(t | x_i))$. Therefore the dummy variable $y_{it,m+1} = 1 - y_{it1} - \ldots - y_{itm}$ has value 1 if individual i does not fail in interval $[a_{t-1}, a_t)$ and $y_{it,m+1} = 0$ if individual i fails in $[a_{t-1}, a_t)$. The likelihood (9.3.6) for the ith observation has the form

$$
\begin{aligned}
L_i &= \prod_{r=1}^{m} \lambda_r(t_i | x_i)^{\delta_i y_{it_i r}} \left(1 - \sum_{r=1}^{m} \lambda_r(t_i | x_i)\right)^{\delta_i y_{it_i,m+1}} \\
&\quad \cdot \prod_{t=1}^{t_i-1} \left\{ \prod_{r=1}^{m} \lambda_r(t | x_i)^{y_{itr}} \right\} \left\{ 1 - \sum_{r=1}^{m} \lambda_r(t | x_i) \right\}^{y_{it,m+1}} \\
&= \prod_{t=1}^{t_i-1+\delta_i} \left\{ \prod_{r=1}^{m} \lambda_r(t | x_i)^{y_{itr}} \right\} \left\{ 1 - \sum_{r=1}^{m} \lambda_r(t | x_i) \right\}^{y_{it,m+1}}.
\end{aligned}
$$

That means the likelihood for the ith observation is the same as that for the $t_i - 1 + \delta_i$ observations $y_{i1}, \ldots, y_{i,t_i-1+\delta_i}$ of the multicategory model $P(Y_{it} = r) = h_r(Z_t \beta)$ where $Y_{it} = r$ if $y_{itr} = 1$. Thus as in the single–cause model ML estimates may be calculated within the framework of generalized linear models. If $\delta_i = 0$ we have the $t_i - 1$ observation vectors $y_{i1}, \ldots, y_{i,t_i-1}$ and if $\delta_i = 1$ we have the t_i observation vectors $y_{i1}, \ldots, y_{i,t_i}$ yielding a blown–up design. For the ith observation response and design matrices are given by

$$
\begin{bmatrix} y_{i1} \\ \vdots \\ y_{i,t_i-1+\delta_i} \end{bmatrix} \quad \begin{bmatrix} Z_1 \\ \vdots \\ Z_{t_i-1+\delta_i} \end{bmatrix}.
$$

The total log–likelihood is given by

time interval

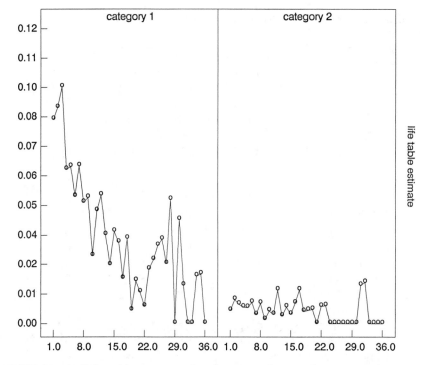

FIGURE 9.3. Life table estimate for duration of unemployment with causes full–time job or part–time job.

$$l = \sum_{i=1}^{n} \sum_{t=1}^{t_i-1+\delta_i} \left(\sum_{r=1}^{m} y_{itr} \log \lambda_r(t|x_i) + y_{it,m+1} \log \left(1 - \sum_{r=1}^{m} \lambda_r(t|x_i) \right) \right)$$

$$= \sum_{t=1}^{q} \sum_{i \in R_t} \left(\sum_{r=1}^{m} y_{itr} \log \lambda_r(t|x_i) + y_{it,m+1} \log \left(1 - \sum_{r=1}^{m} \lambda_r(t|x_i) \right) \right) \quad (9.3.7)$$

where in the latter form $R_t = \{i | t_i - 1 + \delta_i \geq t\}$ is the number of individuals under risk in interval $[a_{t-1}, a_t)$.

Example 9.3: Duration of unemployment (Example 9.2 continued)
The duration data already considered in Example 9.2 are now considered for the case of multiple modes of failure. Instead of the single absorbing event full–time job the causes full–time job or part–time job are considered. Figure 9.3 shows the simple life–time table estimate for these causes. It is seen that the transition from unemployment to full–time job is the

TABLE 9.2. Estimates of cause–specific logistic model for duration of unemployment data.

	Full–time	p–value	Part–time	p–value
1	-3.361	0.000	-5.355	0.000
POLY(1)	0.002	0.944	-0.195	0.168
POLY(2)	-0.004	0.161	0.023	0.112
POLY(3)	0.000	0.167	-0.000	0.101
SEX	0.403	0.000	-0.780	0.000
EDUCATION(1)	-0.150	0.356	-0.472	0.312
EDUCATION(2)	-0.073	0.338	-0.290	0.200
EDUCATION(3)	0.191	0.023	0.276	0.252
ILLNESS	-0.111	0.029	0.059	0.673
NATIONALITY	0.243	0.000	0.886	0.000
LEVEL(1)	0.408	0.000	-0.690	0.001
LEVEL(2)	-0.381	0.027	0.564	0.015
LEVEL(3)	0.484	0.000	0.203	0.503
LEVEL(4)	0.028	0.834	0.334	0.252
COUNTRY(1)	-0.262	0.211	0.210	0.594
COUNTRY(2)	-0.577	0.054	0.686	0.220
COUNTRY(3)	-0.044	0.693	0.249	0.383
COUNTRY(4)	-0.332	0.213	0.321	0.636
COUNTRY(5)	-0.024	0.790	-0.609	0.030
COUNTRY(6)	0.063	0.626	-0.259	0.523
COUNTRY(7)	0.142	0.309	-1.104	0.095
COUNTRY(8)	0.328	0.001	-0.021	0.945
COUNTRY(9)	0.426	0.000	-0.090	0.770
AGE(1)	0.653	0.000	0.456	0.040
AGE(2)	0.361	0.000	0.149	0.566
AGE(3)	0.226	0.015	0.058	0.828
MONTH(1)	0.003	0.959	-0.068	0.732
MONTH(2)	-0.155	0.040	0.223	0.292
MONTH(3)	-0.061	0.369	0.149	0.452

dominant one. Transition to part–time jobs has a rather low value at all times.

Table 9.2 shows the fitted values for the parameters of the logistic model where the polynomial term for the baseline hazard and the explanatory variables have cause–specific weights. Comparison with Table 9.1 shows that the full–time parameters are almost the same as for the duration model considering only the absorbing event full–time job. This is easily explained by looking at the log–likelihood (9.3.7) which for the logistic model may be separated into the sum of log–likelihoods for models specifying each cause separately. However, this holds only for the natural link function underlying the logistic model.

From Table 9.2 it is seen that effects for full–time jobs are often quite different from effects for part–time jobs; sometimes even the sign is different. For example, the effects of sex are 0.403 and -0.780, showing that men have a higher hazard rate than women with respect to full-time jobs but a lower hazard rate with respect to part–time jobs. That might be due to the strong preference of men for full–time jobs. Similar effects are found for the variable "previous level of employment."   □

9.4   Smoothing in discrete survival analysis

Smoothing of hazard rates may be based on various approaches. In this subsection, we consider smoothing based on state space or dynamic modelling techniques, which is closely related to discrete spline smoothing, and smoothing by discrete kernels.

9.4.1   Dynamic discrete time survival models

As already mentioned at the end of Section 9.2.2, "static" models for discrete time which treat baseline hazard coefficients and covariate parameters as "fixed effects" are appropriate if the number of intervals is comparably small. In situations with many intervals, but not enough to apply models for continuous time, such unrestricted modelling and fitting of hazard functions will often lead to nonexistence and divergence of ML estimates due to the large number of parameters. This difficulty in real data problems becomes even more apparent if covariate effects are also assumed to be time–varying, as, for example, the effect of a certain therapy in a medical context or the effect of financial support while unemployed on duration of unemployment.

To avoid such problems, one may try a more parsimonious parameterization by specifying certain functional forms, e.g., (piecewise) polynomials, for time–varying coefficients. However, simply imposing such parametric functions can conceal spikes in the hazard function or other unexpected patterns. A purely nonparametric method to overcome such problems has recently been discussed by Huffer and McKeague (1991), based on Aalen's (1980,1989) additive risk model for continuous time. We describe here another flexible approach by adopting state space techniques for categorical longitudinal data (Section 8.4) to the present context. The dynamic models obtained in this way make simultaneous analysis and smoothing of the baseline hazard function and covariate effects possible. The development is related to a dynamic version of the piecewise exponential model studied by Gamerman (1991), who applied the estimation method of West et al. (1985), and to the ideas in Kiefer (1990).

The data are given as before by observed discrete survival times $t_i = \min(T_i, C_i)$, indicator variables δ_i, possibly time–varying covariates $x_{i1}, \ldots, x_{it}, \ldots, i = 1, \ldots, n$, and the indicator sequences $y_{i1}, \ldots, y_{it}, \ldots$ are defined by

$$y_{it} = \begin{cases} 1, & \text{individual } i \text{ fails in } [a_{t-1}, a_t) \\ 0, & \text{individual } i \text{ survives in } [a_{t-1}, a_t), \end{cases}$$

$t = 1, \ldots, t_i$. We consider the general hazard rate model (9.2.24) in the form

$$\lambda(t|x_i(t)) = F(z'_{it}\alpha_t), \qquad (9.4.1)$$

which will play the role of the observation model (8.4.4). The components of the state vector α_t represent baseline hazard parameters as well as covariate effects, and the design vector z_{it} is a function of covariates. In the simplest case we define $z_{it} = (1, x_{it})$, $\alpha_t = (\gamma_{0t}, \gamma_t)$, and the state vector α_t follows a first–order random walk model

$$\alpha_t = \alpha_{t-1} + \xi_t, \quad \xi_t \sim N(0, Q),$$

$Q = \operatorname{diag}(\sigma_0^2, \sigma_1^2, \ldots, \sigma_p^2)$, $p = \dim(\gamma_t)$. Setting some of the variances $\sigma_1^2, \ldots$ $\ldots, \sigma_p^2$ to zero, means that corresponding covariate effects are assumed to be constant in time. Of course other more complex forms of design vectors and parameter variations as in Chapter 8 are admissible: In the present context the baseline parameters γ_{0t} play the role of the trend component τ_t, while covariate effects are assumed to be constant in time or are modelled by a first–order random walk. The general parameter transition model is again linear and Gaussian:

$$\alpha_t = F_t\alpha_{t-1} + \xi_t, \quad t = 1, 2, \ldots \qquad (9.4.2)$$

with the usual assumptions on $\xi_t \sim N(0, Q)$, $\alpha_0 \sim N(a_0, Q_0)$.

Posterior mode smoothing

In the following all assumptions made for ML estimation in Sections 9.2.2 and 9.2.3 are supposed to hold conditionally, given the sequence $\{\alpha_t\}$ of states. For time–varying covariates we assume that the factors Q_i in (9.2.30) are noninformative. This condition replaces assumption (A2) of Section 8.2.1. Relying on the state space approach, and on the conditional independence assumption (A4) in Section 8.4 and proceeding as in that section, one arrives at the penalized log–likelihood criterion

$$L(\alpha_0, \ldots, \alpha_t, \ldots, \alpha_q) = \sum_{t=1}^{q} \sum_{i \in R_t} l_{it}(\alpha_t) - \frac{1}{2}(\alpha_0 - a_0)' Q_0^{-1}(\alpha_0 - a_0)$$

$$-\frac{1}{2}\sum_{t=1}^{q}(\alpha_t - F_t\alpha_{t-1})'Q_t^{-1}(\alpha_t - F_t\alpha_{t-1}), \tag{9.4.3}$$

where

$$l_{it}(\alpha_t) = y_{it}\log\lambda(t|x_i(t)) + (1-y_{it})\log(1-\lambda(t|x_i(t)),$$

where $\lambda(t|x_i(t)) = F(z_{it}'\alpha_t)$ is the individual log–likelihood contribution as in Sections 9.2.2 and 9.2.3. The risk set $R_t = \{i : t \le t_i - (1 - \delta_i)\}$ can be replaced by $R_t' = \{i : t \le t_i\}$ if censoring is considered to happen at the end of the inverval $[a_{t-1}, a_t)$. As a very simple example, consider a logit model

$$\lambda(t) = \exp(\alpha_t)/(1 + \exp(\alpha_t))$$

for the hazard function without covariates. Together with a first–order random walk model for α_t, the penalized log–likelihood criterion becomes

$$L = \sum_{t=1}^{q}\sum_{i\in R_t} l_{it}(\alpha_t) - \frac{1}{2\sigma_0^2}(\alpha_0 - a_0)^2 - \frac{1}{2\sigma^2}\sum_{t=1}^{q}(\alpha_t - \alpha_{t-1})^2.$$

The last term penalizes large deviations between successive baseline parameters and leads to smoothed estimates. Other stochastic models for the states α_t, such as second–order random walk models, lead to other forms of the penalty term, but with a similar effect.

Comparing with the criterion (8.4.5), we see that (9.4.3) has the same form as (8.4.5) for binary observations with response function $\lambda(t|x_i(t)) = F(z_{it}'\alpha_t)$, with the only difference that the sum of log–likelihood contributions in interval t extends only over individuals i in the risk set R_t. Thus recursive posterior mode estimation of $\{\alpha_t\}$ by the generalized Kalman filter and smoother in Section 8.4.2 requires only a slight modification: Correction steps (3) or (3)* in interval t have to be carried out for $i \in R_t$ only, all other steps remain unchanged. Therefore we do not rewrite the algorithm here. Joint estimation of hyperparameters, usually unknown variances and initial values in the parameter transition model, is again possible by the same EM–type algorithm. For more details see Fahrmeir (1994).

If one is not willing to rely on the Bayesian smoothness priors for γ_{0t} and γ_t, implied by the transition model (9.4.2), one may also start directly from the penalized likelihood criterion. Then the technique may be viewed as discrete time spline smoothing for survival models. The approach is also easily extended to dynamic discrete time models for multiple modes of failure: Binomial models for the hazard function are replaced by corresponding multinomial models, as, for example, in (9.3.1) and (9.3.5), binary indicator sequences by multicategorical indicator sequences

$$y_{itr} = \begin{cases} 1, & \text{failure of type } r \text{ in interval } [a_{t-1}, a_t) \\ 0, & \text{no failure in interval } [a_{t-1}, a_t), \end{cases}$$

and binary likelihood contributions $l_{it}(\alpha_t)$ by corresponding multinomial likelihoods. The resulting penalized likelihood criterion can be maximized by the correspondingly modified generalized Kalman filter and smoother for multicategorical longitudinal data in Section 8.4.

9.4.2   Kernel smoothing

In the simple case of no covariates the life table estimate (9.2.8) is of the simple form

$$\hat{\lambda}(t) = \frac{\text{number of failures in } [a_{t-1}, a_t)}{\text{number of individuals under risk in } [a_{t-1}, a_t)}.\tag{9.4.4}$$

As is seen from Figure 9.1 this estimate may be quite unsteady; in particular for large t where the number of individuals under risk is low there is much noise in the data. The simple relative frequency (9.4.4) is not suited for discovering the underlying form of the hazard function.

A smoothed version of the life table estimate may be based on a polynomial or spline functions fit or on smoothing procedures for discrete responses as considered in Section 5.2. Since the response variable considered in Section 5.2 takes integer values, recoding is appropriate. The indicator variable y_{it} takes values 1 if the individual fails in $[a_{t-1}, a_t)$ and value 0 if the individual survives. Simple recoding yields the variable $\tilde{y}_{it} = 2 - y_{it}$, which is given by

$$\tilde{y}_{it} = \begin{cases} 1, & \text{individual fails in } [a_{t-1}, a_t) \\ 2, & \text{individual survives in } [a_{t-1}, a_t). \end{cases}$$

In Section 9.2.3 (equation (9.2.23)) the risk set R_t indicating the indices of individuals under risk in $[a_{t-1}, a_t)$ has been used. For smoothed estimates that take the response in the neighbourhood into account we define the observations at time t (due to individuals from R_t) by

$$S_t = \{\tilde{y}_{it} | t \leq t_i - (1 - \delta_i)\}.\tag{9.4.5}$$

For this local sample the categorical kernel regression estimate may be directly applied by

$$\hat{\lambda}(t) = \hat{P}(y = 1 | t, \lambda, \delta) = \sum_{\tilde{t}=1}^{q} s_\nu(t, \tilde{t}) \sum_{\tilde{y} \in S_{\tilde{t}}} K_\lambda(1|\tilde{y})\tag{9.4.6}$$

where K is a discrete kernel and $s_\nu(t, \tilde{t})$ is a weight function determining the influence of observations $y_{i\tilde{t}}$ taken at time $\tilde{t}$. The weight function s_ν fulfills $\sum_{\tilde{t}} s_\nu(t, \tilde{t}) = 1$ (see Section 5.2).

For the extreme case $\lambda \to 1$ and ν such that

$$\bar{s}_\nu(t, \tilde{t}) = \begin{cases} 1/q, & \text{if } t = \tilde{t} \\ 0, & \text{otherwise} \end{cases}$$

the kernel regression estimate (9.4.6) becomes the simple life table estimate (9.4.4).

Inclusion of covariates and the extension to the competing risks case is straightforward. Define

$$
\tilde{y}_{it} = \begin{cases} r, & \text{failure of type } r \text{ in interval } [a_{t-1}, a_t) \\ m+1, & \text{no failure in } [a_{t-1}, a_t) \end{cases}
$$

for the multicategorical response. Since now the explanatory variables are t and x the local sample at (t, x) is given by

$$
S_{(t,x)} = \left\{ \tilde{y}_{it} \mid t \leq t_i - (1 - \delta_i), \, x_i = x \right\}.
$$

Then the smoothing estimate based on (5.2.7) has the form

$$
\begin{aligned}
\hat{\lambda}_r(t|x) &= \hat{P}\big(y = r|(t,x), \lambda, \mu\big) \\
&= \sum_{\tilde{t}, \tilde{x}} s_\nu\big((t,x), (\tilde{t}, \tilde{x})\big) \sum_{\tilde{y} \in S_{(\tilde{t}, \tilde{x})}} K_\lambda(r|\tilde{y})
\end{aligned}
$$

where the weight function depends on the target value (t, x) and the observation point $(\tilde{t}, \tilde{x})$. Both variables, time t and covariates x, are treated as explanatory. Since these variables may be quite different in scaling multidimensional smoothing is to be preferred. Let us consider Nadaraya–Watson weights with product kernels

$$
s_\nu\big((t,x), (\tilde{t}, \tilde{x})\big) = \frac{K_{\nu_0}(t|\tilde{t}) \prod_{i=1}^{p} K_{\nu_i}(x_i|\tilde{x}_i)}{\sum_{\tilde{t}, \tilde{x}} K_{\nu_0}(t|\tilde{t}) \prod_{i=1}^{p} K_{\nu_i}(x_i|\tilde{x}_i)}
$$

where $x = (x_1, \ldots, x_p)$ and $\tilde{x} = (\tilde{x}_1, \ldots, \tilde{x}_p)$. The parameters $\nu_1, \ldots, \nu_p$ are the smoothing parameters for the components of x whereas the parameter ν_0 determines the degree of smoothing for observations in the neighbourhood of t.

If time is considered a discrete variable $\nu_0 \to 1$ means faith with the data and $\nu_0 \to 0$ means an ultrasmooth estimate. The same behaviour is found for continuous kernel when the transformed smoothing parameter $\nu_0^* = 1 - \exp(-\nu_0)$ is considered. In discrete survival models the local sample of individuals under risk at time t is given by n_t and is steadily decreasing in t. Therefore smoothing should be stronger for increasing t. This may be accomplished by time–dependent smoothing $\nu_0(t) = \nu_0 n_t/n_1$ where n_t is the number of individuals at risk in $[a_{t-1}, a_t)$. Then $\nu_0(t)$ decreases from $\nu_0(1) = \nu_0$ to $\nu_0(q)$.

Example 9.4: Head and neck cancer
Efron (1988) considers a head and neck cancer study where time is discretized by one–month intervals. Table 9.3 shows the data (from Efron,

TABLE 9.3. Head and neck cancer (Efron, 1988)

Month	Patients at risk	Deaths	With-drawals	Month	Patients at risk	Deaths	With-drawals
1	51	1	0	25	7	0	0
2	50	2	0	26	7	0	0
3	48	5	1	27	7	0	0
4	42	2	0	28	7	0	0
5	40	8	0	29	7	0	0
6	32	7	0	30	7	0	0
7	25	0	1	31	7	0	0
8	24	3	0	32	7	0	0
9	21	2	0	33	7	0	0
10	19	2	1	34	7	0	0
11	16	0	1	35	7	0	0
12	15	0	0	36	7	0	0
13	15	0	0	37	7	1	1
14	15	3	0	38	5	1	0
15	12	1	0	39	4	0	0
16	11	0	0	40	4	0	0
17	11	0	0	41	4	0	1
18	11	1	1	42	3	0	0
19	9	0	0	43	3	0	0
20	9	2	0	44	3	0	0
21	7	0	0	45	3	0	1
22	7	0	0	46	2	0	0
23	7	0	0	47	2	1	1
24	7	0	0				

1988, table 1). Figure 9.4 shows a cubic spline fit following (9.2.16) where the cut–off point is chosen by $t_c = 11$ as suggested by Efron (1988).

Moreover, posterior mode smoothing based on the generalized Kalman filter and smoother is given in Figure 9.4. The underlying model is the simple logistic model $\lambda(t) = \exp(\alpha_t)/(1 + \exp(\alpha_t))$ and first–order random walk $\alpha_t = \alpha_{t-1} + \xi_t$. Alternatively the data are smoothed by the categorical kernel estimate (9.4.6) based on Nadaraya–Watson weights with normal kernels and the Aitchison and Aitken kernel as discrete kernel. The smoothing parameters chosen by cross–validation are $\nu_0^* = 0.5$ for time–dependent smoothing and $\lambda = 0.97$. In Figure 9.5 the life table estimate is also given by full cubicles. The life table estimate is quite jagged, whereas the nonparametric estimate based on cross–validated smoothing parameter is a rather smooth function. It is seen that for this simple data set the esti-

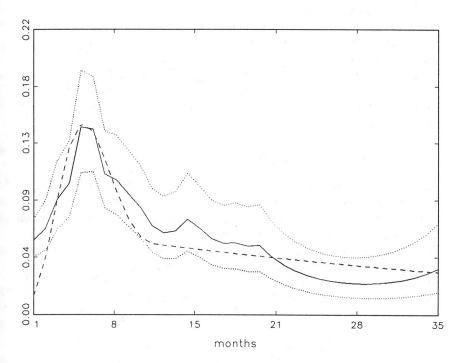

FIGURE 9.4. Cubic–linear spline fit for head and neck cancer data (- - -) and posterior mode smoother (—) with ± standard deviation confidence bands.

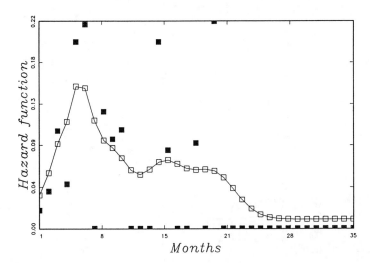

FIGURE 9.5. Smoothed kernel estimate for head and neck cancer data

mates yield rather similar results: after short increase the hazard function is decreasing. Some difference is seen in the decreasing branch. The spline fit is cubic for $t \leq 11$ and linear for $t \geq 11$. Consequently there is a steady decrease for $t \geq 11$. Posterior mode smoothing, in particular the smoother kernel estimate, shows that the decline is very slow for $11 \leq t \leq 18$ and steeper for $t \geq 18$. These estimates are more locally determined and are closer to the relative frequencies which in unsmoothed form give no feeling for the underlying hazard function.   □

9.4.3   Further developments

Posterior mode and penalized likelihood smoothing for dynamic models with multiple modes of failure is described in Fahrmeir and Wagenpfeil (1996), with applications to unemployment data. Full Bayesian analysis with MCMC techniques is developed in Fahrmeir and Knorr–Held (1997b). Other semi- or nonparametric approaches like spline smoothing (e.g. Dannegger, Klinger and Ulm, 1995; Sleeper and Harrington, 1990) or local likelihood (e.g. Wu and Tuma, 1991; Tutz, 1995) can also be adapted to the situation of duration and, more generally, event history data.

Appendix A

A.1 Exponential families and generalized linear models

Simple exponential families

We say that a q–dimensional random variable y, or more exactly its distribution resp. density, belongs to a simple exponential family if its discrete or continuous density with respect to a σ–finite measure has the form

$$f(y|\theta, \lambda) = \exp\{[y'\theta - b(\theta)]/\lambda + c(y, \lambda)\}, \qquad \text{(A.1.1)}$$

with $c(y, \lambda) \geq 0$ and measurable. Jorgenson (1992) calls (A.1.1) an exponential dispersion model. The q–dimensional parameter $\theta \in \Theta \subset \mathbb{R}^q$ is the natural parameter of the family and $\lambda > 0$ is a *nuisance* or *dispersion* parameter. For given λ, we will generally assume that Θ is the natural parameter space, i.e., the set of all θ satisfying $0 < \int \exp\{y'\theta/\lambda + c(y, \lambda)\} dy < \infty$. Then Θ is convex, and in the interior Θ^0—assumed to be nonvoid—all derivatives of $b(\theta)$ and all moments of y exist. In particular

$$E_\theta(y) \;=\; \mu(\theta) = \frac{\partial b(\theta)}{\partial \theta}, \qquad \text{(A.1.2)}$$

$$\text{cov}_\theta(y) \;=\; \Sigma(\theta) = \lambda \frac{\partial^2 b(\theta)}{\partial \theta \partial \theta'}. \qquad \text{(A.1.3)}$$

The covariance matrix $\Sigma(\theta)$ is supposed to be positive definite in Θ^0, so that $\mu : \Theta^0 \to M = \mu(\Theta^0)$ is injective. Inserting the inverse function $\theta(\mu)$ into $\partial^2 b(\theta) / \partial\theta\, \partial\theta'$ one obtains the variance function

$$v(\mu) = \frac{\partial^2 b\big(\theta(\mu)\big)}{\partial\theta\partial\theta'},$$

and

$$\text{cov}(y) = \lambda\, v(\mu)$$

as a function of $\mu = Ey$ and the dispersion parameter λ. As a generalization of (A.1.1) Zhao, Prentice and Self (1992) introduced "partly" exponential models, a very broad class of distributions.

Generalized linear models

Generalized linear models (GLM's) for independent q–dimensional observations $y_1, \ldots, y_n$ and covariates $x_1, \ldots, x_n$ are characterized by the following structure:

(i) The $\{y_i\}$ are independent with densities from simple exponential families

$$f(y_i|\theta_i, \lambda_i), \qquad i = 1, \ldots, n$$

with $\lambda_i = \phi/\omega_i$, where ω_i are *known* weights and ϕ a, possibly unknown, nuisance parameter that is *constant* across observations.

(ii) The covariate vector x_i influences y_i in the form of a q–dimensional *linear predictor*

$$\eta_i = Z_i\beta, \tag{A.1.4}$$

where β is a p–dimensional parameter out of an admissible set $B \subset \mathbb{R}^p$, and $Z_i = Z(x_i)$, the $(q \times p)$–*design matrix*, is a function of the covariates.

(iii) The linear predictor η_i is related to the mean $\mu_i = \mu(\theta_i)$ by the *response function* $h : \mathbb{R}^q \to M$,

$$\mu_i = h(\eta_i) = h(Z_i\beta). \tag{A.1.5}$$

If the inverse $g = h^{-1} : M \to \mathbb{R}^q$ of h exists, then

$$g(\mu_i) = \eta_i = Z_i\beta, \tag{A.1.6}$$

and g is called *link function*.

For some theoretical purposes it is more convenient to relate the linear predictor η_i to the natural parameter θ_i (instead of $\mu_i = \mu(\theta_i)$) by $u = (g \circ \mu)^{-1} = \mu^{-1} \circ h$, i.e.,

$$\theta_i = u(Z_i\beta) = \mu^{-1}(h(Z_i\beta)) \qquad (A.1.7)$$

as in the original definition of Nelder and Wedderburn (1972).

Of special importance are *natural* link functions $g = \mu^{-1}$, and u the identity. Then we obtain a linear model $\theta_i = Z_i\beta$ for the natural parameter. Natural link functions are e.g., the logit function $\log(\mu/(1-\mu))$ for the binomial distribution and the $\log(\mu)$ function for the Poisson distribution.

In the above definitions, covariates x_i and, as a consequence, design matrices Z_i are tacitly assumed to be known constants. For *stochastic* regressors x_i, all definitions have to be understood conditionally: The $y_1, \ldots, y_n$ are conditionally independent, and $f(y_i|\theta_i, \lambda_i), \mu(\theta_i)$, etc., are conditional densities, means, etc.

Inserting (A.1.7) in $\mu(\theta_i)$ gives back (A.1.5), and insertion in $\Sigma(\theta_i)$ yields $cov(y)$ as a function of β. To stress dependence on β we will often write

$$E_\beta y_i = \mu_i(\beta), \quad cov_\beta y_i = \Sigma_i(\beta). \qquad (A.1.8)$$

Log–likelihood, score function, expected and observed information

For the following we assume that

(i) the admissible parameter space B is open,

(ii) $h(Z_i\beta) \in M = \mu(\Theta^0)$, $i = 1, 2, \ldots$, for all $\beta \in B$,

(iii) h, g and u are twice continuously differentiable, $\det(\partial g/\partial \eta) \neq 0$,

(iv) $\sum_{i=1}^{n} Z_i Z_i'$ has full rank for $n \geq n_0$, say.

Condition (ii) is necessary to have a well–defined GLM for all β. Conditions (i) and (iii) guarantee that the second derivatives of the log–likelihood are continuous. The rank condition and $\det(\partial g/\partial \eta) \neq 0$ will ensure that the expected information matrix is positive definite for all $\beta, n \geq n_0$.

The log–likelihood contribution of observation y_i for β is, up to a constant not dependent on β,

$$l_i(\beta) = [y_i'\theta_i - b(\theta_i)]/\lambda_i.$$

The individual score function $s_i(\beta) = \partial l_i(\beta)/\partial \beta$ is obtained by differentiation, using (A.1.2) and (A.1.3):

$$
\begin{aligned}
s_i(\beta) &= \frac{\partial h(Z_i\beta)}{\partial \beta} \Sigma_i^{-1}(\beta)(y_i - \mu_i(\beta)) \\
&= Z_i' D_i(\beta) \Sigma_i^{-1}(\beta)(y_i - \mu_i(\beta)),
\end{aligned}
\qquad (A.1.9)
$$

with $D_i(\beta) = \partial h(\eta)/\partial \eta$ the Jacobian of $h(\eta)$, evaluated at $\eta_i = Z_i\beta$.
An equivalent form is

$$s_i(\beta) = Z_i'W_i(\beta)\frac{\partial g(\mu_i)}{\partial \mu'}(y_i - \mu_i(\beta)), \qquad (A.1.10)$$

with the "weight matrix"

$$W_i(\beta) = \left[\frac{\partial g(\mu_i)}{\partial \mu'}\Sigma_i(\beta)\frac{\partial g(\mu_i)}{\partial \mu}\right]^{-1} = D_i(\beta)\Sigma_i^{-1}(\beta)D_i(\beta)'.$$

From (A.1.9) and (A.1.10) it is easily seen that $E_\beta s_i = 0$, as is common in
ML estimation under regularity conditions.

The contribution of y_i to the Fisher information or expected information
is

$$
\begin{aligned}
F_i(\beta) &= \text{cov}_\beta(s_i(\beta)) = E_\beta(s_i(\beta)s_i'(\beta)) \\
&= Z_i'D_i(\beta)\Sigma_i^{-1}(\beta)D_i(\beta)'Z_i = Z_i'W_i(\beta)Z_i.
\end{aligned}
\qquad (A.1.11)
$$

Further differentiation shows that the observed information of y_i is

$$F_{i,obs}(\beta) = \frac{-\partial^2 l_i(\beta)}{\partial\beta\partial\beta'} = F_i(\beta) - R_i(\beta),$$

with

$$R_i(\beta) = \sum_{r=1}^{q} Z_i'U_{ir}(\beta)Z_i(y_{ir} - \mu_{ir}(\beta)), \qquad (A.1.12)$$

where $U_{ir}(\beta) = \partial^2 u_r(Z_i\beta)/\partial\eta\partial\eta'$, and $u_r(\eta), y_{ir}$ and $\mu_{ir}(\beta)$ are the components of $u(\eta), y_i, \mu_i(\beta)$. It is easily seen that

$$E_\beta(F_{i,obs}(\beta)) = F_i(\beta), \qquad E_\beta(R_i(\beta)) = 0.$$

For natural link functions, where $\theta_i = \eta_i = Z_i\beta$ the expressions for $s_i(\beta)$
and $F_i(\beta)$ simplify, since $D_i(\beta) = \partial^2 b(\theta)/\partial\theta\,\partial\theta'$ evaluated at $\theta_i = Z_i\beta = \Sigma_i(\beta)\omega_i/\phi$. Also expected and observed information matrices coincide.

The formula for the score function on the right side of (A.1.9) was obtained by applying the chain rule for differentiation to $\mu(\beta) = h(Z\beta)$, so
that $\partial\mu/\partial\beta = Z'D(\beta)$ with $D(\beta) = \partial h/\partial\eta$, evaluated at $\eta = Z\beta$. Defining
directly the first derivative of $\mu(\beta)$ by

$$M(\beta) := \partial\mu/\partial\beta,$$

(A.1.9) becomes

$$s_i(\beta) = M_i(\beta)\Sigma_i^{-1}(\beta)(y_i - \mu_i(\beta)), \qquad (A.1.13)$$

and the Fisher information is

$$F_i(\beta) = M_i(\beta)\Sigma_i^{-1}(\beta)M_i'(\beta). \qquad (A.1.14)$$

For generalized linear models these formulas are of course equivalent to (A.1.9) and (A.1.11), since $M(\beta) = Z'D(\beta)$. However, (A.1.13) and (A.1.14) remain valid in the more general case of nonlinear exponential family models, where the assumption of a linear predictor is dropped and the mean is assumed to be a general nonlinear function

$$\mu(\beta) = \mu(x; \beta)$$

of covariates and parameters.

A.2   Basic ideas for asymptotics

In the main text we made repeated use of statements like "under appropriate regularity conditions (quasi–)maximum likelihood estimators are consistent and asymptotically normal," "test statistics are asymptotically χ^2–distributed," etc. This appendix briefly describes the line of arguments that lead to such asymptotic results. It may be of interest for more mathematically oriented readers but is not a necessary requirement for reading the text.

For a compact notation the stochastic versions o_p and O_p of the "Landau" symbols o and O are convenient. Let x_n be a sequence of deterministic vectors and r_n a sequence of positive real numbers. Then

$$x_n = o(r_n) \Leftrightarrow \|x_n\|/r_n < c, n \geq N$$

for all $c > 0$ and sufficiently large N, and

$$x_n = O(r_n) \Leftrightarrow \|x_n\|/r_n < C \quad \text{for all} \quad n \geq N$$

for some constant C and sufficiently large N. Obviously $\|x_n\| = o(r_n)$ is equivalent to $\|x_n\|/r_n \to 0$ for $n \to \infty$. The "p–versions" are:

$$x_n = o_p(r_n) \Leftrightarrow P(\|x_n\|/r_n < c) \geq 1 - \varepsilon, \quad n \geq N$$

for every $c > 0, \varepsilon > 0$ and sufficiently large N, and

$$x_n = O_p(r_n) \Leftrightarrow \quad \text{for every} \quad \varepsilon > 0 \quad \text{there is a} \quad C > 0$$

with

$$P(\|x_n\|/r_n < C) \geq 1 - \varepsilon, \quad n \geq N,$$

for sufficiently large N.

Obviously, $x_n = o_p(r_n)$ is equivalent to $\|x_n\|/r_n \overset{p}{\to} 0$ (in probability), with the special case $x_n \overset{p}{\to} 0$ for $r_n \equiv 1$. If $x_n = O_p(1)$, then x_n is said to be bounded in probability. For equivalent definitions and useful properties of o_p, O_p see, e.g., Prakasa Rao (1987).

Let us first discuss asymptotics of maximum likelihood inference for independent but, in general, not identically distributed observations $y_1, \dots, y_n$. We consider local MLE's $\hat{\beta}_n$ for β in the interior of the parameter space B, i.e., MLE's that are local maximizers of the log–likelihood or, equivalently, corresponding roots of the ML equations. In the following we tacitly assume that the dispersion parameter ϕ is known. However, results remain valid if ϕ is replaced by a consistent estimate. Let

$$\ell_n(\beta) \quad = \quad \sum_{i=1}^{n} \frac{y_i \theta(\mu_i) - b\big(\theta(\mu_i)\big)}{\phi} \omega_i, \quad \mu_i = h(Z_i \beta),$$

$$s_n(\beta) \;=\; \sum_{i=1}^{n} Z_i' \, D_i(\beta) \, \Sigma_i^{-1}(\beta) \, (y_i - \mu_i),$$

$$F_n(\beta) \;=\; \sum_{i=1}^{n} Z_i' \, D_i(\beta) \, \Sigma_i^{-1}(\beta) \, D_i'(\beta) \, Z_i,$$

$$H_n(\beta) \;=\; -\partial^2 \ell_n(\beta)/\partial\beta \, \partial\beta'$$

denote the log–likelihood, score function, expected and observed information matrix of the sample $y_1, \ldots, y_n$. The index n is introduced to make the dependence of likelihoods, MLE's, etc., on sample size n explicit. Note that in this section $\ell_n(\beta), s_n(\beta)$, etc., denote log–likelihoods, score functions of the whole sample $y_1, \ldots, y_n$ and not corresponding individual contributions of observation y_n.

Given a finite sample of size n, there are situations where no maximum of $\ell_n(\beta)$ and no root of $s_n(\beta)$ exists in the interior of B. Furthermore, local and global maxima need not coincide in general. However, for many important models local and global maxima are identical and unique if they exist, see Chapter 2 for references. For asymptotic considerations only *asymptotic existence*, i.e.,

$$P\Big(s_n(\hat{\beta}_n) = 0, \; H_n(\hat{\beta}_n) \text{ p.d.}\Big) \rightarrow 1$$

for $n \rightarrow \infty$ is required.

We give a short outline of standard $n^{1/2}$–asymptotics. For this case typical "regularity assumptions" are weak conditions on third derivatives of $\ell_n(\beta)$, existence of third moments of y, and in particular convergence of $F_n(\beta)/n = \operatorname{cov} s_n(\beta)/n = E\, H_n(\beta)/n$ to a p.d. limit, say

$$F_n(\beta)/n \;\rightarrow\; F(\beta). \qquad (A.2.1)$$

Then the following asymptotic results hold under regularity assumptions: Asymptotic normality of the score function,

$$n^{-1/2} s_n(\beta) \xrightarrow{d} N(0, F(\beta)), \qquad (A.2.2)$$

asymptotic existence and consistency of $\hat{\beta}_n$, and asymptotic normality of the (normed) MLE,

$$n^{1/2}(\hat{\beta} - \beta) \xrightarrow{d} N(0, F(\beta)^{-1}). \qquad (A.2.3)$$

We now give a brief outline of how to prove these statements (compare with McCullagh, 1983). The score function $s_n(\beta)$ is the sum of individual contributions $s_i(\beta)$ with $E\, s_i(\beta) = 0$ and finite variance. The total variance is $\operatorname{cov} s_n(\beta) = F_n(\beta)$. Since $F_n(\beta)/n \rightarrow F(\beta)$ no finite set of individual contributions is dominant, and some central limit theorem implies

$$n^{-1/2} s_n(\beta) \sim N(0, F_n(\beta)/n) + O_p(n^{-1/2})$$

and (A.2.2).

Proofs for consistency can be based on the following: Since $Ey_i = \mu_i(\beta)$ for the "true" parameter β, we have

$$E\, s_n(\beta) = \sum_{i=1}^{n} Z_i'\, D_i(\beta)\, \Sigma_i^{-1}(\beta)\, (Ey_i - \mu_i(\beta)) = 0$$

where the "true" β is a root of the *expected* score function $Es_n(\beta)$. The MLE $\hat{\beta}_n$ is a root of the "observed" score function

$$s_n(\beta) = \sum_{i=1}^{n} Z_i'\, D_i(\beta)\, \Sigma_i^{-1}(\beta)\, (y_i - \mu_i(\beta)).$$

By some law of large numbers

$$[s_n(\beta) - E\, s_n(\beta)]\big/ n \longrightarrow 0$$

in probability. With some additional arguments one obtains $\hat{\beta}_n \to \beta$ in probability. Other proofs make more direct use of the asymptotic behavior of $\ell_n(\beta)$ in small neighbourhoods of β.

Asymptotic normality (A.2.3) of $\hat{\beta}$ is shown as follows: Expansion of the ML equations $s_n(\hat{\beta}_n) = 0$ about β gives

$$s_n(\beta) = \bar{H}_n(\hat{\beta}_n - \beta)$$

where $\bar{H}_n = H_n(\bar{\beta})$ for some $\bar{\beta}$ between $\hat{\beta}_n$ and β, and

$$\hat{\beta}_n - \beta = \bar{H}_n^{-1} s_n(\beta) = F_n^{-1}(\beta)\, s_n(\beta) + O_p(n^{-1}). \qquad (A.2.4)$$

Multiplying both sides by $n^{1/2}$ and using (A.2.1) and (A.2.2) one obtains the $n^{1/2}$–asymptotic normality result (A.2.3).

Asymptotic χ^2–distributions for the (log–)likelihood ratio, the Wald and score statistic are again obtained by expansion of $\ell_n(\beta)$ in a Taylor series about $\hat{\beta}_n$:

$$
\begin{aligned}
\lambda_n &= -2(\ell_n(\beta) - \ell_n(\hat{\beta}_n)) = (\hat{\beta}_n - \beta)'\, \bar{H}_n(\hat{\beta}_n - \beta) \\
&= (\hat{\beta}_n - \beta)'\, \hat{F}_n(\hat{\beta}_n) + O_p(n^{-1/2}) = s_n'(\beta)\hat{F}_n^{-1}(\beta)s_n(\beta) + O_p(n^{-1/2}).
\end{aligned}
$$

Since $n^{1/2}(\hat{\beta} - \beta)$ and $n^{-1/2}s_n(\beta)$ are asymptotically normal, it follows that the likelihood ratio statistic λ_n for testing $H_0 : \beta = \beta_0$ against $H_1 : \beta \neq \beta_0$ is asymptotically χ^2 and asymptotically equivalent to the Wald and score statistic. For composite null hypotheses, $\lambda_n = -2(\ell_n(\tilde{\beta}_n) - \ell_n(\hat{\beta}_n))$ can be expressed as the difference of two quadratic forms, and this difference can be shown to be asymptotically χ^2 with the correct degree of freedoms; see, e.g., Cox and Hinkley (1974) for details.

The standard $n^{1/2}$–approach to first–order asymptotics sketched above mimics corresponding proofs for i.i.d. observations. Apart from certain domination conditions, the convergence assumption $F_n(\beta)/n \to F(\beta)$ in (A.2.1) is the really crucial one. This assumption will hold in the situation of "stochastic regressors," where pairs (y_i, x_i) are independent realisations of (y, x), and $n^{1/2}$–asymptotics are appropriate. However, the convergence condition (A.2.1) will typically not hold in the situation of trending regressors, for example, in planned experiments. Based on the same general ideas but applying them more carefully, asymptotic results can be obtained that allow for considerably more heterogeneity of the observations. In particular, convergence conditions like (A.2.1) can be completely avoided if matrix normalization by square roots of the (expected or observed) information matrix is used instead of $n^{1/2}$–normalization as in (A.2.3). Under rather weak assumptions one obtains asymptotic normality of normed score functions and MLE's in the form

$$
\begin{aligned}
F_n^{-1/2}(\beta)s_n(\beta) &\;\overset{d}{\to}\; N(0, I) \\
F_n^{T/2}(\hat{\beta}_n)(\hat{\beta}_n - \beta) &\;\overset{d}{\to}\; N(0, I),
\end{aligned}
$$

where $F_n^{-1/2}$ is the inverse of a left square root $F_n^{1/2}$ of F_n and $F_n^{T/2} = (F_n^{1/2})^T$ is the corresponding right square root. For details we refer the reader to Fahrmeir and Kaufmann (1985, 1986) and Fahrmeir (1987a, 1988) for results of this kind.

Next let us briefly discuss asymptotics for quasi–likelihood models in Section 2.3 and later in the text, where it is assumed that the mean is still correctly specified, i.e.,

$$
\mu_i = E\,y_i = h(Z_i\beta),
$$

but the variance function $\Sigma_i = \Sigma(\mu_i)$ may be different from the true variance $S_i = \text{cov}\,y_i$. Going through the outline of asymptotics for genuine likelihood inference above, it can be seen that most arguments go through unchanged up to the following modification: The quasi–score function $s_n(\beta)$ is asymptotically normal, but with

$$
\text{cov}\,s_n(\beta) = V_n(\beta) = \sum_{i=1}^{n} Z_i D_i \Sigma_i^{-1} S_i \Sigma_i^{-1} D_i' Z_i'.
$$

The expected "quasi–information" is still given by the common form

$$
F_n(\beta) = \sum_{i=1}^{n} Z_i' D_i \Sigma_i^{-1} D_i' Z_i
$$

so that

$$V_n(\beta) = F_n(\beta) \quad \text{if} \quad \Sigma_i = S_i.$$

If we assume

$$V_n(\beta)/n \to V(\beta) \quad \text{p.d.}$$

then

$$n^{-1/2} s_n(\beta) \xrightarrow{d} N(0, V(\beta))$$

instead of (A.2.2). Inserting this result in (A.2.4), one obtains asymptotic normality

$$n^{1/2}(\hat{\beta}_n - \beta) \xrightarrow{d} N\left(0, \, F(\beta)^{-1} V(\beta) F(\beta)^{-1}\right)$$

of the QLME with an adjusted asymptotic covariance matrix in the form of a sandwich matrix. Replacing the sandwich matrix by an estimate one obtains

$$\hat{\beta}_n \overset{a}{\sim} N\left(\beta, \, \hat{F}_n^{-1} \, \hat{V}_n \, \hat{F}_n^{-1}\right)$$

where $\hat{F}_n = F_n(\hat{\beta}_n)$ and

$$\hat{V}_n = \sum_{i=1}^{n} Z_i' \, \hat{D}_i \, \hat{\Sigma}_i^{-1} \, (y_i - \hat{\mu}_i)(y_i - \hat{\mu}_i)' \, \hat{\Sigma}_i^{-1} \, \hat{D}_i' \, Z_i$$

with $\hat{D}_i = D_i(\hat{\beta}_n)$, $\hat{\Sigma}_i = \Sigma_i(\hat{\beta}_n)$, $\hat{\mu}_i = \mu_i(\hat{\beta}_n)$.

Corresponding results with rigorous proofs for the general approach using matrix normalization can be found in Fahrmeir (1990). For dependent observations as in generalized autoregressive linear models (Section 6.1), the main ideas go through again under the following modification: the sequence $s_i(\beta)$ of individual score function distributions is no longer a sequence of independent random variables, but a martingale difference sequence. Applying laws and limit theorems for martingales, consistency and asymptotic normality of the MLE can be shown under appropriate assumptions; see Kaufmann (1987) for detailed proofs and, e.g., the survey in Fahrmeir (1987b).

A.3   EM–algorithm

The EM–algorithm is a general iterative method to obtain maximum like-lihood estimators in incomplete data situations. It was first proposed by Hartley (1958) and was generalized by Dempster, Laird and Rubin (1977).

Let $y \in \mathbb{R}^n$ denote a vector of observed data and $z \in \mathbb{R}^m$ a vector of unobservable data. Then the hypothetical complete data are given by (y, z) and the incomplete data that were observed are given by y. Furthermore, let $f(y, z; \theta)$ denote the joint density of the complete data depending on an unknown parameter vector $\theta \in \Theta$, and $k(z|y; \theta)$ is the conditional density of the unobservable data z, given the observed data y, which also depends on θ.

To obtain the maximum likelihood estimator (MLE) for θ the marginal log–likelihood

$$l(\theta) = \log \int f(y, z; \theta) dz \qquad (A.3.1)$$

usually is maximized in a direct way. Indirect maximization of (A.3.1) by the EM–algorithm avoids the numerical evaluation of the integral by considering

$$l(\theta) = \log f(y, z; \theta) - \log k(z|y; \theta) \qquad (A.3.2)$$

instead of (A.3.1). The problem in maximizing (A.3.2) is that z is unob-servable. Therefore, expectations are taken on both sides of (A.3.2) with respect to the conditional density $k(z|y, \theta_0), \theta_0 \in \Theta$, so that

$$l(\theta) = M(\theta|\theta_0) - H(\theta|\theta_0)$$

is obtained, where

$$
\begin{aligned}
M(\theta|\theta_0) &= E\{\log f(y, z; \theta)|y; \theta_0\} \\
&= \int \log f(y, z; \theta)\, k(z|y; \theta_0)\, dz, \qquad (A.3.3) \\
H(\theta|\theta_0) &= E\{\log k(z|y; \theta)|y; \theta_0)\} = \int \log k(z|y; \theta)\, k(z|y; \theta_0)\, dz.
\end{aligned}
$$

Then the EM–algorithm maximizes $l(\theta)$ iteratively by maximizing $M(\theta|\theta_0)$ with respect to θ, where θ_0 is given at each cycle of the iteration. In contrast to the integral in (A.3.1) evaluation of the integral in (A.3.3) is straightfor-ward for many applications (see, e.g., Dempster, Laird and Rubin, 1977). If $\theta^{(0)}$ denotes a starting value for θ, the $(p+1)$–th cycle of the EM–algorithm consists of the following two steps for $p = 0, 1, \ldots$:

E(xpectation)–step: Compute the expectation $M(\theta|\theta^{(p)})$ given by (A.3.3)

M(aximizing)–step: Determine $\theta^{(p+1)}$ by $M(\theta|\theta^{(p)}) \to \max_\theta$

The EM–algorithm has the desirable property that the log–likelihood l always increases or stays constant at each cycle: Let $\hat\theta$ maximize $M(\theta|\theta_0)$, given θ_0. Then we have $M(\hat{\theta}|\theta_0) \geq M(\theta|\theta_0)$ for all θ by definition and $H(\theta|\theta_0) \leq H(\theta_0|\theta_0)$ for all θ by Jensen's inequality, so that

$$l(\hat{\theta}) \geq l(\theta_0).$$

Convergence of the log–likelihood sequence $l(\theta^{(p)})$, $p = 0, 1, 2 \ldots$ against a global or local maximum or a stationary point l_* is ensured under weak regularity conditions concerning Θ and $l(\theta)$ (see, e.g., Dempster, Laird and Rubin, 1977). However, if there exists more than one maximum or stationary point, convergence against one of these points depends on the starting value. Moreover, convergence of the log–likelihood sequence $l(\theta^{(p)})$, $p = 0, 1, 2, \ldots$ against l_* does not imply the convergence of $(\theta^{(p)})$ against a point θ_* as has been pointed out by Wu (1983) and Boyles (1983). In general, convergence of $(\theta^{(p)})$ requires stronger regularity conditions, which are ensured in particular for complete data densities $f(y, z; \theta)$ of the simple or curved exponential family. For finite mixtures of densities from the exponential family, see Redner and Walker (1984).

The rate of convergence depends on the relative size of the unobservable information on θ. If the information loss due to the missing of z is a small fraction of the information in the complete data (y, z), the algorithm converges rapidly. On the other hand, the rate of convergence becomes rather slow for parameters θ near the boundary of Θ and an estimator for the variance–covariance matrix of the MLE for θ, e.g., the observed or expected information on θ in the observed data y, is not provided by the EM–algorithm. Newton–Raphson or other gradient methods that maximize (A.3.1) directly are generally faster and yield an estimator for the variance–covariance matrix of the MLE. However, the EM–algorithm is simpler to implement and numerically more stable. An estimate for the variance–covariance matrix of the MLE is obtained if an additional analysis is applied after the last cycle of the EM–algorithm (Louis, 1982). The method can also be used to speed up the EM–algorithm (see also Meilijson, 1989). However, for complex complete data structures the procedure is rather cumbersome.

A.4   Numerical integration

Numerical integration techniques approximate integrals that cannot be solved analytically. Simpson's rule and quadrature methods are prominent among the techniques for univariate integrals (see Davis and Rabinowitz, 1975). These methods are based on reexpressing a regular function $f(x) : \mathbb{R} \to \mathbb{R}$ as the product of a *weight function* $w(x) : \mathbb{R} \to \mathbb{R}_+$ and another function $g(x) : \mathbb{R} \to \mathbb{R}$,

$$f(x) = w(x)g(x).$$

Then, most numerical integration methods approximate an integral by a discrete summation,

$$\int_{\mathbb{R}} f(x)dx \approx \sum_{i=1}^{k} w_i g(x_i), \qquad (A.4.1)$$

where the points x_i are called nodes, the w_i are the weights, and the nodes and weights together constitute an integration rule.

Following Stroud (1971), Davis and Rabinowitz (1984) an integration rule should have at least the following properties to reduce the numerical effort and integration error in (A.4.1) as far as possible:

- Nodes and weights should be easily found and calculated.

- Nodes should lie in the region of integration.

- Weights should all be positive.

Such integration rules are available as long as the weight function $w(x)$ is known.

Univariate Gauss–Hermite integration

For most statistical integration problems the normal density

$$w_N(x; \mu, \sigma^2) = \frac{1}{\sqrt{2\pi}\sigma} \exp\left(-\frac{(x-\mu)^2}{2\sigma^2}\right)$$

with mean μ and variance σ^2 can be deduced as a weight function.

Such a weight function is the basis of the Gauss–Hermite integration rule. If $g(x), x \in \mathbb{R}$, denotes a regular function, the Gauss–Hermite rule approximates integrals of the form

$$\int_{-\infty}^{+\infty} \exp(-x^2)g(x)dx \qquad (A.4.2)$$

by the sum

$$\sum_{i=1}^{k} w_i g(x_i), \tag{A.4.3}$$

where the node x_i is the ith zero of the Hermite polynomial having degree k, $H_k(x)$, and the weight w_i depends on the number k of nodes and the Hermite polynomial $H_{k-1}(x)$ evaluated at x_i. Tables for the nodes and weights can be found in Stroud and Secrest (1966) and Abramowitz and Stegun (1972). As long as $g(x)$ is a polynomial of maximal degree $2k-1$ the sum (A.4.3) delivers the exact value of (A.4.2). That means approximation (A.4.3) becomes arbitrarily accurate when the number k of nodes is increased.

Let us consider now, more generally, the function

$$f(x) = w_N(x; \mu, \sigma^2) g(x). \tag{A.4.4}$$

The simple substitution $x = \sqrt{2}\sigma z + \mu$ yields the identity

$$\int_{-\infty}^{+\infty} f(x)dx = \int_{-\infty}^{+\infty} \pi^{-1/2} \exp(-z^2) g(\sqrt{2}\sigma z + \mu) dz$$

which can be solved by Gauss–Hermite integration in the following way:

$$\int_{-\infty}^{+\infty} f(x)dx \approx \sum_{i=1}^{k} v_i g(\sqrt{2}\sigma x_i + \mu) \tag{A.4.5}$$

where $v_i = \pi^{-1/2} w_i$ is the transformed weight and x_i is the tabulated ith zero of the Hermite polynomial. Thus the integral of a function is well approximated by Gauss–Hermite if $f(x)$ has the form (A.4.4), where $g(x)$ stands for a polynomial in x.

Within this book most integration problems can be traced back to the Bayesian paradigm that can be characterised in the following way (see e.g., Naylor and Smith, 1982; Smith et al., 1985; Shaw, 1988): Suppose we have data y together with a probability model $f(y|\theta)$ indexed by a parameter $\theta \in \mathbb{R}$ with prior density $q(\theta)$. Suppose further that the resulting posterior density

$$p(\theta|y) = \frac{f(y|\theta)q(\theta)}{\int_{\mathbb{R}} f(y|\theta)q(\theta)d\theta} \tag{A.4.6}$$

is analytically intractable, since the integral that also determines the marginal density of y cannot be solved analytically. As a consequence no closed–form solutions are available for the first two posterior moments

$$E(\theta^r|y) = \int_{\mathbb{R}} \theta^r p(\theta|y)d\theta, \quad r = 1, 2 \tag{A.4.7}$$

which are assumed to exist. For the application of the Gauss–Hermite rule to (A.4.6) and (A.4.7) we consider in the following the integral function

$$S(a(\theta)) = \int_{\mathbb{R}} a(\theta) f(y|\theta) q(\theta) d\theta$$

with $a(\theta) = 1, \theta$ or θ^2. $S(a(\theta))$ covers all interesting measures. For example, $S(1)$ corresponds to the marginal density of y and $S(\theta)/S(1)$ denotes the posterior mean.

Application of the Gauss–Hermite integration rule to $S(a(\theta))$ is straight–forward when the prior $q(\theta)$ corresponds to a normal density with known mean μ and known variance σ^2. Since $a(\theta)$ is a polynomial of degree at most 2, the numerical accuracy of the approximation only depends on the polynomial degree of the likelihood $f(y|\theta)$ and the number of nodes used.

If the prior $q(\theta)$ is non–normal iterative use of the Gauss–Hermite rule is recommended by various authors (see, e.g., Naylor and Smith, 1982; Smith et al., 1985). The procedure is based on the Bayesian analogue of the central limit theorem indicating that the posterior may be well approximated by a normal density multiplied by a polynomial in θ. Expanding the posterior kernel by such a normal component with mean μ and variance σ^2, say $w_N(\theta; \mu, \sigma^2)$, yields

$$S(a(\theta)) = \int_{\mathbb{R}} a(\theta) h(\theta) w_N(\theta; \mu, \sigma^2) d\theta$$

with $h(\theta) = f(y|\theta) q(\theta)/w_N(\theta; \mu, \sigma^2)$. Given μ and σ^2, the Gauss–Hermite rule A.4.5) could be applied, provided $a(\theta)h(\theta)$ is at least approximately a polynomial in θ. In most practical situations, however, μ and σ^2 are unknown. A crude, but sometimes satisfactory, procedure is to replace μ by the posterior mode and σ^2 by the posterior curvature. Both are easily estimated by maximizing the log–posterior

$$\ell(\theta) = \log f(y|\theta) + \log q(\theta)$$

with respect to θ. The procedure can be improved by the following iterative process:

1. Choose initial values μ_0 and σ_0^2 for μ and σ^2.

2. For some chosen k, approximate $S(\theta)$ by

$$S_p(a(\theta)) = \sum_{i=1}^{k} a(\theta_{p,i}) h_p(\theta_{p,i}) v_i,$$

with nodes $\theta_{p,i} = \sqrt{2}\sigma_{p-1} x_i + \mu_{p-1}$, weights $v_i = \pi^{-1/2} w_i$ and the function $h_p(\theta) = f(y|\theta) q(\theta)/w_N(\theta; \mu_{p-1}, \sigma_{p-1}^2)$.

3. Calculate updates of μ and σ^2 according to

$$\mu_p = S_p(\theta)/S_p(1) \quad \text{and} \quad \sigma_p^2 = S_p(\theta^2)/S_p(1) - \mu_p^2$$

4. Repeat steps 2 and 3 as long as the changes in updated moments are not small enough.

This procedure gives approximations for the posterior mean and the posterior variance at each cycle p. These approximations are used to construct the nodes $\theta_{p,i}$ and the function h_p for the next cycle. With an increasing number of cycles the number k of nodes should be successively increased to obtain stable values for μ and σ^2 at each size of k.

Multivariate Gauss–Hermite integration

Let us consider now the case of m–dimensional $x = (x_1, \ldots, x_m)$ with a regular function $f(x) : \mathbb{R}^m \to \mathbb{R}$, weight function $w(x) : \mathbb{R}^m \to \mathbb{R}_+$ and another function $g(x) : \mathbb{R}^m \to \mathbb{R}$, so that $f(x) = w(x)g(x)$. A Cartesian product rule based on (A.4.3) may be applied to approximate the m–dimensional integral

$$\int_{\mathbb{R}^m} f(x)dx = \int_{\mathbb{R}} \cdots \int_{\mathbb{R}} w(x_1, \ldots, x_m)g(x_1, \ldots, x_m)dx_1 \ldots dx_m,$$

provided that the weight function is given by

$$w(x) = \exp(-x'x) = \exp(-x_1^2) \cdot \ldots \cdot \exp(-x_m^2).$$

Then the univariate Gauss–Hermite rule applies to each of the components of x in turn. Using k_r nodes in the rth dimension, $r = 1, \ldots, m$, yields the approximation

$$\int_{\mathbb{R}^m} f(x)dx \approx \sum_{i_1=1}^{k_1} w_{i_1}^{(1)} \cdots \sum_{i_m=1}^{k_m} w_{i_m}^{(m)} g\left(x_{i_1}^{(1)}, \ldots, x_{i_m}^{(m)}\right), \qquad (A.4.8)$$

where $x_{i_r}^{(r)}$ is the i_rth zero of the Hermite polynomial with degree k_r and $w_{i_r}^{(r)}$ is the corresponding weight. The Cartesian product rule (A.4.8) has $k = \prod_{r=1}^{m} k_r$ nodes $x_i = \left(x_{i_1}^{(1)}, \ldots, x_{i_m}^{(m)}\right), i = (i_1, \ldots, i_m)$ and is exact as long as g is a polynomial containing terms up to $x_r^{2k_r-1}$ for each dimension $r = 1, \ldots, m$. Unfortunately, the number k of nodes increases exponentially with the number m of dimensions. So Cartesian product rules are less appropriate to approximate high–dimensional integrals. In practice Cartesian product rules work efficiently up to 5– or 6–dimensional integrals.

If we consider, more generally, the multivariate extensions of the Bayesian paradigm that is based on the posterior density

$$p(\theta|y) = \frac{f(y|\theta)q(\theta)}{\int f(y|\theta)q(\theta)d\theta}$$

with parameter $\theta = (\theta_1, \ldots, \theta_m) \in \mathbb{R}^m$, data $y = (y_1, \ldots, y_m)$, likelihood $f(y|\theta)$ and prior density $q(\theta)$, all integration problems can be traced back to

$$S(a(\theta)) = \int_{\mathbb{R}^m} a(\theta)f(y|\theta)q(\theta)d\theta. \tag{A.4.9}$$

For example, posterior mean and covariance are given by

$$E(\theta|y) = S(\theta)/S(1) \quad \text{and} \quad \text{cov}(\theta|y) = S(\theta\theta')/S(1) - E(\theta|y)E(\theta|y)'.$$

To apply the Cartesian Gauss–Hermite product rule to $S(a(\theta))$ we assume, in analogy to the univariate case, that the posterior density can be well approximated by the product of a multivariate normal density and a polynomial in θ. Expansion of the posterior kernel by a multivariate normal density $w_N(\theta; \mu, \Sigma)$ with mean $\mu = (\mu_1, \ldots, \mu_m)$ and variance–covariance matrix Σ yields

$$S(a(\theta)) = \int_{\mathbb{R}^m} a(\theta)h(\theta)w_N(\theta; \mu, \Sigma)d\theta.$$

The substitution

$$\theta = \sqrt{2}\,\Sigma^{1/2}z + \mu$$

with the left Cholesky square root $\Sigma^{1/2}$ has the desirable property that the components of z are nearly orthogonal and that

$$S(a(\theta)) = \int_{\mathbb{R}^m} a(z)h(z)\pi^{-m/2}exp(-z'z)dz$$

has a weight function that is required to use the Cartesian product rule of the Gauss–Hermite type.

Then (A.4.9) can be approximated by an extension of the iterative integration scheme that was proposed for the univariate case. The steps are as follows:

1. Choose initial values μ_0 and Σ_0 for μ and Σ, and set $p = 1$.

2. For some chosen $k_1, \ldots, k_m$ approximate $S(a(\theta))$ by

$$S_p(a(\theta)) = \sum_{i_1=1}^{k_1} v_{i_1}^{(1)} \cdots \sum_{i_m=1}^{k_m} v_{i_m}^{(m)} a(\theta_{p,i})h_p(\theta_{p,i})$$

where for the multiple index $i = (i_1, \ldots, i_m)$ the nodes are given by

$$\theta_{p,i} = \sqrt{2}\, \Sigma_{p-1}^{1/2} z_i + \mu_{p-1}$$

with $z_i = \left(z_{i_1}^{(1)}, \ldots, z_{i_m}^{(m)} \right)$, $z_{i_r}^{(r)}, r = 1, \ldots, m$, denoting the tabled nodes of univariate Gauss–Hermite integration of order k_r. The corresponding weights are given by $v_{i_r}^{(r)} = \pi^{-1/2} w_{i_r}^{(r)}$ and $h_p(\theta) = f(y|\theta) q(\theta)/w_N(\theta; \mu_{p-1}, \Sigma_{p-1})$.

3. Calculate updates of μ and Σ according to $\mu_p = S_p(\theta)/S_p(1)$ and $\Sigma_p = S_p(\theta\theta')/S_p(1) - \mu_p \mu_p'$.

4. Repeat steps 2 and 3 and set $p = p + 1$ as long as the changes of the updated posterior moments are not small enough.

Examples for the efficiency of the above scheme can be found in Naylor and Smith (1982) and Smith et al. (1985), among others. For a prior density $q(\theta)$ with known mean $E(\theta)$ and known variance–covariance matrix $\mathrm{cov}(\theta)$ Hennevogl (1991) recommends using

$$\mu_0 = E(\theta) \quad \text{and} \quad \Sigma_0 = \mathrm{cov}(\theta)$$

as starting values. Alternatively posterior mode and curvature, which are easy and fast to compute, may be used.

A.5   Monte Carlo methods

Importance sampling

The simplest Monte Carlo method for computing integrals of the form

$$I = \int_{-\infty}^{+\infty} g(x)f(x)dx,$$

where g is a continuous function and f is a density, consists in approximating I by the arithmetic mean

$$\hat{I} = \frac{1}{m}\sum_{j=1}^{m} g(x_j),$$

where $x_j, j = 1, \ldots, m$ are i.i.d. drawings from the density $f(x)$. A better and more efficient approximation of I can be obtained by importance sampling: Rewriting I as

$$I = \int_{-\infty}^{+\infty} g(x)\frac{f(x)}{\phi(x)}\phi(x)dx,$$

with the density $\phi(x)$ as the importance function one may also approximate I by

$$\bar{I} = \frac{1}{m}\sum_{j=1}^{m} g(x_j)\frac{f(x_j)}{\phi(x_j)},$$

where the x_i's are now drawn from $\phi(x)$. It can be shown (Ripley, 1987) that this approximation by importance sampling becomes quite good if $g(x)f(x)/\phi(x)$ is nearly constant. An optimal choice would be

$$\phi(x) \propto |g(x)|f(x);$$

however, $\phi(x)$ should also allow fast and simple random number generation. In a Bayesian context, integrals are often of the form

$$I = \int a(x)f(y|x)f(x)dx,$$

where $f(y|x)f(x)$ is proportional to the posterior density $f(x|y)$. Since this posterior becomes approximately normal for larger numbers of observations y, the importance sampling function is often chosen as a (multivariate) normal $N(\mu, \Sigma)$, where μ and Σ are matched to the posterior mean and covariance of $f(x|y)$ or some approximation, such as mode and curvature of $f(x|y)$.

Rejection sampling

Rejection sampling is a technique for generating random numbers from a density $f(x)$ without drawing from $f(x)$ itself directly. A typical application arises in Bayesian inference where $f(x)$ is not available itself but only an un–normalized "density" $g(x)$ (often the nominator in Bayes' theorem):

$$g(x) \propto f(x).$$

If there is a density $h(x)$ available for drawing random numbers and a constant C such that

$$g(x) \leq C \cdot h(x)$$

for all x, then the following rejection algorithm generates random numbers from $f(x)$:

1. Generate a random number x^* from $h(x)$

2. Accept x^* as a random number from $f(x)$ with probability

$$\pi(x^*) = \frac{g(x^*)}{Ch(x^*)}.$$

Step 2 is usually implemented by drawing u from the uniform distribution on $[0, 1]$, and x^* is accepted if $u \leq \pi(x^*)$, rejected otherwise. Repeated application generates a random sample from $f(x)$. In order not to waste too many drawings from $h(x)$, the "envelope" function $C \cdot h(x)$ should be as close as possible to $f(x)$. More details, such as formal proofs or choice of the envelope function, can be found in Devroye (1986) and Ripley (1987).

Gibbs sampler

Suppose we are interested in the joint density $f(u_1, \ldots, u_p)$ or in marginal densities as $f(u_j)$ of a random vector $U = (U_1, \ldots, U_p)$. The Gibbs sampler, introduced by Geman and Geman (1984) allows drawing random samples from this joint density in situations where $f(u_1, \ldots, u_p)$ is unknown or not available, but where the conditional densities

$$f(u_i | u_{-i}), \quad u_{-i} = (u_1, \ldots, u_{i-1}, u_{i+1}, \ldots, u_p)$$

are available for sampling for all $i = 1, \ldots, p$.

The Gibbs sampler consists of the following iterative updating scheme:

0. Choose starting values $u^{(0)} = u_1^{(0)}, \ldots, u_p^{(0)}$.

   For $k = 1, 2, \ldots$

1. Draw $u_1^{(k)}$ from $f(u_1 | u_2^{(k-1)}, \ldots, u_p^{(k-1)})$.

2. Draw $u_2^{(k)}$ from $f(u_2|u_1^{(k)}, u_3^{(k-1)}, \ldots, u_p^{(k-1)})$.

$\vdots$

p. Draw $u_p^{(k)}$ from $f(u_p|u_1^{(k)}, \ldots, u_{p-1}^{(k)})$.

Under mild conditions Geman and Geman (1984) show that

$$U^{(k)} = \left(U_1^{(k)}, \ldots, U_p^{(k)}\right) \to U$$

in distribution as $k \to \infty$. Applying the algorithm m times provides a sample of size m from $f(u_1, \ldots, u_p)$. For m large enough, joint and marginal densities and their moments may be estimated from this sample, e.g., by density smoothers or empirical moments. In practice of course one also has to choose a termination criterion for stopping the Gibbs iterations for some finite k. For this question and more details we refer the reader to Gelfand and Smith (1990), Tanner (1991) and Ritter and Tanner (1992).

Markov chain Monte Carlo (MCMC)

MCMC techniques have revolutionized general Bayesian inference in the last few years. The Gibbs sampling algorithm, probably the most prominent member of MCMC algorithms, iteratively updates all components by samples from their full conditionals. Markov chain theory shows that under very general conditions the sequence of random numbers generated in this way converges to the posterior. However, often these full conditionals are themselves still quite complex, so generation of the required random numbers might be a difficult task. Relief lies in the fact that it is not necessary to sample from the full conditionals; a member of the much more general class of Metropolis–Hastings (MH) algorithms can be used to update the full conditionals. Such a MH step is typically easier to implement and often makes a MCMC algorithm more efficient in terms of CPU time. A MH step proposes a new value for a given component and accepts it with a certain probability. A Gibbs step (i.e., a sample from a full conditional) turns out to be a special case where the proposal is always accepted. There is a great flexibility in the choice of the MH proposal. For details we refer the reader to Tierney (1994), Besag, Green, Higdon and Mengersen (1995) and relevant parts of Gilks, Richardson and Spiegelhalter (1996) as well as the literature cited here.

Appendix B
Software for fitting generalized linear models

In this section software will be descibed that can be used for fitting GLMs and extensions. It includes not only software specialized on GLMs, but also well–known general–purpose statistical packages and programming enviroments with general statistical features. It also includes statistical packages that are generally less well known since they were originally developed for special applications. The following list gives an overview of all packages that will be described:

Software specialized in fitting GLMs
GLIM4 : programming environment for fitting GLMs
GLAMOUR : menu–driven package for fitting GLMs

General–purpose statistical packages
SAS : statistical package for all fields of applications
SPSS/PC+ : statistical package with special features for social sciences
BMDP: statistical package with special features for biometrical applications
GENSTAT 5 : statistical package for all fields of applications

Programming environments with statistical features
S–Plus : programming environment with many built-in functions for statistical analysis (one of which is for fitting GLMs)
GAUSS : programming language with add-on modules for fitting GLMs
XPLORE : programming environment for exploratory regression and data analysis

Specialized packages for certain fields of application
EGRET : package designed for epidemiological applications
LIMDEP : package designed for econometrical applications

In the following, the features for fitting GLMs of all packages listed above will be summarized. We will not compare the available features of these packages, mainly for two reasons:

(1) The packages vary with respect to their extensibility. For packages that are extensible it is possible to achieve a certain feature not available as a standard option by using available programming facilities and thus extending the features of the package. For example, GLIM4 is designed as a programming language and innumerable models may be fitted by using the language though only very few are implemented as standard options. On the other hand, GLAMOUR provides a range of models, but it is menu-driven and does not include any programming facilities.

(2) The descriptions are based on the package version available to the authors at the time of writing. Since features for fitting GLMs are increasingly implemented in recent times new versions are likely to include more features for modelling GLMs. The version of each package or the year of its release will be given in each description.

In the following, a short description of each package and its features for fitting GLMs will be given. Further information can be obtained from the manuals that come with the package or from the distibutor. Distributors' addresses are given after each package description.

GLIM4 (1993)

GLIM4 (Generalized Linear Interactive Modelling) is an interpretive language especially designed for fitting generalized linear models. It was developed at the Centre for Applied Statistics, University of Lancaster, U.K. GLIM4 includes comprehensive facilities of manipulating and displaying data. The interpretive character of the language is adjusted to the needs of iterative statistical model building. GLIM4 supports developing, fitting and checking of statistical models.

GLIM4 is essentially designed for statistical modelling with univariate responses. The following exponential families are included as standard options in GLIM4: normal, binomial, gamma, inverse normal and Poisson. Additionally, it is possible to specify distributions. The multinomial distribution is included, since it can easily be modelled using a relationship between the Poisson and the multinomial distribution. This "Poisson trick" is described, e.g., in Aitkin et al. (1989) or in the GLIM4-manual. GLIM4 includes nine link functions as standard options, but again it is possible to specify link functions.

GLIM4 is extensible in many respects. The user can write macros using the GLIM4 language or include FORTRAN routines. A macro library comes with the package. This library contains, e.g., macros to fit generalized additive models, or others to fit survival models with distributions such as Weibull, extreme value or (log)logistic. Further macros published in the GLIM4 newsletters are available on an archieve server "statlib", which can be reached via electronic mail (address: statlib@lib.stat.cmu.edu) or FTP (ftp to lib.stat.cmu.edu (128.2.241.142) and login with user name statlib).

The user has access to the system settings of GLIM4 and most of the data structures derived in model fitting and calculations. GLIM4 uses a weighted–least–squares algorithm for parameter estimation. It allows user–defined macros to be called at various stages of the fitting algorithm. Additionally, any structure available to the user can be modified during the fit. This allows the user to modify the iteratively reweighted–least–squares algorithm to perform special actions for specialized and nonstandard statistical models.

NAG Inc.
1400 Opus Place, Suite 200
Downers Grove,
Illinois 60515-5702, USA

NAG Ltd.
Wilkinson House
Jordan Hill Road
Oxford OX2 8DR, U.K.

GLAMOUR (Version 3.0, 1992)

GLAMOUR is an interactive package for a wide range of models based on extensions of generalized linear models. It was developed in 1984 at the Universities of Regensburg and Munich, Germany. The user interface of GLAMOUR is controlled by menus and masks. GLAMOUR provides basic features for data transformation, data description and graphics. It includes statistical procedures for parameter estimation, testing of linear hypotheses, variable selection and model checking.

Unidimensional responses can be modelled for normal, binomial, gamma and Poisson distributions. Response functions can be linear, logistic, double–exponential, log-linear or reciprocal.

Multinomial responses can be modelled using linear or logistic models (see Section 3.2). For ordered responses the models available include the cumulative logistic model and the cumulative double exponential model (see Section 3.3.1), the sequential logistic model (see Section 3.3.4) and the two–step cumulative logistic model (see Section 3.3.5). All parameters can be fitted as either category–specific or not, thus leading to extended versions of all models listed above (see, e.g., Section 3.3.2). Time series and panel data can be analysed based on auto-regressive models, since there are comfortable features to enter past values of the response variate into the list of covariates.

GLAMOUR provides a wide range of survival models for discrete time. Survival models can have just one absorbing final state (see Section 9.2) or multiple final states (see Section 9.3). One state model includes logistic, double exponential and exponential models as well as the model of Aranda-Ordaz for the discrete hazard. The baseline hazard function can be approximated by polynomials or splines. Multiple–state or competing risk models include the logistic model for multinomial final states, and cumulative logistic, sequential logistic or two–step cumulative logistic models for ordinal final states. Again, (cause-specific) baseline hazards can be approximated by polynomials or splines. The present version of GLAMOUR cannot fit models with time-varying covariates.

There is an extented version of GLAMOUR, called GAUSS-GLAMOUR, which provides some additional features. In particular it includes: posterior mode and mean approaches for random effects models (Chapter 7), state space modelling for time series and longitudinal data based on posterior mode filtering and smoothing (Chapter 8), and dynamic models for one- and multistate survival models in discrete time. All models are extended versions of univariate and multicategorical response models available in GLAMOUR. Many of the examples in this book have been computed with GAUSS-GLAMOUR. It is available together with GLAMOUR but is less well documented.

Information on GLAMOUR (and GAUSS-GLAMOUR) can be obtained from:

Institut für Statistik
Seminar für Statistik und ihre Anwendungen
(Ludwig Fahrmeir)
Ludwigstr 33/II
80539 München, Germany

SAS (Version 6.08, 1992)

SAS (Statistical Analysis System) is a statistical package with a particularly wide range of facilities for data management and data analysis. The SAS base system includes tools for data management, report writing and descriptive statistics and is rather limited in terms of statistical analysis. Other SAS software products have to be integrated to provide one total analysis system. The most important is the SAS/STAT software, which includes a powerful set of statistical analysis procedures. Other software products include SAS/GRAPH for generating high–quality graphics; SAS/IML, which provides an interactive matrix programming language for writing procedures in SAS; SAS/ETS for econometrical applications; or SAS/QC for statistical methods in quality control.

Metrical responses can be modelled within SAS/STAT using single or multiple linear regression, non-linear regression and general linear models that may include random effects. There are several methods available for variable selection. A new procedure MIXED can fit mixed linear models, i.e. general linear models which include both fixed effects and random effects. It is part of the SAS/STAT software in release 6.07 and later versions. Survival models in SAS/STAT include linear location-scale models (see Section 9.1.2) where the distribution of the noise can be taken from a class of distributions that includes the extreme-value, logistic, the exponential, Weibull, log-normal, log-logistic, and gamma distributions. The data may be left-, right- or interval-censored.

Categorical responses can be fitted using three procedures of SAS/STAT. These procedures are CATMOD, PROBIT and LOGISTIC.

CATMOD is a procedure for CATegorical data MODeling. It includes logit models for multinomial responses and cumulative logit response functions to take into account ordered categories. Additionally, marginal probabilities can be specified as response functions and simple models for repeated measurements can be fitted using a "repeated"-statement. It is possible to define response functions by using transformations and matrix operations within the "response"-statement, or by reading them directly together with their covariance matrix from a data set.

PROBIT can fit models for binary and multinomial responses. Probit, logit and complementary log-log links are available. There is an option to take overdispersion into account. Deviances are not given along with the standard output, but can be easily calculated from quantities given.

LOGISTIC is a new procedure available in SAS/STAT version 6.04 and later. Cumulative logistic models for ordered responses can be fitted. The procedure includes facilities for automatic variable selection and a good range of statistics for model checking.

From version 6.08 a new module 'SAS/INSIGHT' based on the Windows philosophy will include basic generalized models for nutrical, binary and count data. A new procedure GENMOD for fitting Generalized linear Models is planned and will be included as an experimental version in SAS version 6.08.

SAS Institute Inc.
Cary,
North Carolina 27512-8000, USA

SAS Software Ltd.
Wittington House
Henley Road, Medmenham
Marlow SL7 2EB, Bucks, U.K.

SPSS/PC+ (Version 4.0, 1990)

SPSS (Statistical Package for Social Scientists) is a statistical package with a wide range of statistical directives especially designed to meet the needs of social scientists. The base system can be extended by add-on modules. SPSS can perform many analyses that are commonly used in social science. However, there is no programming language included to allow for users' own programs.

Metrical responses can be modelled using simple and multiple regression. There are several methods available for variable selection. Time series can be modelled with procedures that are part of the module "Trends." It includes AR- and ARIMA-models, exponential-smoothing models and spectral-analysis.

Binary responses can be modelled with two procedures that are part of the "Advanced Statistics" module: LOGISTIC REGRESSION and PROBIT. They allow fitting logit and probit models. For LOGISTIC REGRESSION the data must not be grouped. However, there is no such restriction for the PROBIT procedure, and probit and logit models can be fitted using this directive.

SPSS/PC+ provides a good range of facilities for model checking. Predicted values, along with their standard errors, and residuals are easily obtained. However, deviances are not included as an option for the standard output, and there are no facilities to model overdispersion.

SPSS Inc.
444 North Michigan Avenue
Chicago,
Illinois 60611, USA

SPSS UK Ltd.
SPSS House, 5 London Street
Chertsey KT16 8AP, Surrey, U.K.

BMDP (Version 7.0, 1992)

BMDP (BioMeDical Package) is a general–purpose statistical package that consists of a collection of independent programs for statistical analysis, some of which was particulary designed for the needs of biological and medical researchers. It is accompanied by a data manager, but does not include a programming language. The graphic facilities are limited.

Metrical response modelling includes multivariate regression, Ridge–regression nonlinear and polynomial regression. Variable selection is available. Time series can be analysed using ARIMA-models and spectral analysis. Survival models include Cox's proportional hazards model and log-linear models with error-term distributions such as Weibull, exponential, log-normal and log-logistic. Variable selection is available for all models.

The program LR (Logistic Regression) fits logit models for *binary* responses. LR can be used to either fit a specified model or perform variable selection. It provides a good range of diagnostics. Deviances are given and predicted values easily obtained. LR can not take overdispersion into account.

Multicategorical responses can be modelled using the program PR (Polychotomous Logistic Regression). The program includes the logistic model for multinomial responses and the cumulative logit model for ordered responses. The features of PR are similar to LR. In particular, variable selection is also available.

LE is a program for estimating the parameters that maximize likelihood functions and may be used to fit models. LE uses an iterative Newton-Raphson algorithm where the (log)density and initial values for the parameters must be provided. The manual illustrates LE with a logit model for binary responses and a cumulative logistic model for ordered responses.

BMDP Statistical Software Inc.
1440 Sepulveda Blvd.
Los Angeles,
California 90025, USA

BMDP Statistical Software Ltd.
Cork Technology Park
Cork, Ireland

GENSTAT 5 (Release 1.3)

GENSTAT 5 (GENeral STATistical package) is a general–purpose statistical package that supplies a wide range of standard directives, but also includes a flexible command language enabling the user to write programs. FORTRAN programs may also be included via the OWN directive. GENSTAT has a wide range of data–handling facilities and provides high-resolution graphics.

Metrical responses can be modelled based on simple or multiple linear regression, standard nonlinear curves or general nonlinear regression. Time series can be analysed using models such as AR and ARIMA.

The following distributions are included for fitting *generalized linear models*: binomial, Poisson, normal, gamma, inverse normal. Implemented link functions are: identity, log, logit, reciprocal, power, square root and complementary log-log.

GENSTAT provides a good range of facilities for model checking. Deviances are given along with the standard output. Overdispersion can be modelled. Predicted values along with their standard errors can be yielded by a separate command.

There is a link from GENSTAT to GLIM, since GLIM-macros can be translated into GENSTAT-procedures (and vice versa).

Numerical Algorithms Group Inc.
1400 Opus Place, Suite 200
Downers Grove,
Illinois 60515-5702, USA

Numerical Algorithms Group Ltd.
Wilkinson House
Jordan Hill Road
Oxford OX2 8DR, U.K.

S–Plus (Version 3.1 , March 93)

S-PLUS is a programming environment that includes a wide range of built-in functions for statistical analysis. Most of these functions can be easily extended by using the S-PLUS programming language. Additionally, S-PLUS provides an interface to FORTRAN and C, thus increasing the extensibility of S-PLUS. S-PLUS has very flexible graphic facilities.

For *metrical* responses the functions available include multiple regression models, nonlinear models, local regression models, tree-based models and regression models using kernel smoothers and other smoothing techniques. Survival times can be fitted using the Cox-model or a counting process extension to the Cox-model. Time series data can be analysed using AR-models, ARIMA-models or spectral analysis.

Generalized linear models can be fitted using the function glm(). Models can be fitted for the Gaussian, binomial, Poisson, gamma and inverse Gaussian distributions. The family-statement in glm() also includes an option "quasi" for the fitting of quasi-likelihood models. Standard links available are: identity, log, logit, probit, sqrt, inverse and log-log. A function step.glm() allows stepwise variable selection. The glm()-code is available on the archive server statlib (see the section on GLIM4) allowing the user to extend the function to nonstandard likelihoods. Simpler ways of extending the capabilities of the glm() function and examples for its use are given in Chambers and Hastie (1992). This book also describes the use of the function gam(), which fits generalized additive models for the models listed above.

S-PLUS provides a very good range of diagnostics for model checking. Deviances and fitted values are calculated automatically.

There are many S-functions written by researchers that extend the facilities of S-PLUS considerably. Some of these functions are available on the statlib server (see the section on GLIM4). Among those is the function logist, which fits ordinal logistic regression models.

Statistical Sciences
52 Sandfield Road
Headington
Oxford OX3 7RJ, U.K.

GAUSS (Version 3.1)

GAUSS is a matrix programming language including standard matrix operations, but also a wide range of statistical and data–handling procedures. GAUSS provides good graphic facilities. There are interfaces to FORTRAN, C, and PASCAL.

There is an add-on module "Quantal Response", which includes routines to fit the following *generalized linear models*: binomial probit, multinomial logit, ordered cumulative logit, and log Poisson models. Deviances and predicted values are calculated automatically. Other measures for model checking are easy to compute but not included as a standard.

Further facilities for modelling *count* and *duration* data are available in the module "Count." It includes models for count responses such as Poisson, negative binomial, hurdle Poisson and seemingly unrelated Poisson regression models. Models for duration data included are exponential, exponential-gamma and Pareto duration models. All models (count and duration) allow taking into account truncation or censoring.

GAUSS may in general be used to compute sophisticated models that are not available in standard statistical packages. There is a module "Maximum–Likelihood," which includes seven algorithms to maximize user–defined log–likelihoods and may reduce the effort in programming models.

Aptech Systems Inc.
23804 S.E. Kent-Kangley Road
Maple Valley,
Washington 98038, USA

XPLORE (Version 3.1)

XPLORE is a computing environment for explanatory regression and data analysis. It has interactive graphics facilities and allows expansions with user–written macros.

Metrical responses can be analyzed by using a wide range of nonparametric methods like additive models, ACE, projection pursuit regression, average derivative estimation and kernel smoothing.

Binary responses can be modelled with the GLM module of XPLORE, which includes logit and probit models and most of the common univariate GLM's. The ADD module allows generalized additive modelling.

XploRe Systems
W. Härdle
Institut für Statistik und Ökonometrie
FB Wirtschaftswissenschaften
Humboldt Universität zu Berlin
D–1086 Berlin, Germany

EGRET (Version 0.23.25)

EGRET (Epidemiological, Graphics, Estimation and Testing program) is a package specializing in the analysis of epidemiological studies and survival models. There is a module for data definition and a module for data analysis. The facilities for data management and data manipulation within the package are limited.

EGRET includes logit models for binary responses, conditional logit models for modelling data from matched case-control studies (see Collett, 1991, section 7.7.1) and logit models with random effects. Survival models include models with exponential and Weibull noise distribution and Cox's proportional hazards model. Poisson regression can be used for modelling cohort data. Anticipated upgrades of EGRET, which may by now be available, will include logit models for multicategorical data, logit models with nested random effects, Poisson regression models with random effects and random effects regression models with asymmetric mixing distributions.

Statistics & Epidemiology Research Corporation
909 Northeast 43rd Street, Suite 310
Seattle,
Washington 98105, USA

LIMDEP (Version 6.0)

LIMDEP (LIMited DEPendent Variables) is a package originally designed for applications in econometrics. Most of the procedures included are designed for metrical regression and time series analysis. However, there are also many generalized linear models for categorical responses that can be fitted within LIMDEP. The package also includes features for nonlinear optimization, evaluation of integrals and minimizing nonlinear functions.

Metrical responses can be fitted using linear models with fixed and random effects. Further models are Tobit models for censored responses, stochastic frontier models and seemingly unrelated regression models. Time series can be analyzed using AR-, ARIMA- and ARMAX-models. Survival models include models with Weibull, exponential, normal, logistic and gamma noise distributions, but also Cox's proportional hazards model. The data may be censored in various ways.

Important generalized linear models for *categorical* responses included in LIMDEP are the logit model for binomial and multinomial responses, and the cumulative logit model for ordered multicategorical responses. For binary data a probit model can be fitted that can include random effects. Additionally, there is a feature to fit a bivariate probit. Nonparametric models for binary data are also available.

Count data can be fitted using Poisson or negative-binomial models. Censoring or truncation of the data can be taken into account.

Econometric Software Inc.
46 Maple Avenue
Bellport
New York 11713, USA

References

AALEN, O.O. (1989). A Linear Regression Model for the Analysis of Life–Times. *Statistics in Medicine*, 8, 907–925.

AALEN, O.O. (1980). A Model for Nonparametric Regression Analysis of Counting Processes. *Lecture Notes in Statistics*, 2, 1–25. New York: Springer.

ABRAMOWITZ, M., STEGUN, I. (1972). *Handbook of Mathematical Functions*. New York, Dover.

AGRESTI, A. (1984). *Analysis of Ordinal Categorical Data*. New York: Wiley.

AGRESTI, A. (1990). *Categorical Data Analysis*. New York: Wiley.

AGRESTI, A. (1992). A Survey of Exact Inference for Contingency Tables. *Statistical Science*, 7, 131–177.

AGRESTI, A. (1993a). Computing Conditional Maximum Likelihood Estimates for Generalized Rasch Models Using Simple Loglinear models with Diagonal Parameters. *Scand. J. Stat.*, 20, 63–72.

AGRESTI, A. (1993b). Distribution–free Fitting of Logit Models with Random Effects for Repeated Categorical Responses. *Statistics in Medicine*, 12, 1969–1988.

AGRESTI, A. (1997). A model for repeated measurements of a multivariate binary response. *Journal of the American Statistical Association* 92, 315–321.

AGRESTI, A., LANG, J.B. (1993). A Proportional Odds Model with Subject–Specific Effects for Repeated Ordered Categorical Responses. *Biometrika*, 80, 527–534.

AITCHISON, I., AITKEN, C. (1976). Multivariate Binary Discrimination. *Biometrika*, 63, 413–420.

AITKEN, C. (1983). Kernel Methods for the Estimation of Discrete Distributions. *J. Statist. Comput. Simul.*, 16, 189–200.

AITKIN, M., ANDERSON, D., FRANCIS, B., HINDE, J. (1989). *Statistical Modelling in GLIM*. Oxford: Clarendon Press.

AITKIN, M., CLAYTON, D. (1980). The Fitting of Exponential, Weibull and Extreme Value Distributions to Complex Censored Survival Data Using GLIM. *Applied Statistics*, 29, 156–163.

AITKIN, M., LONGFORD, N.T. (1986), Statistical Modelling Issues in School Effectiveness Studies. *Journal of the Royal Statistical Society*, A 149, 1–43.

ALBERT, J.H. (1988). Computational Methods Using a Bayesian Hierarchical Generalized Linear Model. *Journal of the American Statistical Association*, 83, 1037–1044.

ANDERSEN, E.B. (1973). *Conditional Inference and Models for Measuring*. Copenhagen: Mentalhygiejnish Forlag.

ANDERSEN, E.B. (1980). *Discrete Statistics with Social Science Applications*. Amsterdam: North Holland.

ANDERSON, J.A. (1972). Separate Sample Logistic Discrimination. *Biometrika*, 59, 19–35.

ANDERSON, J.A. (1984). Regression and Ordered Categorical Variables. *Journal of the Royal Statistical Society*, B 46, 1–30.

ANDERSON, B., MOORE, J. (1979). *Optimal Filtering*. Englewood Cliffs, N.J.: Prentice Hall.

ANDERSON, D.A., AITKIN, M. (1985). Variance Component Models With Binary Response: Interviewer Variability. *Journal of the Royal Statistical Society*, B 47, 203–210.

ANDERSON, D.A., HINDE, J. (1988). Random Effects in Generalized Linear Models and the EM Algorithm. *Comm. Statist.–Theor. Meth.*, 17, 3847–3856.

ANDREWS, D.W.K. (1987). Asymptotic Results for Generalized Wald Tests. *Econometric Theory*, 3, 348-358.

ANDREWS, D.W.K. (1988). Chi-Square Diagnostic Tests for Econometric Models: Theory. *Econometrica*, 56, 1419-1453.

AOKI, M. (1987). *State Space Modelling of Time Series*. Heidelberg: Springer.

ARANDA-ORDAZ, F.J. (1983). An Extension of the Proportional-Hazard-Model for Grouped Data. *Biometrics* 39, 109-118.

ARMINGER, G., KÜSTERS, U. (1985). Simultaneous Equation Systems with Categorical Observed Variables. In: Gilchrist, R., Francis, B., Whittaker, J. (Eds.). *Generalized Linear Models. Lecture Notes in Statistics*. Berlin: Springer.

ARMINGER, G., SCHOENBERG, R.J. (1989). Pseudo Maximum Likelihood Estimation and a Test for Misspecification in Mean and Covariance Structure Models. *Psychometrika*, 54, 409-425.

ARMINGER, G., SOBEL, M. (1990). Pseudo Maximum Likelihood Estimation of Mean- and Covariance Structures with Missing Data. *Journal of the American Statitstical Association*, 85, 195-203.

ARMSTRONG, B.G., SLOAN, M. (1989). Ordinal Regression Models for Epidemiologic Data. *American Journal of Epidemiology*, 129, 191-204.

ASHBY, M., NEUHAUS, J.M., HAUCK, W.W., BACCHETTI, P., HEILBRON, D.C. JEWELL, N.P., SEGAL, M.R., FUSARO, R.E. (1992). An Annotated Bibliography of Methods for Analysing Correlated Categorical Data. *Statistics in Medicine*, 11, 67-99.

AZZALINI, A. (1983). Maximum Likelihood Estimation of Order m for Stationary Stochastic Processes. *Biometrika*, 70, 381-387.

AZZALINI, A., BOWMAN, A.W. (1993). On the use of nonparametric regression for checking linear relationships. *Journal of the Royal Statistical Society*, B, 549-557.

AZZALINI, A., BOWMAN, A.W., HÄRDLE, W. (1989). On the Use of Nonparametric Regression for Model Checking. *Biometrika*, 76, 1-11.

BARTHOLOMEW. D.J. (1980). Factor Analysis for Categorical Data. *Journal of the Royal Statistical Society*, B 42, 293-321.

BELSLEY, D.A., KUH, E., WELSCH, R.E. (1980). *Regression Diagnostics*. New York: Wiley

BEN–AKIVA, M., LERMAN, S. (1985). *Discrete Choice Analysis: Theory and Application to Travel Demand.* Cambridge, Mass.: MIT Press.

BENEDETTI, J.K. (1977). On the Nonparametric Estimation of Regression Functions. *Journal of the Royal Statistical Society*, B, 39, 248–253.

BESAG, J., GREEN, P.J., HIGDON, D., MENGERSEN, K. (1995). Bayesian Computation and Stochastic Systems. *Statistical Science*, 10, 3–66.

BHAPKAR, V.P. (1980). ANOVA and MANOVA: Models for Categorical Data. In: Krishnaiah, P. (Ed.), *Handbook of Statistics* 1, 343–387. Amsterdam: North Holland.

BILLER, C. (1997). Posterior Mode Estimation in Dynamic Generalized Linear Mixed Models. Discussion Paper 70, Sonderforschungsbereich 386, Ludwig–Maximilians–Universität München.

BILLER, C., FAHRMEIR, L. (1997). Bayesian Spline-type smoothing in Generalized Regression Models. *Computational Statistics*, 12, 135–151.

BIRCH, M.W. (1963). Maximum Likelihood in Three–Way Contingency Tables. *Journal of the Royal Statistical Society*, B 25, 220–233.

BISHOP, Y.M.M., FIENBERG, S.E., HOLLAND, P.W. (1975) *Discrete Multivariate Analysis. Theory and Practice.* Cambridge, Mass.: MIT Press.

BLOSSFELD, H.–P., HAMERLE, A., MAYER, K.U. (1989). *Event History Analysis.* Hillsdale, N.J.: Lawrence Erlbaum Associates.

BÖRSCH–SUPAN, A. (1990). On the Compatibility of Nested Logit Models with Utility Maximization. *Journal of Econometrics* 43, 373–388.

BONNEY, G.E. (1987). Logistic Regression for Dependent Binary Observations. *Biometrics*, 43, 951–973.

BOWMAN. A.W. (1980). A Note on Consistency of the Kernel Method for the Analysis of Categorical Data. *Biometrika*, 67, 682–684.

BOWMAN, A.W., HALL, P., TITTERINGTON, D.M. (1984). Cross–Validation in Nonparametric Estimation of Probability Densities. *Biometrika*, 71, 341–351.

BOYLES, R.A. (1983). On the Covergence of the EM Algorithm. *Journal of the Royal Statistical Society*, B 45, 47–50.

BRESLOW, N.E. (1984), Extra–Poisson Variation in Log–Linear Models. *Applied Statistics*, 33, 38–44.

BRESLOW, N.E., CLAYTON, D.G. (1993). Approximate Inference in Generalized Linear Mixed Models. *Journal of the American Statistical Association*, 88, 9–25.

BRILLINGER, D.R., PREISLER, M.K. (1983). Maximum Likelihood Estimation in a Latent Variable Problem. In: Karlin, S., Amemiya, T., Goodman, L.A. (Eds.). *Studies in Econometrics, Time Series and Multivariate Statistics*, 31–65. New York: Academic Press.

BROEMELING, L.D. (1985). *Bayesian Analysis of Linear Models*. New York: Marcel Dekker.

BROWN, P.J., PAYNE, C.D. (1986). Aggregate Data. Ecological Regression and Voting Transitions. *Journal of the American Statistical Association*, 81, 452–460.

BUSE, A. (1982). The Likelihood Ratio, Wald and Lagrange Multiplier Test: An Expository Note. *The American Statistician*, 36, 153–157.

CAMERON, A.D., TRIVEDI, P.K. (1986). Econometric Models Based on Count Data: Comparisons and Applications of Some Estimators and Tests. *Journal of Applied Econometrics*, 1, 29–53.

CAREY, V.C., ZEGER, S.L., DIGGLE, P.F. (1993). Modelling Multivariate Binary Data with Alternating Logistic Regressions. *Biometrika*, 80, 517–526.

CARLIN, B.P., POLSON, N.G., STOFFER, D.S. (1992). A Monte Carlo Approach to Nonnormal and Nonlinear State–Space Modelling. *Journal of the American Statistical Association*, 87, 493–500.

CARROLL, R.J. (1992). Approaches to Estimation with Errors in Predictors. In: *GLIM and Statistical Modelling*, Springer Lecture Notes in Statistics, 78, 40–47.

CARTER, C.K., KOHN, R. (1994). On Gibbs sampling for state space models. *Biometrika*, 81, 541–553.

CARTER, C.K., KOHN, R. (1996). Robust Bayesian nonparametric regression. In: Härdle, W., Schimek, M.G. (Eds.). *Statistical Theory and Computational Aspects of Smoothing*. Heidelberg: Physica–Verlag, 128–148.

CHAMBERS, J.M., HASTIE, T.J. (1992). *Statistical Models in S*. Pacific Grove: Wadsworth, Brooks/Cole.

CHRISTENSEN, R. (1990). *Log–Linear Models*. New York: Springer.

CLEVELAND, W.S. (1979). Robust Locally Weighted Regression and Smoothing Scatterplots. *Journal of the American Statistical Association* 74, 829–836.

CLOGG, C. (1982). Some Models for the Analysis of Association in Multiway Crossclassifications Having Ordered Categories. *Journal of the American Statistical Association*, 77, 803–815.

COLLETT, D. (1991). *Modelling Binary Data.* London: Chapman and Hall.

CONAWAY, M.R. (1989). Analysis of Repeated Categorical Measurements with Conditional Likelihood Methods. *Journal of the American Statistical Association*, 84, 53–62.

CONAWAY, M.R. (1990). A Random Effects Model for Binary Data. *Biometrics*, 46, 317–328.

CONOLLY, M.A., LIANG, K.Y. (1988). Conditional Logistic Regression Models for Correlated Binary Data. *Biometrika*, 75, 501–506.

COOK, R.D. (1977). Detection of Influential Observations in Linear Regression. *Technometrics* 19, 15–18.

COOK, R.D., WEISBERG, S. (1982). *Residuals and Influence in Regression.* London: Chapman and Hall.

COPAS, J.B. (1983). Plotting p Against x. *Applied Statistics*, 32, 25–31.

CORBEIL, R.R., SEARLE, S.R. (1976). A Comparison of Variance Component Estimators. *Biometrics*, 32, 779–791.

COX, D.R. (1961). Tests of Separate Families of Hypotheses. *Proc. of the 4th Berkeley Symposium on Mathematical Statistics and Probability,* Vol. 1. Berkeley: University of California Press.

COX, D.R. (1962). Further Results on Tests of Separate Families of Hypotheses. *Journal of the Royal Statistical Society*, B 24, 406–424.

COX, D.R. (1970). *The Analysis of Binary Data.* London: Chapman and Hall.

COX, D.R. (1972). Regression Models and Life Tables (with discussion). *Journal of the Royal Statistical Society*, B 34, 187–220.

COX, D.R. (1975). Partial Likelihood. *Biometrika*, 62, 269–275.

COX, D.R., HINKLEY, D.V. (1974). *Theoretical Statistics.* London: Chapman and Hall.

CRAVEN, P. WAHBA, G. (1979) Smoothing Noisy Data with Spline Functions. *Numerische Mathematik*, 31, 377–403.

CRESSIE, N., READ, I. (1984). Multinomial Goodness–of–Fit Tests. *Journal of the Royal Statistical Society*, B 46, 440–464.

CZADO, C. (1992). On Link Selection in Generalized Linear Models. In: *Advances in GLIM and Statistical Modelling*. Springer Lecture Notes in Statistics, 78, 60–65.

DAGANZO, C. (1980). *Multinomial Probit*. New York: Academic Press.

DALE, J. (1986). Asymptotic Normality of Goodnes–of–Fit Statistics for Sparse Product Multinomials. *Journal of Royal Statistical Society*, B 48, 48–59.

DANNEGGER, F., KLINGER, A., ULM, K. (1995). Identification of Prognostic Factors with Censored Data. Discussion Paper 11, Sonder-forschungsbereich 386, Ludwig–Maximilians–Universität München.

DAVIDSON, R., MC KINNON, J.G. (1980). On a Simple Procedure for Testing Non–nested Regression Models. *Economics Letters*, 5, 45–48.

DAVIS, C.S. (1991). Semi–parametric and Non–parametric Methods for the Analysis of Repeated Measurement with Applications to Clinical Trials. *Statistics in Medicine*, 10, 1959–1980.

DAVIS, P.J., RABINOWITZ, P. (1975). *Numerical Integration*. Waltham, Mass.: Blaisdell.

DAVIS, P.J., RABINOWITZ, P. (1984). *Methods of Numerical Integration*. Orlando, F.: Academic Press.

DEAN, C.B. (1992). Testing for Overdispersion in Poisson and Binomial Regression Models. *Journal of the American Statistical Association*, 87, 451–457.

DE BOOR, C. (1978). *A Practical Guide to Splines*. New York: Springer.

DECARLI, A., FRANCIS, B.J., GILCHRIST, R. SEEBER, G.H. (1989). Statistical Modelling, *Springer Lecture Notes in Statistics*, 57.

DE JONG, P. (1989). Smoothing and Interpolation with the State Space Model. *Journal of the American Statistical Association*, 84, 1085–1088.

DEMPSTER, A.P., LAIRD, N.M., RUBIN, D.B. (1977). Maximum Likelihood from Incomplete Data via the EM–Algorithm. *Journal of the Royal Statistical Society*, B 39, 1–38.

DEVROYE, L. (1986). Non–uniform Random Variate Generation. New York: Springer.

DIELMAN, T.E. (1989). *Pooled Cross–Sectional and Time Series Analysis.* New York: Marcel Dekker.

DIGGLE, P.J., LIANG, K.Y., ZEGER, S.L. (1994). *Analysis of Longitudinal Data.* London: Chapman and Hall (to appear).

DOBSON, A.J. (1989). *Introduction to Statistical Modelling.* London: Chapman and Hall.

DOKSUM, K.A., GASKO, M. (1990). On a Correspondence Between Models in Binary Regression Analysis and in Survival Analysis. *International Statistical Review*, 58, 243–252.

DRUM, M.L., MC CULLAGH, P. (1993). REML estimation with exact covariance in the logistic mixed model. *Biometrics* 49, 677–689.

DUFFY, D.E., SANTNER, T.J. (1989). On the Small Sample Properties of Restricted Maximum Likelihood Estimators for Logistic Regression Models. *Comm. Statist.–Theor. Meth.*, 18, 959–989.

DURBIN, J., KOOPMAN, S.J. (1993). Filtering, Smoothing and Estimation of Time Series Models when the Observations come from Exponential Family Distributions. Discussion paper, London School of Economics.

EDWARDS, D., HAVRANEK, T. (1987). A Fast Model Selection Procedure for Large Families of Models. *Journal of the American Statistical Association*, 82, 205–213.

EDWARDS, A.L., THURSTONE, L.L. (1952). An Internal Consistency Check for Scale Values Determined by the Method of Successive Intervals. *Psychometrika*, 17, 169–180.

EFRON, B. (1986). Double Exponential Families and Their Use in Generalized Linear Regression. *Journal of the American Statistical Association*, 81, 709–721.

EFRON, B. (1988). Logistic Regression, Survival Analysis, and the Kaplan–Meier–Curve. *Journal of the Amercican Statistical Association*, 83, 414–425.

ENGLE, R.F. (1982). Autoregressive Conditional Heteroscedasticity with Estimates of United Kingdom Inflation. *Econometrica*, 50, 987–1007.

ENGLE, R.F., GRANGER, W.J., RICE, J., WEISS, A. (1986). Semiparametric Estimates of the Relation Between Weather and Electricity Sales. *Journal of the American Statistical Association*, 81, 310–320.

EPANECHNIKOV, V. (1969). Nonparametric Estimates of a Multivariate Probability Density. *Theory of Probability and Its Applications*, 14, 153–158.

EUBANK, R.L. (1988). *Smoothing Splines and Nonparametric Regression.* New York: Dekker.

EUBANK, R.L., HART, J. (1993). Testing goodness of fit via nonparametric function estimation techniques. *Communications in Statistics, Theory and Methods* 22, 3327–3354.

FAHRMEIR, L. (1987a). Asymptotic Testing Theory for Generalized Linear Models. *Statistics*, 18, 65–76.

FAHRMEIR, L. (1987b). Asymptotic Likelihood Inference for Nonhomogeneous Observations. *Statistical Papers*, 28, 81–116.

FAHRMEIR, L. (1988). A Note on Asymptotic Testing Theory for Nonhomogenous Observation. *Stochastic Processes and Its Applications*, 28, 267–273.

FAHRMEIR, L. (1990). Maximum Likelihood Estimation in Misspecified Generalized Linear Models. *Statistics*, 21, 487–502.

FAHRMEIR, L. (1992a). Posterior Mode Estimation by Extended Kalman Filtering for Multivariate Dynamic Generalized Linear Models. *Journal of the American Statistical Association*, 87, 501–509.

FAHRMEIR, L. (1992b). State Space Modeling and Conditional Mode Estimation for Categorical Time Series. *New Directions in Time Series Analysis*, Brillinger, D., et. al.(Eds.), 87–110, New York: Springer.

FAHRMEIR, L. (1994). Dynamic Modeling and Penalized Likelihood Estimation for Discrete Time Survival Data. *Biometrika*, 81, 317–330.

FAHRMEIR, L., FRANCIS, B., GILCHRIST, R., TUTZ, G. (1992). Advances in GLIM and Statistical Modelling. *Lecture Notes in Statistics*, 78.

FAHRMEIR, L., FROST, H. (1992). On Stepwise Variable Selection in Generalized Linear Regression and Time Series Models. *Computional Statistics*, 7, 137–154.

FAHRMEIR, L., GOSS, M. (1992). On Filtering and Smoothing in Dynamic Models for Categorical Longitudinal Data. In: Heijden, P. v.d., Jansen, W., Francis, B., Seeber ,G. (Eds.), *Statistical Modelling*, 85–94. Amsterdam: North Holland.

FAHRMEIR, L., HAMERLE, A. (1984, Eds.). *Multivariate statistische Verfahren.* Berlin: De Gruyter.

FAHRMEIR, L., HENNEVOGL, W., KLEMME, K., (1992). Smoothing in Dynamic Generalized Linear Models by Gibbs Sampling. In: Fahrmeir, L., Francis, B., Gilchrist, R., Tutz, G. (Eds.). *Advances in GLIM and Statistical Modelling*, 85–90. Heidelberg: Springer.

FAHRMEIR, L., KAUFMANN, H. (1985). Consistency and Asymptotic Normality of the Maximum Likelihood Estimator in Generalized Linear Models. *The Annals of Statistics*, 13, 342–368.

FAHRMEIR, L., KAUFMANN, H. (1986). Asymptotic Inference in Discrete Response Models. *Statistical Papers*, 27, 179–205.

FAHRMEIR, L., KAUFMANN, H. (1987). Regression Model for Nonstationary Categorical Time Series. *Journal of Time Series Analysis*, 8, 147–160.

FAHRMEIR, L., KAUFMANN, H. (1991). On Kalman Filtering, Posterior Mode Estimation and Fisher Scoring in Dynamic Exponential Family Regression. *Metrika*, 38, 37–60.

FAHRMEIR, L., KAUFMANN, H., MORAWITZ, B. (1989). Varying Parameter Models for Panel Data with Aplications to Business Test Data. *Regensburger Beiträge zur Statistik und Ökonometrie*. Universität Regensburg.

FAHRMEIR, L., KNORR–HELD, L. (1997a). Dynamic and Semiparametric Models. Discussion Paper 57, Sonderforschungsbereich 386, Ludwig–Maximilians–Universität München.

FAHRMEIR, L., KNORR–HELD, L. (1997b). Dynamic Discrete–Time Duration Models: Estimation via Markov Chain Monte Carlo. *Sociological Methodology* (to appear).

FAHRMEIR, L., KREDLER, CH. (1984). Verallgemeinerte Lineare Modelle. In: Fahrmeir, L., Hamerle, A. (Eds.). *Multivariate statistische Verfahren*. Berlin: De Gruyter.

FAHRMEIR, L., PRITSCHER, L. (1996). Regression Analysis of Forest Damage by Marginal Models for Correlated Ordinal Responses. *Environmental and Ecological Statistics*, 3, 257–268.

FAHRMEIR, L., WAGENPFEIL, S. (1996). Smoothing Hazard Functions and Time-Varying Effects in Discrete Duration and Competing Risks Models. *Journal of the American Statistical Association*, 91, 1584-1594.

FAN, J. (1992). Design–adaptive Nonparametric Regression. *Journal of the American Statistical Association* 87, 998–1004.

FAN, J. (1993). Local Linear Regression Smoothers and their Minimax Efficiencies. *Annals of Statistics* 21, 196–216.

FAN, J., GIJBELS, I. (1996). *Local Polynomial Modelling and its Applications*. London: Chapman and Hall.

FAN, J., HECKMAN, N.E., WAND, M.P. (1995). Local polynomial kernel regression for generalized linear models and quasi–likelihood functions. *Journal of the American Statistical Association* 90, 141–150.

FIENBERG, S.E., HOLLAND, P.W. (1973). Simultaneous Estimation of Multinomial Cell Probabilities. *Journal of the American Statistical Association*, 68, 683–691.

FINKELSTEIN, D. (1986). A Proportional Hazard Model for Interval–Censored Failure Time Data. *Biometrics*, 42, 845–854.

FINNEY, D.J. (1947). The Estimation from Individual Records of the Relationship Between Close and Quantal Response. *Biometrika*, 34, 320–334.

FIRTH, D. (1991). Generalized Linear Models. In: Hinkley, D.V., Reid, N., Snell, E.J. (Eds.), *Statistical Theory and Modeling*. London: Chapman and Hall.

FIRTH, D., GLOSUP, J., HINKLEY, D.V. (1991). Model checking with nonparametric curves, *Biometrika* 78, 245–252.

FITZMAURICE, G.M., LAIRD, N.M. (1993). A Likelihood-based Method for Analysing Longitudinal Binary Responses. *Biometrika*, 80, 141–151.

FITZMAURICE, G.M., LAIRD, N.M., ROTNITZKY, A.G. (1993). Regression Models for Discrete Longitudinal Responses. *Statistical Science*, 8, 284–309.

FITZMAURICE, G.M., MOLENBERGHS, G., LIPSITZ, S.R. (1995). Regression Models for Longitudinal Binary Responses with Informative Drop–Outs. *Journal of the Royal Statistical Society*, B 57, 691–704.

FORTHOFER, R.N., LEHNEN, R.G. (1981). *Public Program Analysis. A New Categorical Data Approach*. Belmont, Calif.: Lifetime Learning Publications.

FOUTZ, R.V., SRIVASTAVA, R.C. (1977). The Performance of the Likelihood Ratio Test when the Model is Incorrect. *Annals of Statistics*, 5, 1183–1194.

FOLKS, J.L., CHHIKARA, R.S. (1978). The Inverse Gaussian Distribution and Its Statistical Application, A Review (with Discussion). *Journal of the Royal Statistical Society*, B 40, 263–289.

FOWLKES, E.B. (1987). Some Diagnostics for Binary Logistic Regression via Smoothing. *Biometrika*, 74, 593–515.

FRIEDL, H. (1991). Verallgemeinerte logistische Modelle in der Analyse von Zervix–Karzinomen. Dissertation, Universität Graz.

FRIEDMAN, J., SILVERMAN, B. (1989). Flexible Parsimonious Smoothing and Additive Modelling (with discussion). *Technometrics*, 31, 3–39.

FROST, H. (1991). Modelltests in generalisierten linearen Modellen. Dissertation, Universität Regensburg.

FRÜHWIRTH–SCHNATTER, S. (1991). Monitoring von ökologischen und biometrischen Prozessen mit statistischen Filtern. In: Minder, Ch., Seeber, G. (Eds.) *Multivariate Modelle: Neue Ansätze für biometrische Anwendungen*. Heidelberg: Springer Lecture Notes M/S.

FRÜHWIRTH–SCHNATTER, S. (1994). Data Augmentation and Dynamic Linear Models. *Journal of Time Series Analysis*, 15, 183–202.

FURNIVAL, G.M., WILSON, R.W. (1974). Regression by Leaps and Bounds. *Technometrics*, 16, 499–511.

GAMERMAN, D. (1991). Dynamic Bayesian Models for Survival Data. *Applied Statistics*, 40, 63–79.

GAMERMAN, D. (1995). Monte Carlo Markov Chains for Dynamic Generalized Linear Models. Discussion Paper: Instituto de Matemática, Universidade Federal do Rio de Janeiro.

GARBER, A.M. (1989). A Discrete–Time Model of the Acquisition of Antibiotic–Resistant Infections in Hospitalized Patients. *Biometrics*, 45, 797–816.

GASSER, T., MÜLLER, H.–G. (1979). Kernel Estimation of Regression Functions. In: Gasser, T., Rosenblatt, H. (Eds.). *Smoothing Techniques for Curve Estimation*. Heidelberg: Springer.

GASSER, T., MÜLLER, H.–G. (1984). Estimating Regression Functions and Their Derivatives by the Kernel Method. *Scand. J. of Statistics*, 11, 171–185.

GASSER, T., MÜLLER, H.–G., MAMMITZSCH, V. (1985). Kernels for Nonparametric Curve Estimation. *Journal of the Royal Statistical Society*, B 47, 238–252.

GAY, D.M., WELSCH, R.E. (1988). Maximum Likelihood and Quasi–Likelihood for Nonlinear Exponential Family Regression Models. *Journal of the American Statistical Association*, 83, 990–998.

GELFAND, A.E., SMITH, A.F.M. (1990). Sampling–Based Approaches to Calculating Marginal Densities. *Journal of the American Statistical Association*, 85, 398–409.

GEMAN, S., GEMAN, D. (1984). Stochastic Relaxation, Gibbs Distributions, and the Bayesian Restoration of Images. *IEEE Transactions on Pattern Analysis and Machine Intelligence* PAMI–6, 721–741.

GENTER, F.C., FAREWELL, V.T. (1985). Goodness–of–Link Testing in Ordinal Regression Models. *The Canadian Journal of Statistics*, 13, 37–44.

GERSCH, W., KITAGAWA, G. (1988). Smoothness Priors in Time Series. In: Spall, J.C. (Ed.). *Bayesian Analysis of Time Series and Dynamic Models*. New York: Dekker.

GIEGER, C. (1997). Non– and Semiparametric Marginal Regression Models for Ordinal Response. Discussion Paper 71, Sonderforschungsbereich 386, Ludwig–Maximilians–Universität München.

GIESBRECHT, F.G., BURNS, J.C. (1985). Two–Stage Analysis Based on a Mixed–Model: Large–Sample Asymptotic Theory and Small–Sample Simulation Results. *Biometrics*, 41, 477–486.

GIGLI, A. (1992). Bootstrap Importance Sampling in Regression. In: Heijden, P. v.d., Jansen, W., Francis, B., Seeber, G. (Eds.), *Statistical Modelling*, 95–104. Amsterdam: North Holland.

GILCHRIST, R., FRANCIS, B. WHITTAKER, J. (1985). Generalized Linear Models, Proceedings, Lancaster 1985. Springer Lecture Notes. New York: Springer.

GILKS, W.R., RICHARDSON, S., SPIEGELHALTER, D.J. (1996). Markov Chain Monte Carlo in Practice. Chapman and Hall, London.

GLONEK, G.V.F. (1996). A Class of Regression Models for Multivariate Categorical Responses. *Biometrika*, 83, 15–28.

GLONEK, G.V.F., MC CULLAGH, P. (1995). Multivariate Logistic Models. *Journal of the Royal Statistical Society*, B 57, 533–546.

GOLDSTEIN, H. (1986). Multilevel Mixed Linear Model Analysis Using Iterative Generalized Least Squares. *Biometrika*, 73, 43–56.

GOLDSTEIN, H. (1989). Restricted Unbiased Iterative Generalized Least Squares Estimation. *Biometrika*, 76, 622–623.

GOODMAN, L.A. (1979). Simple Models for the Analysis of Association in Cross–Classification Having Ordered Categories. *Journal of the American Statistical Association*, 74, 537–552.

GOODMAN, L.A. (1981a). Association Models and Canonical Correlation in the Analysis of Cross–Classifications Having Ordered Categories. *Journal of the American Statistical Association*, 76, 320–324.

GOODMAN, L.A. (1981b). Association Models and the Bivariate Normal for Contingency Tables with Ordered Categories. *Biometrika*, 68, 347–355.

GORDON, K. (1986). The Multi State Kalman Filter in Medical Monitoring. *Computer Methods and Programs in Biomedicine*, 23, 147–154.

GOSS, M. (1990). *Schätzung und Identifikation von Struktur– und Hyperstrukturparametern in dynamischen generalisierten linearen Modellen*. Dissertation. Universität Regensburg.

GOULD, A., LAWLESS, J.F. (1988). Estimation Efficiency in Lifetime Regression Models when Responses are Censored or Grouped. *Comm. Statist. Simul.*, 17, 689–712.

GOURIEROUX, C. (1985). Asymptotic Comparison of Tests for Non–nested Hypotheses by Bahadur's A.R.E. In: Florens, J.–P., *Model Choice*. Proceedings of the 4th Franco–Belgian Meeting of Statisticians, Facultes Univ.: Saint–Louis, Bruxelles.

GOURIEROUX, C., HOLLY, A., MONTFORT, A. (1982). Likelihood Ratio Test, Wald Test and Kuhn–Tucker Test in Linear Models with Inequality Constraints on the Regression Parameters. *Econometrica*, 50, 63–80.

GOURIEROUX, C., MONTFORT, A. (1989). Simulation Based Inference with Heterogeneity, Document de Travail INSEE/ENSAE, No. 8902.

GOURIEROUX, C., MONTFORT, A., TROGNON, A. (1983a). Estimation and Test in Probit Models with Serial Correlation. CEPREMAP Discussion paper No.8220.

GOURIEROUX, C., MONTFORT, A., TROGNON, A. (1983b). Testing Nested or Non–nested Hypotheses. *Journal of Econometrics*, 21, 83–115.

GOURIEROUX, C., MONTFORT, A., TROGNON, A. (1984). Pseudo Maximum Likelihood Methods: Theory. *Econometrica*, 52, 681–700.

GREEN, P. (1984). Heratively Reweighted Least Squares for Maximum Likelihood Estimation, and Some Robust and Resistant Alternatives (with Discussion). *Royal Statistical Society*, B 46, 149–192.

GREEN, P. (1987). Penalized Likelihood for General Semi–Parametric Regression Models. *International Statistical Review*, 55, 245–259.

GREEN, P. (1989). Generalized Linear Models and Some Extensions. Geometry and Algorithms. In: Decarli, A., Francis, B.J., Gilchrist, R., Seeber, G.U.H. (Eds.). *Statistical Modelling*, 26–36. Heidelberg: Springer Lecture Notes in Statistics 57.

GREEN, P., SILVERMAN, B.W. (1994). *Nonparametric Regression and Generalized Linear Models: A Roughness Penalty Approach*. London: Chapman and Hall.

GREEN, P., YANDELL, B. (1985). Semi–Parametric Generalized Linear Models. In: Gilchrist, R., Francis, B., Whittaker, J. (Eds.) *Generalized Linear Models*. Heidelberg: Springer Lecture Notes.

GREENWOOD, M. (1926). The Natural Duration of Cancer. *Reports of Public Health and Medical Subjects*, 33. London: Her Majesty's Stationary Office.

GRETHER, D.M., MADDALA, G.S. (1982). A Time Series Model with Qualitative Variables. In: Deistler, M., et al (Eds.). *Games, Economic Dynamics and Time Series Analysis*, 291–305. Wien: Physica.

GU, C., (1990). Adaptive Spline Smoothing in Non–Gaussian Regression Models. *Journal of the American Statistical Association*, 85, 801–807.

GUAN, D., YUAN, L. (1991). The Integer–Valued Autoregressive (INAR(p)) Model. *J. Time Series Analysis*, 12, 129–142.

HABBEMA, J.D.F., HERMANS, J., REMME, J. (1978). Variable Kernel Density Estimation in Discriminant Analysis. In: Corstner, L.C.A., Hermands, J. (Eds.). *Compstat*, Vienna: Physica, 1978, 178–185.

HABERMAN, S.J. (1974). *The Analysis of Frequency Data*. Chicago: University of Chicago Press.

HABERMAN, S.J. (1977). Maximum Likelihood Estimates in Exponential Response Models. *Annals of Statistics*, 5, 815–841.

HABERMAN, S.J. (1978). *Analysis of Qualitative Data, Vol. 1*. New York: Academic Press.

HALL, P. (1981). On Nonparametric Multivariate Binary Discrimination. *Biometrika*, 68, 287–294.

HALL, P., HEYDE, C.C. (1980). *Martingale Limit Theory and Its Applications*. New York: Academic Press.

HAMERLE, A., NAGL, W. (1988). Misspecification in Models for Discrete Panel Data: Applications and Comparisons of Some Estimators. Diskussionsbeitrag Nr. 105. Universität Konstanz.

HAMERLE, A., RONNING, G. (1992) Panel Anylysis for Qualitative Variables. In: Arminger, G., Clogg, C.C., Sobel, M.E. (Eds.). *A Handbook for Statistical Modelling in the Social and Behavioral Sciences.* New York: Plenum.

HAMERLE, A., TUTZ, G. (1989). *Diskrete Modelle zur Analyse von Verweildauer und Lebenszeiten.* Frankfurt/New York: Campus.

HANEFELD, U. (1987). *Das sozio-ökonomische Panel.* Frankfurt: Campus.

HÄRDLE, W. (1990a). *Applied Nonparametric Regression.* Cambridge: Cambridge University Press.

HÄRDLE, W. (1990b). *Smoothing Techniques.* With Implementation in S. New York: Springer Verlag.

HÄRDLE, W., TURLACH, B. (1992). Nonparametric Approaches to Generalized Linear Models. *Springer Lecture Notes,* 78, 213–225.

HÄRDLE, W., HALL, P., MARRON, J.S., (1988). How Far are Automatically Chosen Regresssion Smoothing Parameters from Their Optimum? (with discussion) *Journal of the American Statistical Association,* 83, 86–99.

HARRISON, P.J., STEVENS, C.F. (1976). Bayesian Forecasting (with Discussion). *Journal of the Royal Statistical Society,* B 38, 205–247.

HARTIGAN, J.A. (1969). Linear Bayesian Methods. *Journal of the Royal Statistical Society,* B 31, 446–454.

HARTLEY, H. (1958). Maximum Likelihood Estimation from Incomplete Data. *Biometrics,* 14, 174–194.

HARVEY, A.C. (1989). *Forecasting, Structural Time Series Models and the Kalman Filter.* Cambridge: Cambridge University Press.

HARVEY, A.C., FERNANDES, C. (1988). Time Series Models for Count or Qualitative Observations. *Journal of Business and Economic Statistics,* 7, 407–422.

HARVILLE, D.A. (1976). Extension of the Gauss–Markov Theorem to Include the Estimation of Random Effects. *Annals of Statistics,* 4, 384–395.

HARVILLE, D.A. (1977). Maximum Likelihood Approaches to Variance Component Estimation and to Related Problems. *Journal of the American Statistical Association*, 72, 320–338.

HARVILLE, D.A., MEE, R.W. (1984). A Mixed Model Procedure for Analyzing Ordered Categorical Data. *Biometrics*, 40, 393–408.

HASTIE, T., LOADER, C. (1993). Local Regression: Automatic Kernel Carpentry. *Statistical Science* 8, 120–143.

HASTIE, T., TIBSHIRANI, R. (1985). Generalized Additive Models (with discussion). *Statist. Sci.*, 1, 297–318.

HASTIE, T., TIBSHIRANI, R. (1986). Generalized Additive Models: Some Applications. *Proceedings 2nd International GLIM Conference*. Lancaster. Berlin, Heidelberg: Springer Lecture Notes in Statistics, 32.

HASTIE, T., TIBSHIRANI, R. (1987). Generalized Additive Models: Some Applications. *Journal of the American Statistical Association*, 82, 371–386.

HASTIE, T., TIBSHIRANI, R. (1990). *Generalized Additive Models*. London: Chapman and Hall.

HASTIE, T., TIBSHIRANI, R. (1993). Varying–coefficient Models. *Journal of the Royal Statistical Society*, B 55, 757–796.

HAUSMAN, J.A. (1978). Specification Tests in Econometrics. *Econometrica*, 46, 1251–1271.

HAUSMAN, J. HALL, B.H., GRILICHES, Z. (1984). Econometric Models for Count Data with an Application to the Patents–R&D Relationship. *Econometrica*, 52, 909–938.

HAUSMAN, J.A., TAYLOR, W.E. (1981). A General Specification Test. *Economics Letters*, 8, 239–245.

HAUSMAN, J.A., WISE, D.A. (1978). A Conditional Probit Model for Qualitative Choice: Discrete Decisions Recognizing Interdependence and Heterogeneous Preference. *Econometrica*, 46, 403–426.

HEAGERTY, P.J., ZEGER, S.L. (1996). Marginal Regression Models for Clustered Ordinal Measurements. *Journal of the American Statistical Association*, 91, 1024–1036.

HEAGERTY, P.J., ZEGER, S.L. (1997). Lorelogram: A Regression Approach to Exploring Dependence in Longitudinal Categorical Responses. *Journal of the American Statistical Association* (to appear).

HEBBEL, H., HEILER, S. (1987). Trend and Seasonal Decomposition in Discrete Time. *Statistical Papers*, 28, 133–158.

HECKMAN, J.J. (1981). Dynamic Discrete Probability Models. In: Manski, C.F., McFadden, D. (Eds.). *Structural Analysis of Discrete Data with Econometric Applications*, 114–195. Cambridge, Mass.: MIT Press.

HENNEVOGL, W. (1991). *Schätzung generalisierter Regressions- und Zeitreihenmodelle mit variierenden Parametern*. Dissertation. Universität Regensburg.

HENNEVOGL, W., KRANERT, T. (1988). Resiudal and Influence Analysis for Multicategorical Response Models. *Regensburger Beiträge zur Statistik und Ökonometrie* 5.

HEUMANN, C. (1996). *Likelihoodbasierte marginale Regressionsmodelle für korrelierte kategoriale Daten*. Dissertation, Ludwig-Maximilians-Universität München.

HINDE, J. (1982). Compound Poisson Regression Models. In: Gilchrist, R. (Ed.). GLIM 82. *Intern. Conf. Generalized Linear Models*, 109–121. New York: Springer.

HINDE, J. (1992). Choosing Between Non–Nested Models: A Simulation Approach. In: *Advances in GLIM and Statistical Modelling*. Springer Lecture Notes in Statistics, 78, 119–124. New York: Springer.

HOAGLIN, D.C., WELSCH, R. (1978). The Hat Matrix in Regression and ANOVA. *American Statistician* 32, 17–22.

HOCKING, R.R. (1976). The Analysis and Selection of Variables in Linear Regression. *Biometrics*, 32, 1–49.

HOLLAND, P.W., WELSCH, R.E. (1977). Robust Regression Using Iteratively Reweighted Least Squares. *Communications in Statistics A, Theory and Methods*, 6, 813–827.

HOLLY, A. (1982). A Remark on Hausman's Specification Test. *Econometrica*, 50, 749–759.

HOLMES, M.C., WILLIAMS, R. (1954). The Distribution of Carriers of Streptococcus Pyogenes among 2413 Healthy Children. *J. Hyg. Camb.*, 52, 165–179.

HOLTBRÜGGE, W., SCHUHMACHER, M. (1991). A Comparison of Regression Models for the Analysis of Ordered Categorical Data. *Applied Statistics*, 40, 249–259.

HOPPER, J.L., YOUNG, G.P. (1989). A Random Walk Model for Evaluating Clinical Trials Involving Serial Observations. *Statistics in Medicine*, 7, 581–590.

HOROWITZ, J., HÄRDLE, W. (1992). Testing a Parametric Model Against a Semiparametric Alternative. Discussion paper 9304, Humboldt–Universität Berlin.

HSIAO, C. (1986). *Analysis of Panel Data*. Cambridge: Cambridge University Press.

HUBER, P. (1981). *Robust Statistics*. New York: John Wiley.

HUFFER, F.W., MC KEAGUE, I.W. (1991). Weigthed Least Squares Estimation for Aalen's Additive Risk Model. *Journal of the American Statistical Association*, 86, 114–129.

HUTCHINSON, C.E. (1984). The Kalman Filter Applied to Aerospace and Electronic Systems. *IEEE Transactions Aero. Elect. Systems*, AES–20, 500–504.

IM, S., GIANOLA, D. (1988). Mixed Models for Bionomial Data with an Application to Lamb Mortality. *Applied Statistics*, 37, 196–204.

JACOBS, P.A., LEWIS, P.A.W. (1983). Stationary Discrete Autoregressive Moving–Average Time Series Generated by Mixtures. *J. Time Series Analysis*, 4, 19–36.

JANSEN, J. (1990). On the Statistical Analysis of Ordinal Data When Extravariation is Present. *Applied Statistics*, 39, 74–85.

JAZWINSKI, A. (1970). *Stochastic Processes and Filtering*. New York: Academic Press.

JONES, R.H. (1993). *Longitudinal Data with Serial Correlation: A State–Space Approach*. London: Chapman and Hall.

JORGENSEN, B. (1982). *Statistical Properties of the Generalized Inverse Gaussian Distribution*. Heidelberg: Springer Lecture Notes.

JORGENSEN, B. (1983). Maximum Likelihood Estimation and Large–Sample Inference for Generalized Linear and Nonlinear Regression Models. *Biometrika*, 70, 19–28.

JORGENSEN, B. (1992). Exponential Dispersion Models and Extension: A Review. *International Statistical Review*, 60, 5–20.

KALBFLEISCH, J., PRENTICE, R. (1973). Marginal Likelihoods Based on Cox's Regression and Life Model. *Biometrika*, 60, 256–278.

KALBFLEISCH, J., PRENTICE, R. (1980). *The Statistical Analysis of Failure Time Data.* New York: Wiley.

KAUERMANN, G., TUTZ, G. (1995). Local likelihood estimation and bias reduction in varying coefficient models. *Technical report*, 95-9, Technical University of Berlin, Institute of Quantitative methods.

KAUFMANN, H. (1987). Regression Models for Nonstationary Categorical Time Series: Asymptotic Estimation Theory. *Annals of Statistics*, 15, 79–98.

KAUFMANN, H. (1988). On Existence and Uniqueness of Maximum Likelihood Estimates in Quantal and Ordinal Response Models. *Metrika*, 35, 291–313.

KAUFMANN, H. (1989). On Likelihood Ratio Tests For and Against Convex Cones in the Linear Model. *Regensburger Beiträge zu Statistik und Ökonometrie*, 17.

KIEFER, N.M. (1990). Ecometric Methods for Grouped Duration Data. In: Hartog, J, Ridder, G., Theeuwes, J. (Eds.). *Panel Data and Labor Market Studies.* Amsterdam: Elsevier Science Publishers.

KIRCHEN, A. (1988). Schätzung zeitveränderlicher Strukturparameter in ökonometrischen Prognosemodellen. *Mathematical Systems in Economics*, 111, Frankfurt a.M.: Athenäum.

KITAGAWA, G. (1987). Non–Gaussian State–Space Modelling of Nonstationary Time Series (with Comments). *Journal of the American Statistical Association*, 82, 1032–1063.

KITAGAWA, G., GERSCH, W. (1984). Smoothness Priors, State Space Modelling of Time Series with Trend and Seasonality. *Journal of the American Statistical Association*, 79, 378–389.

KNORR–HELD, L. (1993). Schätzen von Zustandsmodellen durch Gibbs Sampling. Diplomarbeit, Institut für Statistik, Universität München.

KNORR–HELD, L. (1996). Hierarchical Modelling of Discrete Longitudinal Data. Applications of Markov Chain Monte Carlo. Dissertation, Universität München.

KOHN, R., ANSLEY, C. (1987). A New Algorithm for Spline Smoothing Based on Smoothing a Stochastic Process. *SIAM Journal Sci. Stat. Comp.*, 8, 33–48.

KOHN, R., ANSLEY, C. (1989). A Fast Algorithm for Signal Extraction, Influence and Cross–Validation in State–Space Models. *Biometrika*, 76, 65–79.

KÖNIG, H., NERLOVE, M., OUDIZ, G. (1981). On the Formation of Price Expectations, An Analysis of Business Test Data by Log–Linear Probability Models. *European Economic Review*, 16, 421-433.

KOOPMAN, S.J. (1993). Disturbance Smoother for State Space Models. *Biometrika*, 80, 117-126.

KRÄMER, W. (1986). Bemerkungen zum Hausman–Test. *Allgemeines Statitisches Archiv*, 70, 170–179.

KRANTZ, D.H., LUCE, R.D., SUPPES, P., TVERSKY, A. (1971). *Foundations of Measurement*, Vol.1. New York: Academic Press.

KREDLER, CH. (1984). Selection of Variables in Certain Nonlinear Regression Models. *Comp. Stat. Quarterly* 1, 13–27.

KÜSTERS, U. (1987). *Hierarchische Mittelwert– und Kovarianzstrukturmodelle mit nichtmetrischen endogenen Variablen.* Heidelberg: Physica.

KUK, A.Y. (1995). Asymptotically unbiased estimation in generalized linear models with random effects. *Journal of the Royal Statistical Society* 57, 395-407.

KULLBACK, S. (1959). *Information Theory and Statistics.* New York: Wiley.

KULLBACK, S. (1985). Minimum Discrimination Information (MDI) Estimation. In: Kotz, S., Johnson N.L. (Eds.). *Encyclopedia of Statistical Sciences*, Vol. S, 527–529, New York: Wiley.

LÄÄRÄ, E. MATTHEWS, J.N. (1985). The Equivalence of Two Models for Ordinal Data. *Biometrika*, 72, 206–207.

LAI, S.L. (1977). Large Sample Properties of K–Nearest Neighbour Procedures. Ph.D. dissertation. Dept. Mathematics, UCLA, Los Angeles.

LAIRD, N.M. (1978). Empirical Bayes Methods for Two–Way Contingency Tables. *Biometrika*, 65, 581–590.

LAIRD, N.M., BECK, G.J., WARE, J.H. (1984). Mixed Models for Serial Categorical Response. Quoted in: Ekholm, A. (1991). *Maximum Likelihood for Many Short Binary Time Series* (preprint).

LAIRD, N.M., WARE, J.H. (1982). Random Effects Models for Longitudinal Data. *Biometrics*, 38, 963–974.

LANCASTER, T. (1990). *The Econometric Analysis of Transition Data.* Cambridge: Cambridge University Press.

LANDWEHR, J.M., PREGIBON, D., SHOEMAKER, A.C. (1984), Graphical Methods for Assessing Logistic Regression Models. *Journal of the American Statistical Association*, 79, 61–71.

LAUDER, I.J. (1983). Discrete Kernel Assessment of Diagnostic Probabilities. *Biometrika*, 70, 251–256.

LAURITZEN, S.L., WERMUTH, N. (1989). Graphical Models for Association Between Variables, Some of which are Qualitative and Some Quantitative. *Annals of Statistics*, 17, 31–57.

LAWLESS, J.F. (1982). *Statistical Models and Methods for Lifetime Data*. New York: Wiley.

LAWLESS, J.F. (1987), Negative Binomial and Mixed Poisson Regression. *The Canadian Journal of Statistics* 15, 209–225.

LAWLESS, J.F., SINGHAL, K. (1978). Efficient Screening of Nonnormal Regression Models. *Biometrics*, 34, 318–327.

LAWLESS, J.F., SINGHAL, K. (1987). ISMOD: An All–Subsets Regression Program for Generalized Linear Models. *Computer Methods and Programs in Biomedicine*, 24, 117–134.

LE CESSIE, S., VAN HOUWELINGEN, J.C. (1991). A goodness-of-fit test for binary regression models, based on smoothing mehtods. *Biometrics* 47, 1267–1282.

LEE, A.H. (1988). Assessing Partial Influence in Generalized Linear Models. *Biometrics*, 44, 71–77.

LEONARD, T. (1972). Bayesian Methods for Biomial Data. *Biometrika*, 59, 581–589.

LEONARD, T. (1977). A Bayesian Approach to Some Multinomial and Pretesting Problems. *Journal of the American Statistical Association*, 72, 869–874.

LEONARD, T., NOVICK, M.R. (1986). Bayesian Full Rank Marginalization for Two–Way Contingency Tables. *Journal of Educational Statistics*, 11, 33–56.

LERMAN, S., MANSKI, C. (1981). On the Use of Simulated Frequencies to Approximate Choice Probabilities. In: Manski, C., Fadden, D. (Eds.). *Structural Analysis of Discrete Data*. Cambridge, Mass.: MIT Press.

LESAFFRE, E., ALBERT, A. (1989). Multiple–Group Logistic Regression Diagnostics. *Applied Statistics*, 38, 425–440.

LESAFFRE, E., MOLENBERGHS, G., DEWULF, L. (1996). Effect of Dropouts in a Longitudinal Study: An Application of a Repeated Ordinal Model. *Statistics in Medicine*, 15, 1123–1141.

LEVINE, D. (1983). A Remark on Serial Correlation in Maximum Likelihood. *Journal of Econometrics*, 23, 337–342.

LI, K.C., DUAN, N. (1989). Regression Analysis under Link Violation. *The Annals of Statistics*, 17, 1009–1052.

LIANG, K.Y., ZEGER, S. (1986). Longitudinal Data Analysis Using Generalized Linear Models. *Biometrika*, 73, 13–22.

LIANG, K.Y., ZEGER, S. (1989). A Class of Logistic Regression Models for Multivariate Binary Time Series. *Journal of the American Statistical Association*, 84, 447–451.

LIANG, K.Y., ZEGER, S.L., QAQISH, B. (1992). Multivariate Regression Analysis for Categorical Data. *Journal of the Royal Statistical Society*, B 54, 3–40.

LINDSEY, J.K. (1993). *Models for Repeated Measurements*. Oxford: Oxford University Press.

LINDSTROM, M., BATES, D. (1990). Nonlinear Mixed Effects Models for Repeated Measures Data. *Biometrics*, 46, 673–687.

LIPSITZ, S., LAIRD, N., HARRINGTON, D. (1991). Generalized Estimation Equations for Correlated Binary Data: Using the Odds Ratio as a Measure of Association. *Biometrika*, 78, 153–160.

LONGFORD, N.T. (1987). A Fast Scoring Algorithm for Maximum Likelihood Estimation in Unbalanced Mixed Models with Nested Random Effects. *Biometrika*, 74, 817–827.

LONGFORD, N.T. (1994). *Random Coefficient Models*. In: Arminger, G., Clogg, C., Sobel, M. (Eds.). *Handbook of Statistical Modelling for the Behavioural Sciences*. New York: Plenum.

LOS, C. (1984). Econometrics of Models with Evolutionary Parameter Structures. Ph.D. Dissertation, Columbia University, New York, N.Y.

LOUIS, T.A. (1982). Finding the Observed Information Matrix when Using the EM Algorithm. *Journal of the Royal Statistical Society*, B 44, 226–233.

MACK, Y.P. (1981). Local properties of k–NN Regression Estimates. *SIAM J. Alg. Disc. Meth.*, 2, 311–323.

MACK, Y.P., SILVERMAN, B.W. (1982). Weak and Strong Uniform Consistency of Kernel Regression Estimates. *Zeitschrift für Wahrscheinlichkeitstheorie und verwandte Gebiete*, 61, 405–461.

MADDALA, G.S. (1983). Limited–Dependent and Qualitative Variables in Econometrics. Cambridge: Cambridge University Press.

MAGNUS, J.R., NEUDECKER, H. (1988). *Matrix Differential Calculus with Applications in Statistics and Econometrics*. London: John Wiley.

MANTEL, N., HANKEY, B.F. (1978). A Logistic Regression Analysis of Response Time Data Where the Hazard Function is Time Dependent. *Comm. Statist.*, A 7, 333–347.

MARTIN, R. (1979). Approximate Conditional–Mean Type Smoothers and Interpolators. In: Gasser, T., Rosenblatt, M. (Eds.). *Smoothing Techniques for Curve Estimation*, 117–143. Berlin: Springer.

MARTIN, R., RAFTERY, A. (1987). Robustness, Computation, and Non-Euclidian Models (Comment). *Journal of the American Statistical Society*, 82, 1044–1050.

MARX, B., SMITH, E. (1990). Principal Component Estimation for Generalized Linear Regression. *Biometrika*, 77, 23–31.

MARX, B., EILERS, P, SMITH, E. (1992). Ridge Likelihood Estimation for Generalized Linear Regression. In: Heijden, P. v.d., Jansen, W., Francis, B., Seeber, G. (Eds.), *Statistical Modelling*, 227–238. Amsterdam: North Holland.

MASTERS, G.N. (1982). A Rasch Model for Partial Credit Scoring. *Psychometrika*, 47, 149–174.

MC CULLAGH, P. (1980). Regression Model for Ordinal Data (with discussion). *Journal of the Royal Statistical Society*, B 42, 109–127.

MC CULLAGH, P. (1983). Quasi–Likelihood Functions. *Annals of Statistics*, 11, 59–67.

MC CULLAGH, P., NELDER, J.A. (1983, 1989 2d ed.). *Generalized Linear Models*. New York: Chapman and Hall.

MC CULLOCH, C.E. (1997). Maximum likelihood algorithms for generalized linear mixed models. *Journal of the American Statistical Association* 92, 162–170.

MC DONALD, B.W. (1993). Estimating Logistic Regression Parameters for Bivariate Binary Data. *Journal of the Royal Statistical Society*, B 55, 391–397.

MC FADDEN, D. (1973). Conditional Logit Analysis of Qualitative Choice Behaviour. In: Zarembka, P. (Ed.). *Frontiers in Econometrics*. New York: Academic Press.

MC FADDEN, D. (1978). Modeling the Choice of Residential Location. In: Karlquist, A., et al (Eds.), *Spatial Interaction Theory and Residential Location*. Amsterdam: North–Holland.

MC FADDEN, D. (1981). Econometric Models of Probabilistic Choice. In: Manski, C., McFadden, D. (Eds.). *Structural Analysis of Discrete Data*. Cambridge, Mass.: MIT Press.

MC FADDEN, D. (1984). Qualiative Response Model. In: Grilicher, M.D. Intrilligator (Eds.), *Handbook of Econometrics*, Cambridge, Mass.: MIT Press.

MC GILCHRIST, C.A. (1994). Estimation in generalized mixed models. *Journal of the Royal Statistical Society*, B 55, 945–955.

MC KINNON, J.G. (1983). Model Specification Tests Against Non–nested Alternatives. *Econometric Reviews*, 2, 85–110.

MEHTA, C.R., PATEL, N.R., TSIATIS, A.A. (1984). Exact Significance Testing to Establish Treatment Equivalence with Ordered Categorical Data, *Biometrics*, 40, 819–825.

MEILIJSON, I. (1989). A Fast Improvement to the EM–Alogorithm on Its Own Terms. *Journal of the Royal Statistical Society*, B 51, 127–138.

MILLER, A.J. (1984). Selection of Subsets and Regression Variables (with discussion). *Journal of the Royal Statistical Society*, A 147, 389–429.

MILLER, A.J. (1989). *Subset Selection in Regression*. London: Chapman and Hall.

MILLER, M.E., DAVIS, C.S., LANDIS, R.J. (1993). The Analysis of Longitudinal Polytomous Data: Generalized Estimating Equations and Connections with Weighted Least Squares. *Biometrics*, 49, 1033–1044.

MOLENBERGHS, G. (1992). A Full Maximum Likelihood Method for the Analysis of Multivariate Ordered Categorical Data. Doctoral thesis, University of Antwerpen.

MOLENBERGHS, G., LESAFFRE, E. (1992). Marginal Modelling of Correlated Ordinal Data Using an n–way Plackett Distribution. In: Advances in GLIM and Statistical Modelling, *Springer Lecture Notes in Statistics* 78, 139–144.

MOLENBERGHS, G., LESAFFRE, E. (1994). Marginal Modelling of Correlated Ordinal Data Using a Multivariate Plackett Distribution. *Journal of the American Statistical Association*, 89, 633-644.

MOORE, D.F. (1987) Modelling the Extraneous Variance in the Presence of Extra–Bionomial Variation. *Applied Statistics*, 36, 8-14.

MORAWITZ, B., TUTZ, G. (1990). Alternative Parametrizations in Business Tendency Surveys. *ZOR–Methods and Models of Operations Research*, 34, 143-156.

MORGAN, B.J.T. (1985). The Cubic Logistic Model for Quantal Assay Data. *Applied Statistics*, 34, 105-113.

MORRIS, C.N. (1982). Natural Exponential Families with Quadratic Variance Functions. *Annals of Statistics*, 10, 65-80.

MOULTON, L., ZEGER, S. (1989). Analysing Repeated Measures in Generalized Linear Models Via the Bootstrap. *Biometrics*, 45, 381-394.

MOULTON, L. ZEGER, S. (1991). Bootstrapping Generalized Linear Models. *Computational Statistics and Data Analysis*, 11, 53-63.

MÜLLER, H.-G. (1984). Smooth Optimum Kernel Estimators of Densities, Regression Curves and Modes. *The Annals of Statistics*, 12, 766-774.

MÜLLER, H.-G., SCHMITT, T. (1988). Kernel and Probit Estimates in Quantal Bioassay. *Journal of the American Statistical Association*, 83, 750-759.

MÜLLER, H.-G., STADTMÜLLER, U. (1987). Estimation of Heteroscedasticity in Regression Analysis. *Annals of Statistics*, 15, 221-232.

MUTHÉN, B. (1984). A General Structural Equation Model with Dichotomous Categorical, and Continuous Latent Variable Indicators. *Psychometrika*, 49, 115-132.

NADARAYA, E.A. (1964). On Estimation Regression. *Theory Prob. Appl.*, 9, 141-142.

NAYLOR, J.C., SMITH, A.F.M. (1982). Applications of a Method for the Efficient Computation of Posterior Distributions. *Applied Statistics*, 31, 214-225.

NELDER, J.A. (1992). Joint Modelling of Mean and Dispersion. In: Heijden, P. v.d., Jansen, W., Francis, B., Seeber. G. (Eds.), *Statistical Modelling*, 263-272. Amsterdam: North Holland.

NELDER, J.A., PREGIBON, D. (1987). An Extended Quasi–Likelihood Function. *Biometrika*, 74, 221–232.

NELDER, J.A., WEDDERBURN, R.W.M. (1972). Generalized Linear Models. *Journal of Royal Statistical Society*, A 135, 370–384.

NERLOVE, M. (1983). Expectations, Plans and Realizations in Theory and Practice. *Econometrica*, 51, 1251–1279.

NEUHAUS, J., HAUCK, W., KALBFLEISCH, J. (1991). The Effects of Mixture Distribution. Misspecification when Fitting Mixed Effect Logistic Models. Department of Epidemiology and Biostatistics University of California, San Francisco, Technical Report 16.

NEUHAUS, J.M., KALBFLEISCH, J.D., HAUCK, W.W. (1991). A Comparison of Cluster–Specific and Population–Averaged Approaches for Analyzing Correlated Binary Data. *International Statistical Review*, 59, 25–35.

NEYMAN, J. (1949). Contribution to the Theory of the Chi–Square–Test. Proceedings of the Berkeley Symposium on Mathematical Statistics and Probability, Berkeley.

OSIUS, G., ROJEK, D. (1992). Normal Goodness–of–Fit Tests for Parametric Multinomial Models with Large Degrees of Freedom. *Journal of the American Statistical Association*, 87, 1145–1152.

O'SULLIVAN, F., YANDELL, B., RAYNOR, W. (1986). Automatic Smoothing of Regression Functions in Generalized Linear Models. *Journal of the American Statistical Association*, 81, 96–103.

PARR, W.C. (1981). Minimum Distance Estimation. A Bibliography. *Communications in Statistics, Theory and Methods*, 10, 1205–1224.

PAULY, R. (1989). A General Structural Model for Decomposing Time Series and Its Analysis as a Generalized Regression Model. *Statistical Papers*, 30, 245–261.

PEPE, M.S., ANDERSON, G.L. (1994). A Cautionary Note on Inference for Marginal Regression Models with Longitudinal Data and General Correlated Response Data. *Communications in Statistics, Part B–Simulation and Computation*, 23, 939–951.

PIEGORSCH, W.W., WEINBERG, C.R., MARGOLIN, B.H. (1988). Exploring Simple Independent Action in Multifactor Tables of Proportions. *Biometrics*, 44, 595–603.

PIERCE, D.A., SCHAFER, D.W. (1986). Residuals in Generalized Linear Models. *Journal of the American Statistical Association*, 81, 977–986.

PRAKASA RAO, B.L.S. (1987). *Asymptotic Theory of Statistical Inference*. New York: Wiley.

PREGIBON, D. (1980). Goodness of Link Tests for Generalized Linear Models. *Applied Statistics*, 29, 15–24.

PREGIBON, D. (1981). Logistic Regression Diagnostics. *Annals of Statistics*, 9, 705–724.

PREGIBON, D. (1982). Resistant Fits for some Commonly Used Logistic Models with Medical Applications. *Biometrics*, 38, 485–498.

PREGIBON, D. (1984). Review of Generalized Linear Models by McCullagh and Nelder. *American Statistician*, 12, 1589–1596.

PREISLER, H.K. (1989). Analysis of a Toxicological Experiment Using a Generalized Linear Model with Nested Random Effects. *International Statistical Review*, 57, 145–159.

PRENTICE, R.L. (1976). A Generalization of the Probit and Logit Methods for Dose Response Curves. *Biometrics*, 32, 761–768.

PRENTICE, R.L. (1988). Correlated Binary Regression with Covariates Specific to Each Binary Observation. *Biometrics* 44, 1033–1084.

PRENTICE, R.L., ZHAO, L.P. (1991). Estimating Equations for Parameters in Mean and Covariances of Multivariate Discrete and Continuous Responses. *Biometrics*, 47, 825–839.

PRIESTLEY, M.B., CHAO, M.T. (1972). Nonparametric Function Fitting. *Journal of the Royal Statistical Society*, B 34, 385–392.

PRUSCHA, H. (1993). Categorical Time Series with a Recursive Scheme and with Covariates. *Statistics*, 24, 43–57.

PUDNEY, S. (1989). *Modelling Individual Choice. The Econometrics of Corners, Kinks and Holes*. London: Basil Blackwell.

QU, Y., WILLIAMS, G.W., BECK, G.J., GOORMASTIC, M. (1987). A Generalized Model of Logistic Regression for Clustered Data. *Communications in Statistics, A Theory and Methods*, 16, 3447–3476.

RANDALL, J.H. (1989). The Analysis of Sensory Data by Generalized Linear Models. *Biom. Journal*, 31, 781–793.

RAO, C., KLEFFE, J. (1988). *Estimation of Variance Components and Applications*. Amsterdam: North Holland.

RASCH, G. (1961). On General Laws and the Meaning of Measurement in Psychology. In: Neyman, J. (Ed.), *Proceedings of the Fourth Berkeley Symposium on Mathematical Statistics and Probability*. Berkeley.

READ, I. CRESSIE, N. (1988). *Goodness–of–Fit Statistics for Discrete Multivariate Data.* New York: Springer Verlag.

REDNER, R.A., WALKER, H.F. (1984). Mixture Densities, Maximum Likelihood and the EM Algorithm. *SIAM Review,* 26, 195–239.

REINSCH, C. (1967). Smoothing by Spline Functions. *Numerische Mathematik,* 10, 177–183.

RIPLEY, B. (1987). *Stochastic Simulation.* New York: Wiley.

RITTER, C., TANNER, M.A. (1992). Facilitating the Gibbs Sampler: The Gibbs Stopper and the Griddy–Gibbs Sampler. *Journal of the American Statistical Association,* 87, 861–868.

ROBERTS, F. (1979). *Measurement Theory.* Reading, Mass.: Addison Wesley.

ROBINS, J.M., ROTNITZKY, A.G., ZHAO, L.P. (1995). Analysis of Semiparametric Regression Models for Repeated Outcomes in the Presence of Missing Data. *Journal of the American Statistical Association,* 90, 106–120.

ROJEK, D. (1989). *Asymptotik für Anpassungstests in Produkt–Multinomial–Modellen bei wachsendem Freiheitsgrad.* Ph.D. Thesis, Universität Bremen.

RONNING, G. (1980). Logit, Tobit and Markov Chains. The Different Approaches of the Analysis of Aggregate Tendency Data. In: Strigel, W. (Ed.). *Business Cycle Analysis.* Farnborough: Westmed.

RONNING, G. (1987). The Informal Content of Responses from Business Surveys. Diskussion Paper 96/s., University of Konstanz.

RONNING, G. (1991). *MikroÖkonometrie.* Berlin: Springer.

RONNING, G., JUNG, R. (1992). Estimation of a First Order Autoregressive Process with Poisson Marginals for Count Data. In: Fahrmeir, L., et al (Eds). Advantages in GLIM and Statistical Modelling, *Lecture Notes in Statistics,* 78, 188–194. Berlin: Springer.

ROSENBERG, B. (1973). Random Coefficient Models. *Annals of Economic and Social Measurement,* 4, 399–428.

ROSNER, B. (1984). Multivariate Methods in Ophthalmology with Applications to Other Paired–Data Situations. *Biometrics,* 40, 1025–1035.

SAGE, A., MELSA, J. (1971). *Estimation Theory, with Applications to Communications and Control.* New York: McGraw Hill.

SANTNER, T.J., DUFFY, D.E. (1989). *The Statistical Analysis of Discrete Data*. New York: Springer.

SCALLAN, A. GILCHRIST, R., GREEN, M. (1984). Fitting Parametric Link Functions in Generalised Linear Models. *Comp. Statistics and Data Analysis*, 2, 37–49.

SCHALL, R. (1991). Estimation in generalised linear models with random effects. *Biometrika* 78, 719–727.

SCHLICHT, E. (1981). A Seasonal Adjustment Principle and Seasonal Adjustment Method Derived from this Principle. *Journal of the American Statistical Association*, 76, 374–378.

SCHNATTER, S. (1992). Integration–Based Kalman–Filtering for a Dynamic Generalized Linear Trend Model. *Computational Statistics and Data Analysis*, 13, 447–459.

SCHNEIDER, W. (1986). *Der Kalmanfilter als Instrument zur Diagnose und Schätzung variabler Parameterstrukturen in Ökonometrischen Modellen*. Heidelberg: Physica.

SCHWARZ, G. (1978). Estimating the Dimension of a Model. *Annals of Statistics*. 6, 461–464.

SEBER, G. (1977). *Linear Regression Analysis*. New York: Wiley.

SEEBER, G. (1989). Statistisches Modellieren in Exponentialfamilien. Habilitationsschrift (unpublished). Innsbruck.

SEIFERT, B., GASSER, T. (1996). Variance properties of local polynomials and ensuring modifications. In: Härdle, W., Schimek, M.G. (Eds.). *Statistical Theory and Computational Aspects of Smoothing*. Heidelberg: Physica–Verlag.

SHAW, J.E.H. (1988). Aspects of Numerical Integration and Summarisation. *Bayesian Statistics*, 3, 411–428.

SHEPHARD, N. (1994). Partial non–Gaussian State Space. *Biometrika*, 81, 115–131.

SHEPHARD, N., Pitt, M.K. (1995). Parameter–Driven Exponential Family Models. Unpublished manuscript, Nuffield College, Oxford, UK.

SILVERMAN, B.W. (1984). Spline Smoothing: The Equivalent Variable Kernel Method. *Annals of Statistics*, 12, 898–916.

SILVAPULLE, M.J. (1981). On the Existence of Maximum Likelihood Estimates for the Bionomial Response Models. *Journal of the Royal Statistical Society*, B43, 310–313.

SIMONOFF, J. (1983). A Penalty Function Approach to Smoothing Large Sparse Contingency Tables. *Am. Statist.*, 11, 208–218.

SIMONOFF, J. (1996). *Smoothing Methods in Statistics.* New York: Springer–Verlag.

SIMONOFF, J., TSAI, C. (1991). Assessing the Influence of Individual Observation on a Goodness–of–fit Test Based on Nonparametric Regression. *Statistics & Probability Letters*, 12, 9–17.

SLEEPER, L.A., HARRINGTON,D.P. (1990). Regression splines in the Cox model with application to covariate effects in liver disease. *Journal of the American Statistical Association* 85, 941–949.

SMALL, K.A. (1987). A Discrete Choice Model for Ordered Alternatives. *Ecomometrica*, 55, 409–424.

SMITH, A., WEST, M. (1983). Monitoring Renal Transplants: An Application of the Multi–Process Kalman Filter. *Biometrics*, 39, 867–878.

SMITH, A.F.M., SKENE, A.M., SHAW, J.E.H., NAYLOR, J.C., DRANSFIELD, M. (1985). The Implementation of the Bayesian Paradigm. *Comm. Statist. Theor. Meth.*, 14, 1079–1102.

SMITH, R., MILLER, J. (1986). A Non–Gaussian State Space Model and Application to Prediction of Records. *Journal of the Royal Statistical Society*, B 48, 79–88.

STANISWALIS, J.G. (1989). Local Bandwidth Selection for Kernel Estimates. *Journal of the American Statistical Association*, 84, 284–288.

STANISWALIS, J.G., COOPER, V.D. (1988). Kernel Estimates of Dose–Responses. *Biometrics*, 44, 1103–1119.

STANISWALIS, J.G., SEVERINI, T.A. (1991). Diagnostics for assessing regression models, *Journal of the American Statistical Association* 86, 684–692.

STEVENS, S.S. (1951). Mathematics, Measurement and Psychophysics. In: Stevens, S.S. (Ed.): *Handbook of Experimental Psychology.* New York: Wiley.

STIRATELLI, R., LAIRD, N., WARE J.H. (1984). Random–Effects Models for Serial Observation with Binary Response. *Biometrics*, 40, 961–971.

STONE, C.J. (1977). Consistent Nonparametric Regression (with discussion). *Annals of Statistics*, 5, 595–645.

STRAM, D.O., WEI, L.J. (1988). Analyzing Repeated Measurements with Possibly Missing Observations by Modelling Marginal Distributions. *Statistics in Medicine*, 7, 139–148.

STRAM, D.O., WEI, L.J., WARE, J.H. (1988). Analysis of Reported Categorical Outcomes with Possibly Missing Observations and Time–Dependent Covariates. *Journal of the American Statistical Association*, 83, 631–637.

STROUD, A. (1971). *Approximate Calculation of Multiple Integrals*. Englewood Cliffs, N.J.: Prentice Hall.

STROUD, A.H., SECREST. D. (1966). *Gaussian Quadrature Formulas*. New York: Prentice Hall.

STUKEL, T.A. (1988). Generalized Logistic Models. *Journal of the American Statistical Assiciation*, 83, 426–431.

STUTE, W. (1984). Asymptotic Normality of Nearest Neighborhood Regression Function Estimates. *Annals of Statistics*, 12, 917–926.

TANNER, M.A. (1991). Tools for Statistical Inference. *Lecture Notes in Statistics*, 67. New York: Springer.

TERZA, J.V. (1985). Ordinal Probit: A Generalization. *Commun. Statist. Theor. Meth.*, 14, 1–11.

THALL, P.F., VAIL, S.C. (1990). Some Covariance Models for Longitudinal Count Data with Overdispersion. *Biometrics*, 46, 657–671.

THIELE, T. (1980). Sur la Compensation de Quelques Erreurs Quasi–Systematiques par la Methode des Moindres Carrees. Copenhagen: Reitzel.

THOMPSON, R. (1981). Survival Data and GLIM (Letter to the Editors). *Applied Statistics*, 30, 310.

THOMPSON, R., BAKER, R.J. (1981). Composite Link Functions in Generalized Linear Models. *Applied Statistics*, 30, 125–131.

THOMPSON, W.A., Jr. (1977): On the Treatment of Grouped Observations in Life Studies. *Biometrics*, 33, 463–470.

TIBSHIRANI, R., CIAMPI, A. (1983). A Family of Proportional– and Additive–Hazards Models for Survival Data. *Biometrics*, 39, 141–147.

TIELSCH, J.M., SOMMER, A., KATZ, J., EZRENE, S. (1989). Sociodemographic Risk Factors for Blindness and Visual Impairment. *The Baltimore Eye Survey*. Archives of Ophtalmology.

TIERNEY, L. (1994). Markov Chains for exploring Posterior Distributions. *Annals of Statistics*, 22, 1701–1762.

TITTERINGTON, D.M. (1980). A Comperative Study of Smoothing Procedures for Ordered Categorical Data. *Technometrics*, 22, 291–342.

TITTERINGTON, D.M. (1985). Common Structure of Smoothing Techniques in Statistics. *International Statistical Review*, 52, 141–170.

TITTERINGTON, D.M., BOWMAN, A.W. (1985). A Comperative Study of Smoothing Procedures for Ordered Categorical Data. *J. Statist. Comput. Simul.*, 21, 291–312.

TSIATIS, A.A. (1981). A Large Sample Study of Cox's Regressions Model. *Annals of Statistics*, 9, 93–108.

TSUTAKAWA, R.K. (1988). Mixed Model for Analyzing Geographic Variability in Mortality Rates. *Journal of the American Statistical Association*, 83, 37–42.

TUTZ, G. (1989a). On Cross–Validation for Discrete Kernel Estimates in Discrimination. *Communications in Statistics, Theory and Methods*, 11, 4145–4162.

TUTZ, G. (1989b). Compound Regression Models for Categorical Ordinal Data. *Biometrical Journal*, 31, 259–272.

TUTZ, G. (1990a). *Modelle für kategoriale Daten mit ordinalem Skalenniveau, parametrische und nonparametrische Ansätze.* GÖttingen: Vandenhoeck & Ruprecht Verlag.

TUTZ, G. (1990b). Smoothed Categorical Regression Based on Direct Kernel Estimates. *Journal of Statistical Computation and Simulation*, 36, 139–156.

TUTZ, G. (1990c). Sequential Item Response Models with an Ordered Response. *British Journal of Statistical and Mathematical Psychology*, 43, 39–55.

TUTZ, G. (1991a). Sequential Models in Ordinal Regression. *Computational Statistics and Data Analysis*, 11, 275–295.

TUTZ, G. (1991b). Choice of Smoothing Parameters for Direct Kernels in Discrimination. *Biometrical Journal*, 33, 519–527.

TUTZ, G. (1991c). Consistency of Cross–Validatory Choice of Smoothing Parameters for Direct Kernel Estimates. *Computational Statistics Quarterly*, 4, 295–314.

TUTZ, G. (1993). Invariance Principles and Scale Information in Regression Models. *Methodika* (in print).

TUTZ, G. (1995). Dynamic modelling of discrete duration data: a local likelihood approach. Report 95-15, Institut für Quantitative Methoden, Tech. Univ. Berlin.

TUTZ, G., GROSS, H. (1994). Discrete Kernels, Loss Functions and Parametric Models in Discrete Discrimination. *ZOR–Methods and Models of Operations Research* (to appear).

TUTZ, G., KAUERMANN, G. (1997). Local estimators in multivariate generalized linear models with varying coefficients. *Computational Statistics* 12, 193–208.

VAN DER HEIJDEN, P., JANSEN, W., FRANCIS, B., SEEBER, G.U.H. (1992). *Statistical Modelling.* Amsterdam: North Holland.

VAN DEUSEN, P. (1989). A Model–Based Approach to Tree Ring Analysis. *Biometrics*, 45, 763–779.

WACLAWIW, M., LIANG, K.Y. (1993). Prediction of Random Effects in the Generalized Linear Model. *Journal of the American Statistical Association*, 88, 171–178.

WACLAWIW, M., LIANG, K.Y. (1994). Empirical Bayes estimation and inference for the random effects model with binary response. *Statistics in Medicine* 13, 541–551.

WAHBA, G. (1978). Improper Prior, Spline Smoothing and the Problem of Guarding Against Model Errors in Regression. *Journal of the Royal Statistical Society*, B 44, 364–372.

WANG, M.-CH., VAN RYZIN, J. (1981). A Class of Smooth Estimators for Discrete Distributions. *Biometrika*, 68, 301–309.

WANG, P.C. (1985). Adding a Variable in Generalized Linear Models *Technometrics*, 27, 273–276.

WANG, P.C. (1987). Residual Plots for Detecting Non–Linearity in Generalized Linear Models. *Technometrics*, 29, 435–438.

WARE, J.H., LIPSITZ, S., SPEIZER, F.E. (1988). Issues in the Analysis of Repeated Categorical Outcomes. *Statistics in Medicine*, 7, 95–107.

WATSON, G.S. (1964). Smooth Regression Analysis. *Synkhyā*, A, 26, 359–372.

WECKER, W., ANSLEY, C. (1983). The Signal Extraction Approach to Nonlinear Regression and Spline Smoothing. *Journal of the American Statistical Association*, 78, 81–89.

WEDDERBURN, R.W.M. (1974). Quasi–Likelihood Functions, General-ized Linear Models, and the Gauss–Newton Method. *Biometrika*, 61, 439–447.

WEDDERBURN, R.W.M. (1976). On the Existence and Uniqueness of the Maximum Likelihood Estimates for Certain Generalized Linear Models. *Biometrika*, 63, 27–32.

WERMUTH, N., LAURITZEN, S.L. (1990). On Substantive Research Hypotheses, Conditional Independence Graphs and Graphical Chain Models. *Journal of the Royal Statistical Society*, B, 52, 21–50.

WEST, M. (1981). Robust Sequential Approximate Bayesian Estimation. *Journal of the Royal Statistical Society*, B 43, 157–166.

WEST, M., HARRISON, P.J., MIGON, M. (1985). Dynamic Generalized Linear Models and Bayesian Forecasting. *Journal of the American Statistical Association*, 80, 73–97.

WEST, M., HARRISON, P.J. (1989). *Bayesian Forecasting and Dynamic Models*. New York: Springer.

WHITE, H. (1981). Consequences and Detection of Misspecified Nonlinear Regression Models. *Journal of the American Statistical Association*, 76, 419–433.

WHITE, H. (1982). Maximum Likelihood Estimation of Misspecified Mod-els. *Econometrica*, 50, 1–25.

WHITE, H. (1984). Maximum Likelihood Estimation of Misspecified Dy-namic Models. In: Dijlestra, T. (Ed.), *Misspecification Analysis*. Berlin: Springer.

WHITTAKER, E.T. (1923). On a New Method of Graduation. *Proc. Ed-inborough Math. Assoc.*, 78, 81–89.

WHITTAKER, J. (1990). *Graphical Models in Applied Multivariate Statis-tics*. Chichester: Wiley.

WILLIAMS, D.A. (1982). Extra Bionomial Variation in Logistic Linear Models. *Applied Statistics*, 31, 144–148.

WILLIAMS, D.A. (1987). Generalized Linear Model Diagnostics Using the Deviance and Single Ease Deletions. *Applied Statistics*, 36, 181–191.

WILLIAMS, O.D., GRIZZLE, J.E. (1972). Analysis of Contingency Ta-bles Having Ordered Response Categories. *Journal of the American Statistical Association*, 67, 55–63.

WILLIAMSON, J.M., KIM, K., LIPSITZ, S.R. (1995). Analyzing Bivariate Ordinal Data using a Global Odds Ratio. *Journal of the American Statistical Association*, 90, 1432–1437.

WILSON, J., KOEHLER, K. (1991). Hierarchical Models for Cross–Classified Overdispersed Multinomial Data. *J. Bus. & Econ. St.*, 9, 103–110.

WOLAK, F. (1987). An Exact Test for Multiple Inequality and Equality Constraints in the Linear Regression Model. *Journal of the American Statistical Association*, 82, 782–793.

WOLAK, F. (1989). Local and Global Testing of Linear and Nonlinear Inequality Constraints in Nonlinear Econometric Models. *Econometric Theory*, 5, 1–35.

WONG, G.Y., MASON, W.M. (1985). The Hierarchical Logistic Regression Model for Multilevel Analysis. *Journal of the American Statistical Association*, 80, 513–524.

WONG, W.H. (1986). Theory of Partial Likelihood. *Annals of Statistics*, 14, 88–123.

WU, J.C.F. (1983). On the Covergence Properties of the EM–Algorithm. *The Annals of Statistics*, 11, 95–103.

WU, L.L., TUMA, N.B. (1991). Assessing bias and fit of global and local hazard models. *Sociological Methods and Research* 19, 354–387.

YANG, S. (1981). Linear Funtions of Concomitants of Order Statistics with Application to Nonparametric Estimation of a Regression Function. *Journal of the American Statistical Association*, 76, 658–662.

YELLOT, J.I. (1977). The Relationship Between Luce's Choice Axiom, Thurstone's Theory of Comparative Judgement, and the Double Exponential Distribution. *Journal of Mathematical Psychology*, 15, 109–144.

ZEGER, S.L. (1988a). A Regression Model for Time Series of Counts. *Biometrika*, 75, 621–629.

ZEGER, S.L. (1988b). Commentary. *Statistics in Medicine*, 7, 161–168.

ZEGER, S.L., DIGGLE, P.J., YASUI, Y. (1990). Marginal Regression Models for Time Series. Inst. for Mathematics and Its Application, Time Series Workshop, Preprint.

ZEGER, S.L., KARIM, M.R. (1991). Generalized Linear Models with Random Effects; A Gibbs' Sampling Approach. *Journal of the American Statistical Association*, 86, 79–95.

ZEGER, S., LIANG, K.-Y. (1986). Longitudinal Data Analysis for Discrete and Continuous Outcomes. *Biometrics*, 42, 121–130.

ZEGER, S.L., LIANG, K.-Y. (1989). A Class of Logistic Regression Models for Multivariate Binary Time Series. *Journal of the American Statistical Association*, 84, 447–451.

ZEGER, S.L., LIANG, K.-Y., ALBERT, P.S. (1988). Models for Longitudinal Data: A Generalized Estimating Equation Approach. *Biometrics*, 44, 1049–1060.

ZEGER, S., LIANG, K.-Y., SELF, S. (1985). The Analysis of Binary Longitudinal Data with Time–Independent Covariates. *Biometrika*, 72, 31–38.

ZEGER, S.L., QAQISH, B. (1988). Markov Regression Models for Time Series: A Quasi–Likelihood Approach. *Biometrics*, 44, 1019–1031.

ZELLNER, A., ROSSI, P.E. (1984). Bayesian Analysis of Dichotomous Quantal Response Models. *Journal of Econometrics*, 25, 365–393.

ZHAO, L.P., PRENTICE, R.L. (1990). Correlated Binary Regression using a Quadratic Exponential Model. *Biometrika* 77, 642–648.

ZHAO, L.P., PRENTICE, R.L., SELF, S. (1992). Multivariate Mean Parameter Estimation by Using a Partly Exponential Model. *Journal of the Royal Statistical Society*, B 54, 805–811.

Author Index

Aalen, 337
Abramowitz, 358
Agresti, 13, 35, 74, 86, 96, 104, 209, 220, 254
Aitchison, 162, 163, 165, 168, 342
Aitken, 162, 163, 165, 168, 342
Aitkin, 15, 220, 224, 230, 244, 312–314
Albert, 140, 220, 221, 252–254
Andersen, 208
Anderson, 15, 33, 74, 95, 218, 220, 230, 244, 258, 264, 265, 269, 274
Andrews, 145, 146
Ansley, 262, 265, 266
Aoki, 268
Aranda–Ordaz, 320
Aranda-Ordaz, 79
Arminger, 105, 144, 145
Armstrong, 91
Ashby, 105
Azzalini, 141, 185, 197

Baker, 60
Bartholomew, 105

Bates, 221
Beck, 10
Belsley, 124
Ben-Akiva, 73
Benedetti, 157
Besag, 365
Bhapkar, 102
Biller, 303
Birch, 103
Bishop, 13, 35
Blossfeld, 305
Börsch-Supan, 71
Bonney, 106, 107
Bowman, 141, 163, 165, 185
Boyles, 356
Breslow, 35, 255
Brillinger, 230, 244, 246
Broemeling, 224
Brown, 255
Burns, 227
Buse, 47

Cameron, 35
Carey, 116, 118
Carlin, 287, 288

Carroll, 62
Carter, 303
Chao, 157
Chhikara, 23
Christensen, 13
Clayton, 255, 312–314
Cleveland, 185
Clogg, 107
Collett, 15, 34
Conaway, 208, 254
Conolly, 109
Cook, 124, 125, 138
Cooper, 162, 168
Copas, 162, 168
Corbeil, 226
Cox, 146, 188, 189, 309, 321, 352
Craven, 161
Cressie, 96, 101–103, 165
Czado, 61, 79, 141

Daganzo, 71
Dale, 103
Dannegger, 344
Davidson, 146
Davis, 118, 213, 357
de Boor, 154–156
De Jong, 265
Dean, 34
Decarli, 13
Dempster, 355, 356
Devroye, 364
Dewulf, 218
Dielman, 204, 221
Diggle, 13, 116, 187, 201, 204, 218
Dobson, 15
Doksum, 310
Drum, 255
Duan, 143
Duffy, 13, 58
Durbin, 280

Edwards, 75, 120
Efron, 55, 321, 341, 342
Eilers, 62
Engle, 158, 192

Epanechnikov, 156
Eubank, 151, 185

Fahrmeir, 3, 13, 15, 35, 41, 44, 46,
        54, 118, 120, 122–124, 142,
        191, 196, 271, 277, 279,
        280, 287, 295, 298, 303,
        339, 344, 353, 354
Fan, 185, 186
Farewell, 79
Fernandes, 271
Fienberg, 13, 163
Finkelstein, 315
Finney, 127
Firth, 15, 185
Fitzmaurice, 118, 218
Folks, 23
Forthofer, 5, 74
Foutz, 55
Fowlkes, 168
Francis, 13, 15
Friedl, 79
Friedman, 156
Frost, 120, 122–124, 145, 148
Frühwirth–Schnatter, 259, 268, 290,
        303
Furnival, 120, 122

Gamerman, 303, 337
Garber, 190, 207
Gasko, 310
Gasser, 157, 158, 185
Gay, 55, 61
Gelfand, 58, 365
Geman, 289, 364, 365
Genter, 79
Gersch, 259, 262
Gianola, 244
Giesbrecht, 227
Gigli, 62
Gijbels, 185, 186
Gilchrist, 13, 61
Gilks, 365
Glonek, 118
Glosup, 185

Goldstein, 224
Goodman, 96
Gordon, 259
Goss, 281, 295
Gould, 320
Gourieroux, 48, 53, 142, 147, 194, 249
Green, 61, 175, 180, 186, 277, 365
Greenwood, 317
Grether, 194
Griliches, 255
Grizzle, 96
Groß, 173
Gu, 180
Guan, 194

Habbema, 163, 165
Haberman, 41, 44, 165
Härdle, 13, 141, 151, 159, 160, 162, 185
Hall, 160, 165, 196, 255
Hamerle, 3, 35, 124, 208, 213, 215, 254, 305, 331, 333
Hanefeld, 12, 170, 317
Hankey, 321
Harrington, 113, 344
Harrison, 57, 257–259, 261, 290
Hart, 185
Hartigan, 277
Hartley, 355
Harvey, 12, 257–262, 264, 268, 269, 271
Harville, 220, 225, 226, 233
Hastie, 13, 151, 175, 177, 180, 181, 183, 185, 267
Hauck, 104, 220, 230, 254
Hausman, 71, 144, 255
Havranek, 120
Heagerty, 118
Hebbel, 266
Heckman, 185, 194, 206
Heiler, 266
Hennevogl, 138, 287, 291, 362
Hermans, 163
Heumann, 118

Heyde, 196
Higdon, 365
Hinde, 15, 62, 220, 230, 244, 246
Hinkley, 185, 352
Hoaglin, 126
Hocking, 120
Holland, 13, 61, 163
Holly, 48, 144
Holmes, 85
Holtbrügge, 95
Hopper, 208
Horowitz, 141
Hsiao, 204, 221, 223, 225, 293
Huber, 61
Huffer, 337
Hutchinson, 259

Im, 244

Jacobs, 194
Jansen, 13, 220, 233, 244, 246
Jazwinski, 269
Jones, 13, 221, 226, 257, 258, 293, 294
Jorgensen, 23
Jorgenson, 61, 345
Jung, 194

Kalbfleisch, 78, 104, 220, 230, 254, 305, 309, 310, 319, 323, 327, 331, 333
Karim, 248, 249, 254
Kauermann, 186
Kaufmann, 41, 44, 48, 191, 196, 271, 277, 279, 295, 298, 353, 354
Kiefer, 337
Kim, 118
Kirchen, 269
Kitagawa, 8, 259, 262, 270, 282, 283, 290, 293
Kleffe, 221, 225, 227
Klemme, 287
Klinger, 344
Knorr–Held, 287, 290, 303, 344

Koehler, 255
König, 209, 298
Kohn, 262, 265, 266, 303
Koopman, 265, 280
Krämer, 145
Kranert, 138
Krantz, 73
Kredler, 15, 41, 124
Küsters, 105
Kuh, 124
Kuk, 255
Kullback, 102

Läärä, 87
Lai, 159
Laird, 10, 58, 113, 118, 215, 220,
        221, 226, 230, 233, 355,
        356
Lancaster, 305
Landis, 118
Landwehr, 125
Lang, 209, 254
Lauder, 168
Lauritzen, 105
Lawless, 120, 122, 305, 306, 308,
        311, 317, 320
Le Cessie, 185
Lee, 140
Lehnen, 5, 74
Leonard, 58, 163
Lerman, 71, 73
Lesaffre, 118, 140, 218
Levine, 197
Lewis, 194
Li, 143
Liang, 7, 11, 13, 108–110, 112, 113,
        115, 116, 118, 187, 204,
        205, 211–213, 215, 218,
        220, 221, 252–255
Lindsey, 13, 221, 271
Lindstrom, 221
Lipsitz, 113, 118, 208, 218
Loader, 185
Longford, 224
Los, 269

Louis, 356

Mack, 158, 159
Maddala, 73, 194
Magnus, 237, 239
Mammitzsch, 157
Manski, 71
Mantel, 321
Margolin, 4
Marron, 160
Martin, 257, 270
Marx, 62
Mason, 230, 233
Masters, 96
Matthews, 87
Mayer, 305
McCullagh, 13, 15, 35, 48, 52, 55,
        78, 85, 90, 118, 119, 132,
        141, 255
McCulloch, 255
McDonald, 113, 215
McFadden, 71
McGilchrist, 255
McKeague, 337
McKinnon, 146
Mee, 220, 233
Mehta, 91
Meilijson, 356
Melsa, 258, 269, 274, 277
Mengersen, 365
Migon, 257
Miller, 118, 120, 271
Molenberghs, 118, 218
Montfort, 48, 53, 249
Moore, 255, 258, 264, 265, 269,
        274
Morawitz, 94, 209, 295
Morgan, 79
Morris, 52
Moulton, 62, 213
Müller, 157, 158, 162, 168
Muthén, 105

Nadaraya, 157, 158, 342
Nagl, 213, 215

Naylor, 57, 358, 359, 362
Nelder, 1, 13, 15, 35, 48, 52, 55,
        119, 132, 141, 347
Nerlove, 209, 298
Neudecker, 237, 239
Neuhaus, 104, 220, 230, 254
Neyman, 102

O'Sullivan, 177, 180, 277
Osius, 103, 104
Oudiz, 209, 298

Parr, 102
Patel, 91
Pauly, 266, 268
Payne, 255
Pepe, 218
Piegorsch, 4
Pierce, 132, 133
Pitt, 303
Polson, 287, 288
Prakasa Rao, 350
Pregibon, 52, 55, 61, 79, 125, 127,
        128, 139–141
Preisler, 230, 244, 246
Prentice, 78, 79, 109, 112, 115,
        118, 305, 309, 310, 319,
        323, 327, 331, 333, 346
Priestley, 157
Pritscher, 118
Pudney, 73

Qaqish, 7, 108, 111, 190–192, 197
Qu, 109

Rabinowitz, 357
Raftery, 257, 270
Randall, 228
Rao, 221, 225, 227
Rasch, 208
Raynor, 177, 180, 277
Read, 96, 101–103, 165
Redner, 356
Reinsch, 154, 155, 266
Remme, 163
Richardson, 365

Ripley, 363, 364
Ritter, 365
Roberts, 73
Robins, 218
Rojek, 103, 104
Ronning, 73, 194, 208, 210, 254
Rosenberg, 293, 294
Rosner, 109
Rossi, 57, 58, 248
Rotnitzky, 118, 218
Rubin, 355, 356

Sage, 258, 269, 274, 277
Santner, 13, 58
Scallan, 61
Schafer, 132, 133
Schall, 255
Schlicht, 266
Schmitt, 162, 168
Schnatter, 290–292
Schneider, 258, 264, 269
Schoenberg, 144, 145
Schuhmacher, 95
Searle, 226
Secrest, 358
Seeber, 13, 61, 120
Seifert, 185
Self, 215, 309, 346
Severini, 185
Shaw, 358
Shephard, 303
Shoemaker, 125
Silvapulle, 41
Silverman, 156, 158, 186
Simonoff, 133, 163, 186
Singhal, 120, 122
Sleeper, 344
Sloan, 91
Small, 71
Smith, 57, 58, 62, 259, 271, 358,
        359, 362, 365
Sobel, 105
Speizer, 208
Spiegelhalter, 365
Srivastava, 55

Stadtmüller, 158
Staniswalis, 158, 162, 168, 185
Stegun, 358
Stevens, 261
Stiratelli, 58, 220, 230, 233
Stoffer, 287, 288
Stone, 158, 185
Stram, 206, 213, 299
Stroud, 357, 358
Stukel, 79, 141
Stute, 158

Tanner, 365
Taylor, 144
Terza, 79
Thall, 213, 215
Thiele, 265
Thompson, 60, 321
Thurstone, 75
Tibshirani, 13, 151, 175, 177, 180,
    181, 183, 185, 267
Tielsch, 108
Tierney, 365
Titterington, 163, 165
Trivedi, 35
Trognon, 53
Tsai, 133
Tsiatis, 91, 309
Tsutakawa, 255
Tuma, 344
Turlach, 151
Tutz, 13, 73, 87, 91, 94–96, 168,
    173, 175, 186, 209, 331,
    333, 344

Ulm, 344

Vail, 213, 215
Van der Heijden, 13
Van Deusen, 259
Van Houwelingen, 185
Van Ryzin, 163

Waclawiw, 255
Wagenpfeil, 344
Wahba, 161, 262, 266

Wald, 196, 215
Walker, 356
Wand, 185
Wang, 125, 163
Ware, 10, 58, 208, 213, 220, 221,
    226, 230, 233
Watson, 157, 158, 342
Wecker, 262
Wedderburn, 1, 41, 52, 53, 347
Wei, 213
Weinberg, 4
Weisberg, 124, 125
Welsch, 55, 61, 124, 126
Wermuth, 105
West, 57, 257–259, 270, 271, 283–
    285, 290, 337
White, 55, 140, 142, 145, 197, 208
Whittaker, 13, 105, 266
Williams, 85, 96, 140, 255
Williamson, 118
Wilson, 120, 122, 255
Wise, 71
Wolak, 48
Wong, 190, 230, 233
Wu, 344, 356

Yandell, 175, 177, 180, 277
Yang, 158
Yasui, 201
Yellott, 71
Young, 208
Yuan, 194

Zeger, 7, 9, 11, 13, 62, 108–110,
    112, 115, 116, 118, 187,
    190–192, 197, 201, 203–
    205, 211–213, 215, 218,
    220, 221, 248, 249, 252–
    254
Zellner, 57, 58, 248
Zhao, 116, 118, 218, 346

Subject Index

AIC, 122
Aitchison and Aitken kernel, 163
Akaikes Information Criterion, 122
Alternative–specific, 72
ARCH models, 192
Asymptotics, 46, 48, 103, 207, 215, 350
    consistency, 42, 215
    existence, 351
    existence and uniqueness, 42
    normality, 42, 58, 351, 352
Auto–covariance function, 201
Autocorrelation function, 203
Autoregressive models, 188
Average predictive squared error, 160
Average squared error, 160

Backfitting, 182
Bandwidth, 157
Bayes models, 57
Bayes risk, 174
Bias–variance trade–off, 158
Binary response, 24
Binary time series, 189, 202

Binomial response, 24
Binomial time series, 189
BMDP, 367, 372

Case deletion, 138
Categorical time series, 191, 271
Censoring, 310
    random, 310
    Type I, 311
Cholesky square root, 361
Cluster, 111
Cluster–specific, 221, 229
Coding
    dummy, 16
    effect, 16
    of covariates, 16
Conditional covariance, 195
Conditional expectation, 195
Conditional gamma models, 192
Conditional independence, 205
Conditional information, 195, 207
Conditional likelihood, 208
Conditional means, 205
Conditional model, 188, 205
    for logit data, 205

statistical inference, 194
Conjugate prior–posterior, 275
Consistency, 42, 214, 351
Cook distance, 138
Correction step, 264
Correlation matrix
    working, 112
Count data, 190
Counts, 192
Covariance matrix
    working, 112
Covariate
    categorical, 20
    external, 327
    internal, 327
    metrical, 20
    time–varying, 325
Cross–validation, 161, 165, 173, 174

DARMA processes, 194
Data
    count, 35, 190, 192
    grouped, 17, 22, 65
    longitudinal, 187, 204, 293
    panel, 293
    ungrouped, 17, 22
Data–driven, 106
Density
    marginal posterior, 57
    posterior, 57
    prior, 57
Design matrix, 66, 69, 71, 80, 94, 258
Design vector, 19, 20
Deviance, 48, 99
Direct kernel estimate, 167
Discrete choice, 73
Discriminant loss function, 173
Dispersion parameter, 19, 214, 345
Distribution
    χ^2, 46
    binomial, 20, 24
    exponential, 23, 306
    extreme value, 71

extreme–maximal–value, 79
extreme–minimal–value, 27, 78
gamma, 20, 23
inverse Gaussian, 20, 23
mixing, 221
multinomial, 64
normal, 20, 22
Poisson, 20, 35
scaled binomial, 25
scaled multinomial, 64
Weibull, 307
Dynamic models, 206, 258, 270, 272, 337

EGRET, 368, 376
EM–algorithm, 226, 227, 268, 281, 355
EM–type algorithm, 59, 237, 281
Empirical Bayes, 240
Epanechnikov kernel, 156
Equicorrelation, 112
Estimating equation, 202
Estimation, 238
    by integration techniques, 238
    categorical kernel regression
        estimate, 167
    direct kernel estimate, 167
    existence, 41
    generalized estimating equa-
        tion, 53, 114, 213
    hierarchical Bayes, 60
    hyperparameter, 281
    marginal, 252
    maximum likelihood, 37, 97,
        238, 322, 333, 350, 355
    MINQUE, 225
    MIVQUE, 225
    of marginal models, 202
    posterior mean, 247, 275, 286,
        295
    posterior mode, 58, 265, 275,
        276, 295
    posterior modes, 233
    quasi–maximum likelihood, 53,
        60, 112, 197, 202

RMLE, 225
  under misspecification, 142
  uniqueness, 41
Exponential family, 18, 345
  simple, 345
Extended Kalman filter, 297

Filter, 263, 286
  Fisher–scoring, 279
  Gauss–Newton, 279
  longitudinal data, 295
  non–normal, 275
Fisher matrix
  expected, 39, 97, 348
Fisher scoring, 40, 99, 181, 235
  for generalized spline smooth-
    ing, 176
Fixed parameter models, 187

Gasser–Müller weight, 157, 167
GAUSS, 367, 375
Gauss–Hermite integration, 241,
    290
  multivariate, 360
  univariate, 357
Gauss–Seidel algorithm, 182
Gaussian kernel, 156
Generalized additive models, 180
Generalized autoregressive mod-
    els, 205
Generalized cross–validation, 161,
    177, 183
Generalized estimating equation,
    114, 202, 213
Generalized Kalman filter, 296
Generalized linear model
  dynamic, 270
  multivariate, 68, 345
  software, 367
  univariate, 18, 345
Generalized sequential model, 87
GENSTAT5, 367, 373
Gibbs sampling, 248, 287, 289, 303,
    364
GLAMOUR, 367, 369

GLIM4, 367, 368
Global, 72
Goodness–of–fit, 45, 48, 99, 130

Hat matrix, 126
  generalized, 126
Hazard
  baseline, 321
Hazard function, 306
  cause–specific, 331
  discrete, 314
  overall, 331
Heterogeneity
  unobserved, 34
Hyperparameter, 58, 259, 267, 268,
    281

Importance function, 291
Importance sampling, 248, 291, 363
Independence
  working, 112
Inference, 206, 213
Information matrix, 38, 145
  expected, 39, 97, 347, 348, 351
  observed, 39, 347, 348, 351
Integrated squared error, 160
Integration techniques, 238
Iteratively weighted least–squares,
    40, 99

k–nearest neighbourhood, 153
Kalman filter, 258, 263
  generalized extended, 277
Kalman gain, 264, 278, 297
Kernel
  Aitchison and Aitken, 163, 168
  discrete, 163
  Epanechnikov, 156
  Gaussian, 156
  ordinal, 163
  product, 163
Kernel smoothing, 162
Kullback–Leibler distance, 164
Kullback–Leibler loss, 174

Landau symbol, 350

Leaving–one–out estimate, 161, 165
Life table, 315
Likelihood
    marginal, 235, 252
    maximum likelihood estima-
        tion, 37, 97, 238, 322, 333,
        350, 355
    penalized, 58, 175, 338
    quasi–maximum, 53, 60, 112,
        197, 202
Likelihood ratio statistic, 45, 196
LIMDEP, 368, 377
Link function, 19, 20, 67, 80, 94,
        346
    log–link, 23
    natural, 20, 347
    violation, 141
Local average, 152
Local linear trend, 260
Log–likelihood, 38, 97, 347, 350
Logit model
    dynamic, 271
Longitudinal data, 187, 204, 257
Loss functions, 173

Marginal cumulative response model,
        213
Marginal estimation, 252
Marginal likelihood, 235
Marginal logit model, 202, 211
Marginal model, 110, 200, 202, 211
Marginal models, 202
Markov chain Monte Carlo, 303,
        365
Markov model, 189, 205
Markovian process, 259
Maximum random utility, 70
Mean average squared error, 160
Metropolis–Hastings algorithm, 303
MINQUE, 225
MIVQUE, 225
Mixing distribution, 221
Model
    Aranda–Ordaz, 320
    ARCH, 192

asymmetric, 105
autoregressive, 188
Bayes, 57
binary regression, 309
competing risks, 331
complementary log–log, 26
compound cumulative, 92
conditional, 105, 188, 205
conditional gamma, 192
conditional Gaussian, 259
cumulative, 75, 76, 79
discrete time, 314
dispersion, 345
dynamic cumulative models,
        272
dynamic discrete time survival,
        337
dynamic generalized linear, 270
dynamic logit model, 271
dynamic multivariate logistic,
        272
dynamic Poisson model, 271
exponential, 312
for categorical time series, 187,
        191
for correlated responses, 104
for longitudinal data, 187
for marginal cumulative re-
        sponse, 213
for nominal responses, 70
for non–normal time series,
        188
for nonexponential family time
        series, 193
for nonlinear time series, 193
for ordinal responses, 73
general state space, 274
generalized autoregressive, 205
generalized sequential, 87
grouped Cox, 78
grouped proportional hazards,
        318
linear Poisson, 35
linear probability, 25
linear transformation, 309

location–scale, 308
log–linear Poisson, 35, 190
logistic, 29, 321
logit, 26
marginal, 110, 200, 202, 211
Markov, 189, 205
multicategorical logit, 66
multicategorical response, 64
multilevel, 224
multinomial logit, 71
multiple modes of failure, 331
multivariate dynamic, 271
non–normal, 269
non–normal state space, 269
nonexponential family, 274
nonexponential family regression, 60
nonhomogeneous for transition probabilities, 190
nonlinear, 269
nonlinear exponential family, 349
nonlinear family, 274
nonlinear regresssion, 60
probit, 25
proportional hazards, 78, 308
proportional odds, 76, 309
quasi–likelihood, 52, 191, 205, 353
quasi–likelihood Markov, 191
random effects, 221, 228
sequential, 84, 321
state space, 258
stereotype regression, 95
survival, 314
symmetric, 108
threshold, 27, 75
transition, 106, 207
two–step, 91
two–step cumulative, 92
variance components, 223
Weibull, 313
Monte Carlo methods, 239, 242, 287, 289, 363
Multilevel models, 224

Nadaraya–Watson weight, 157, 167, 341
Nearest neighbourhood, 158
Non–normal time series, 188
Nonexponential family regression models, 60
Nonexponential family time series, 193
Nonlinear time series, 193
Nonstationary Markov chains, 189
Numerical integration, 357

Observation equation, 258
Odds ratio, 29, 66, 112, 113
One–step estimate, 138
Overdispersion, 34

Panel waves, 205, 293
Parameter
    category–specific, 72
    choice of smoothing parameter, 177
    dispersion, 19, 214, 345
    global, 73
    natural, 19, 345
    nonconstant dispersion, 55
    overdispersion, 44
    scale, 19
    smoothness, 160, 266
Partial likelihood, 194
Pearson statistic, 48, 132
Penalized, 276
Penalized least–squares, 154, 266
Penalized log–likelihood, 175, 338
Penalty matrix, 267
Plots
    added variable, 125
    partial residual, 125
Poisson distribution, 35
Poisson model, 35, 190
    dynamic, 271
Population–specific, 221
Posterior covariance matrix, 57, 241
Posterior curvatures, 234, 237

Posterior distribution, 263
Posterior mean, 57, 240, 263, 267
Posterior mean estimation, 247
Posterior mode, 58, 233, 234, 237,
        265, 267
Power–divergence family, 101
Prediction, 263, 286
Prediction step, 264
Projection matrix, 126

Quadratic loss, 164, 173, 174
Quasi–information, 353
Quasi–likelihood, 53, 109
Quasi–likelihood Markov models,
        191
Quasi–likelihood model, 52, 191,
        196, 197, 202, 353
Quasi–likelihood models, 205
Quasi–score function, 53, 109, 202,
        353

Random effects, 221, 223, 228
        two–stage, 221
Random intercepts, 222
Random slopes, 223
Random walks, 260
Regression splines, 155
Rejection sampling, 248, 288, 364
Reliability function, 306
Repeated observations, 204
Residual
        Anscombe, 132
        deviance, 133
        Pearson, 131
        studentized, 132
Response
        binary, 24, 27
        multicategorical, 271
        multinomial, 271
        smoothing techniques for con-
                tinuous response, 152
Response function, 19, 20, 69, 346
        exponential, 23
        violation, 141
RMLE, 225

S–PLUS, 367, 374
SAS, 367, 370
SC, 122
Scatterplot smoother, 152
Schwarz' criterion, 122
Score function, 38, 97, 98, 347,
        351
Score statistic, 45, 54, 121, 196,
        352
Seasonal component, 260
Selection
        all–subsets, 122
        criteria, 120
        stepwise, 122
        variable, 119
Semiparametric, 262
Simple exponential family, 345
Smoothing, 152, 258, 263, 265, 286,
        296, 337
        choice of smoothing parame-
                ter, 172, 177
        cubic spline smoothing, 175
        Fisher–scoring, 279
        for longitudinal data, 295, 296
        Gauss–Newton, 277, 279
        generalized extended, 277
        kernel, 156, 340
        non–normal, 275
        of a trend component, 266
        penalized least squares crite-
                rion, 266
        posterior mode, 276, 338
        running–line smoother, 153
        simple neighbourhood, 152
        smoothed categorical regres-
                sion, 167
        spline, 153
        spline smoothing in general-
                ized linear models, 175
Splines, 153, 158, 321
        cubic smoothing splines, 153,
                180
        regression splines, 155
SPSS/PC+, 367, 372
State space model, 258, 259, 293

Bayesian interpretation, 266
Statistic
    (log–)likelihood ratio, 352
    Freeman–Tukey, 102
    Kullback, 102
    likelihood ratio, 196
    Neyman, 102
    Pearson, 99
    Pearson goodness–of–fit, 130
    power–divergence, 101
    score, 196, 352
    Wald, 196, 352
Strict stochastic ordering, 78, 90
Structural time series, 259
Subject–specific approaches, 208
Survivor function, 306
    discrete, 314

Test
    for non–nested hypotheses, 146
    generalized score, 147
    generalized Wald, 147
    goodness–of–fit, 45, 48, 99, 130
    Hausman, 144
    information matrix, 145
    likelihood ratio, 45, 47, 99,
        121
    misspecification, 140
    modified likelihood ratio, 121
    modified Wald, 54
    quasi–likelihood ratio, 55
    score, 45, 54, 121, 196
    Wald, 45, 47, 121, 196
Time series, 187, 204, 257
Time–varying covariate, 262
Transition equation, 258
    nonlinear, 274
Transition model, 207, 294
Transition probabilities, 190
Trend component, 260
Two–stage random effects, 221

Variance components, 223, 235
Variance function, 20, 22, 55
    working, 53

Wald statistic, 196, 352
Working correlation matrix, 212
Working covariances, 211

XPLORE, 367, 376

Springer Series in Statistics

(continued from p. ii)

Le Cam/Yang: Asymptotics in Statistics: Some Basic Concepts.

Longford: Models for Uncertainty in Educational Testing.

Manoukian: Modern Concepts and Theorems of Mathematical Statistics.

Miller, Jr.: Simultaneous Statistical Inference, 2nd edition.

Mosteller/Wallace: Applied Bayesian and Classical Inference: The Case of *The Federalist Papers.*

Pollard: Convergence of Stochastic Processes.

Pratt/Gibbons: Concepts of Nonparametric Theory.

Ramsay/Silverman: Functional Data Analysis.

Read/Cressie: Goodness-of-Fit Statistics for Discrete Multivariate Data.

Reinsel: Elements of Multivariate Time Series Analysis, 2nd edition.

Reiss: A Course on Point Processes.

Reiss: Approximate Distributions of Order Statistics: With Applications to Non-parametric Statistics.

Rieder: Robust Asymptotic Statistics.

Rosenbaum: Observational Studies.

Ross: Nonlinear Estimation.

Sachs: Applied Statistics: A Handbook of Techniques, 2nd edition.

Särndal/Swensson/Wretman: Model Assisted Survey Sampling.

Schervish: Theory of Statistics.

Seneta: Non-Negative Matrices and Markov Chains, 2nd edition.

Shao/Tu: The Jackknife and Bootstrap.

Siegmund: Sequential Analysis: Tests and Confidence Intervals.

Simonoff: Smoothing Methods in Statistics.

Small: The Statistical Theory of Shape.

Tanner: Tools for Statistical Inference: Methods for the Exploration of Posterior Distributions and Likelihood Functions, 3rd edition.

Tong: The Multivariate Normal Distribution.

van der Vaart/Wellner: Weak Convergence and Empirical Processes: With Applications to Statistics.

Vapnik: Estimation of Dependences Based on Empirical Data.

Weerahandi: Exact Statistical Methods for Data Analysis.

West/Harrison: Bayesian Forecasting and Dynamic Models, 2nd edition.

Wolter: Introduction to Variance Estimation.

Yaglom: Correlation Theory of Stationary and Related Random Functions I: Basic Results.

Yaglom: Correlation Theory of Stationary and Related Random Functions II: Supplementary Notes and References.